Kumbakonam Rajagopal

Lecture Notes in Engineering

Also of Interest

Modern Signal Processing
Xian-Da Zhang, 2023
ISBN 978-3-11-047555-5, e-ISBN 978-3-11-047556-2

Vehicle Technology
Technical foundations of current and future motor vehicles
Dieter Schramm, Benjamin Hesse, Niko Maas and Michael Unterreiner, 2020
ISBN 978-3-11-059569-7, e-ISBN 978-3-11-059570-3

Electrical Engineering
Fundamentals
Viktor Hacker and Christof Sumereder, 2020
ISBN 978-3-11-052102-3, e-ISBN 978-3-11-052111-5

Electrical Machines
A Practical Approach
Satish Kumar Peddapelli and Sridhar Gaddam, 2020
ISBN 978-3-11-068195-6, e-ISBN 978-3-11-068227-4

Kumbakonam Rajagopal

Lecture Notes in Engineering

Introduction to Linearized Elasticity

DE GRUYTER

Author
Prof. Kumbakonam Rajagopal
Texas A&M University
Department of Mechanical Engineering
100 Mechanical Engineering Office Building
College Station 77843-3123
United States of America
krajagopal@tamu.edu

ISBN 978-3-11-078942-3
e-ISBN (PDF) 978-3-11-078951-5
e-ISBN (EPUB) 978-3-11-078963-8

Library of Congress Control Number: 2023938204

Bibliographic information published by the Deutsche Nationalbibliothek
The Deutsche Nationalbibliothek lists this publication in the Deutsche Nationalbibliografie; detailed bibliographic data are available on the Internet at http://dnb.dnb.de.

Typesetting: VTeX UAB, Lithuania
Printing and binding: CPI books GmbH, Leck

www.degruyter.com

Preface

In a delightful poem, one of America's 19th century Romantic poets, James Russell Lowell, describes the glorious fountain as "Fresh, changeful, constant" and flowing ever "upward." In a similar vein, I like to think of textbooks as fountains of knowledge and information from which the students are expected to imbibe and quench their intellectual thirst. This knowledge, like the emanation from Lowell's fountain, is ever changing though textbooks invariably treat the subject matter to be truths set in stone. It is not entirely clear whether the "truths" contained in textbooks are "truths" by general consensus or whether they are necessary truths. Undergraduate students cannot be expected to grasp and understand this important distinction, and textbooks sadly never mention this critical difference. That textbooks deal solely with facts that are unquestionable is an erroneous though popular misconception.

Kac, Rota and Schawartz [1] in the essay "Mathematical Tensions" remark that *"The late W. A. Hurwitz used to say that in teaching at an elementary level, one must tell the truth, nothing but the truth, but not the whole truth"*. And herein lies the rub, as the preface to this very same collection of essays begins with the statement, *"In mathematics, as anywhere today, it is more and more difficult to tell the truth."* It is also true that some "truths" of the past became fodder for future ridicule. An appropriate modification of Hurwitz might read thus: when teaching at an elementary level, one must tell what one believes to be the truth and nothing other than that one believes to be the truth, and accept with great humility that one's belief as to what is the truth is not necessarily the truth, let alone the whole truth.

A well-written textbook should therefore convey the sense that the material being advanced for consumption by the student, while not being the whole truth, is nonetheless sufficiently close enough to the truth to be applicable for its intended use. While not dogmatic it should nevertheless be unambiguous and provide a sense of assurance without giving a false sense of confidence. Above all, it should inculcate in the student a thirst for the subject matter. This might seem like a tall order for a textbook but falls far short of what is expected of the teacher. As Rabbi Heschel remarked (see Abraham Joshua Heschel, A Spiritual Anthology, Samuel H. Dresner (editor), Crossroads Publishing Company (1983)):

> Everything depends on the person who stands in front of the classroom. The teacher is not an automatic fountain from which intellectual beverages may be obtained. The teacher is either a witness or a stranger. To guide a pupil into the promised land, the teacher must have been there themselves. When asking themselves: Do I stand for what I teach? Do I believe what I say? the teacher must be able to answer in the affirmative. What we need more than anything else is not textbooks, but text people. It is the personality of the teacher which is the text that the pupils read: the text that they will never forget.

These lecture notes are meant to come as close as possible to the "text people" that Heschel describes: They should have the feel of a teacher instructing and educating a stu-

https://doi.org/10.1515/9783110789515-201

dent. It is meant to have a conversational and colloquial feel and inject a discussion of the ambiguities and uncertainties underlying the subject matter. The emphasis should be on conveying concepts and their consequences with the clear understanding that the concepts on which the edifice is erected will be able to withstand some perturbations to it. While mathematical rigor is neither disdained nor abandoned, there is no overwhelming preoccupation with it, for mathematics itself is the product of the interplay of antithetical notions with its axioms based purely and firmly on intuition, and its theorems a consequence of rigorous logic sans any appeal to intuition.

These lecture notes are meant for students majoring in engineering who possess a basic knowledge of undergraduate mathematics. Here, we first develop the basic framework for the response of nonlinear elastic bodies within the context of Cauchy elasticity, and then carry out a linearization based on the displacement gradient being small, which leads to the classical approximate constitutive relation for linearized elastic response. Most textbooks on elasticity present results as though the points of view being espoused are universally agreed upon. This is far from true. As an example, it is shown that it is possible that bodies could exhibit elastic response and not belong to the class of Cauchy elastic bodies: these bodies can be described by implicit constitutive relations between the stress and the stretch, and linearization of such a general theory leads to the possibility of a nonlinear relationship between the linearized strain and the stress. The student is made aware that even within the context of response, which is considered as being well-understood, there is great deal that has evaded comprehension. Conventional wisdom concerning elasticity is greatly misplaced.

A course using these lecture notes can be completed in one semester. Few exercises are provided but each chapter ends with a list of suggested readings, albeit incomplete, where such exercises may be found. A glaring lacuna in the lecture notes is the complete absence of numerical solutions to problems involving elastic bodies. Since the equations in question are linear and powerful codes have been developed that are available commercially, and more importantly since courses devoted to such numerical studies are offered regularly at most institutions of higher learning, this is not a part of the lecture notes. Such material can be found in several of the references that are provided in the lecture notes. Furthermore, these lecture notes are not meant to be a textbook. The student is expected to refer to books that are provided as references to fill in details that are not provided as the i's are not dotted and the t's are not crossed in the material that is presented, ensuring the participation of the student in learning the material thoroughly.

This lecture notes is an outgrowth of a first-year graduate course that is offered regularly to mechanical engineering students at Texas A&M University. While teaching the course in Fall 2020 in the midst of the dreaded Covid 19 pandemic, I could not have remotely imagined that publishing them as a textbook could potentially be an outcome of my pedagogical efforts. Jacob Rogers, a highly motivated and hardworking graduate student, had diligently taken notes during the course, and I was pleasantly surprised when he presented me with a set of type-written notes at the end of the semester to serve for my future use.

The choice of the subject matter and selection, of what to include and what to omit in this publication, is a reflection of my idiosyncrasies and biases. I have been greatly influenced by those from whom I have learnt, both as a student and as a colleague. I thank Jacob for his concerted effort in meticulously transcribing my lecture notes. He was the sounding board reflecting the interests of the students to whom the book is addressed. Unfortunately, Jacob was unable to see the lecture notes to its fruition due to other demands on his time. I was fortunate to obtain the assistance of Dr. Gokulnath Chinnasamy, a young faculty member at IIT Palakkad to fill in the gaps that existed and complete the lecture notes. I appreciate his patience and poise in incorporating the myriad of rewrites and corrections that I made, trying to achieve clarity, which invariably evaded my efforts. I thank Profs. Ponnalagu Alagappan, Krishna Kannan, Umakanthan Saravanan and Alan Wineman for proof reading the manuscript and making helpful suggestions to improve the presentation of the material in these lecture notes. While most if not all that is worthwhile in the notes, is attributable to those from whom I have learnt, I bear sole responsibility for all errors and omissions.

These lecture notes would not have been possible, if not for the patience, forbearance and encouragement of my wife, Chandrika Rajagopal to whom I owe immeasurable gratitude.

References

[1] M. Kac, G-C. Rota and J. T. Schwartz. Discrete thoughts: Essays on Mathematics. Science and Philosophy, Springer Science, New York, 1992.

Contents

Nomenclature

Scalar Fields

γ	Extensional strain
λ	Lamé's modulus
$\lambda_{\mathbf{n}}$	Stretch along the direction **n**
μ, G	Shear modulus
ν	Poisson's ratio
Φ	Prandtl stress function
ϕ	Airy stress function
ρ	Mass density
τ	Shear stress
φ	Warping function
c	Arbitrary constant
E	Young's modulus
k	Bulk modulus
t	Current time
$\mathrm{d}A$	Infinitesimal area element in the reference (material) configuration
$\mathrm{d}a$	Infinitesimal area element in the current (spatial) configuration
$\mathrm{d}V$	Infinitesimal volume element in the reference (material) configuration
$\mathrm{d}\upsilon$	Infinitesimal volume element in the current (spatial) configuration

General

$\mathbb{R}$	Real numbers
$\mathcal{B}$	Abstract body
$\mathcal{E}$	Euclidean space
$\mathcal{P}$	Abstract material point
$\mathcal{O}$	Set of Orthogonal transformations
κ	Placer

Operators

$:$	Double contraction
$\cdot$	Scalar product
∇	Gradient in the material configuration
∇^2	Laplacian
∇^4	Biharmonic Operator
$\otimes$	Tensor product
div	Divergence
$\times$	Cross product
$\lvert \cdot \rvert$	Euclidean norm
$\lvert\lvert \cdot \rvert\rvert$	Frobenius norm
det	Determinant
grad	Gradient in the spatial configuration
tr	Trace

https://doi.org/10.1515/9783110789515-202

Tensor Fields

$\mathcal{C}$	Elasticity tensor
$\boldsymbol{\omega}$	Linearized rotation
$\boldsymbol{\varepsilon}$	Linearized strain
$\boldsymbol{e}$	Almansi-Hamel strain
δ_{ij}	Kronecker delta
$\mathbf{B}$	Left Cauchy-Green tensor
$\mathbf{C}$	Right Cauchy-Green tensor
$\mathbf{E}$	Green-St. Venant strain
$\mathbf{F}$	Deformation gradient
$\mathbf{I}$	Identity map
$\mathbf{L}$	Velocity gradient
$\mathbf{Q}$	Orthogonal transformation
$\mathbf{R}$	Rotation or inversion
$\mathbf{T}$	Cauchy stress
$\mathbf{U}$	Right stretch
$\mathbf{V}$	Left stretch

Vector Fields

χ	Motion map
$\mathbf{a}$	Acceleration of a particle
$\mathbf{b}$	Specific Body force
$\mathbf{e}$	Base vector
$\mathbf{N}$	Unit normal in the material configuration
$\mathbf{n}$	Unit normal in the spatial configuration
$\mathbf{t}$	Traction
$\mathbf{u}$	Displacement
$\boldsymbol{\upsilon}$	Velocity of a particle
$\mathbf{X}$	Position in the reference (material) configuration
$\mathbf{x}$	Position in the current (spatial) configuration
$\mathrm{d}\mathbf{X}$	Infinitesimal filament in the reference (material) configuration
$\mathrm{d}\mathbf{x}$	Infinitesimal filament in the current (spatial) configuration

1 Introduction

1.1 Introductory remarks

The term elasticity is used in a variety of subjects. In economics we come across the usage *'elasticity of demand'*, *'elasticity of supply'*, and *'price elasticity'*; in psychology *'elasticity of thought'*, *'elasticity of mind'*; in livestock production *'elasticities in livestock production'*, and in the physical sciences *'elastic body'* and *'elasticity of response'*. It has been used to signify *'buoyancy of mind or character'*, *'capacity for overcoming depression'*, and Jane Austen [1] mentions a *'village of wonderful Elasticity'* (see the Oxford English Dictionary [2] for the various meanings ascribed to the word elasticity).

Turning our attention to the concept of elasticity relevant to this monograph, according to Truesdell [26], inchoate description of elasticity can be found in the works of Philon, who lived in the fourth century BC. In the Prologue to the Rational Mechanics of Flexible or Elastic bodies, Truesdell [26] provides a history of elasticity and vibration, as it pertains to strings, thin structures like beams, arches, etc. Therein, he remarks,

> The earliest known descriptions of elasticity, and in particular of the elasticity of metals, are found in PHILON'S work. He advises that the bow cord be stretched so tight "that when the machine is drawn, the diameter is lessened by a third part." He mentions the fatigue of the cord as a result of use and advises against the common practice of trying to regain the tension by twisting the cord until it is tight again. He recommends "tightening all the strings of the bow cord at once, in their natural straight position," so as to avoid weakening them by twisting. He claims the invention of bronze leaf springs and describes their fabrication. His innovation appears to have aroused some doubts: "...many persons... say that it is impossible that curved bands [i. e. springs] when straightened out by the force of the bow will not remain straight thereafter but will instead regain their original curvature. While indeed by its nature horn has this property, and some kinds of woods (and bows are made of such), bronze on the contrary is hard and stiff in its nature, as is iron, so that when bent... it cannot straighten itself out. Let these persons be forgiven for holding such an opinion without trying the details."

Truesdell [26] later in the same article goes on to say

> How great a proportion of mediaeval work survives, and how much of that is now available, I do not know. The only writing of value on deformable bodies that I have been able to see is the fourth book of JORDAN DE NEMORE's Theory of Weight (13th Century), and remarkable it is, Western in spirit, ambitious beyond anything in the Greek or Arab tradition'. The seventeen propositions on fluid flow, resistance, fracture, and elasticity are all original. While the style is mathematical, it would be unfair to expect what Jordan brings forward as "proofs" to be more than plausible reasoning alleged in favor of assertions drawn from experience and conjecture by a scientist well trained in the ancient mathematical statics. Only two of the propositions concern our present subject. In Prop. 12 we are told that the coherence of a beam hung up by its two ends may or may not suffice to keep it from breaking in the middle. The beam, whether supported in this way or at one end only, is to be regarded as a lever. Greater bending is produced by a body striking the beam than by the same body resting upon it. [This is the earliest distinction between static and dynamic loading in respect to deformation.]

https://doi.org/10.1515/9783110789515-001

> Prop. 13 reads, "When the middle is held back, the ends are more easily curved." The "body" is taken as a line fixed at its midpoint; the ends are supposed to receive an impulse. ceé "... since the ends yield more easily, while the other parts follow more easily insofar as they lie closer to them, it turns out that the whole body is curved into a circle." [This is the earliest statement of the problem of the elastic curve or elastica]. JORDAN asserts, in modern terms, that a band clamped at one end and struck by a weight falling upon the other assumes the form of a circular arc.
> The reasoning is vague, qualitative, and insufficient if not erroneous, but the attempt at a precise argument to prove a concrete result in a domain never previously entered is of splendid daring. [This work of the thirteenth century is better than many to be published by learned academies in the seventeenth and even the early eighteenth.]

In neither of the above comments by Philon and Jordan (Jordanus) do we find a clear statement as to what is meant by an "elastic body", but it would be unreasonable to expect a clear statement at such time. For instance, the comments of Philon which Truesdell quotes is concerned with avoiding fatigue by instantaneously tightening the bow cord, which on using our hindsight based on our current understanding of elastic response may allow us to attribute to Philon a hazy understanding of elasticity. The terminology "elasticity" and "elastical" came into usage much later. However, given that Philon's work was in the fourth century BC and that of Jordan in the thirteenth century, their remarks show singular observational proficiency and remarkable perspicacity.

These lecture notes are restricted to a discussion of the elastic response of bodies, but even this from a very limited perspective, namely a discussion of the linearized response of elastic solids. Looking at textbooks devoted to elasticity including the acclaimed book by Love [3], one would get a distorted, in fact perverted, outlook that elastic bodies refers to solid bodies, but nothing could be further from the truth. In fact, the terminology "elasticity" or "elastical" was used in the mid-seventeenth century to describe gases. In 1653, Jean Pecquet [4], in his New Anatomical Experiments mentions *'The spontaneous dilatation [of the air] enerveth the power of the Elastic (impulsive) faculty'* and in 1664 Henry Power [5] states *'The external and internal AYR were come of the same...Elasticity'*. He also remarks about the *'Elastical pressure of the external Ayr upon the surface of Quicksilver in the vessel'*. Robert Boyle [6] in an essay concerning the behavior of air remarks that *'There is a spring or elastical power in the air in which we live'*. As discussed below, Hooke [7] erroneously thought his constitutive relation applied to solids as well as gases. William Petty [8], in 1674, while discussing the property of springs, comments *'Elasticity is the power of recovering the figure, upon removal of such Force'*. We observe that while the above comments by Pecquet, Power and Petty provide characteristics of elastic response, namely "spontaneity of response", "springiness", "recovery of shape", *etc.*, they did not provide a clear response relationship between cause and effect.

According to Love [3], Hooke was aware of his law in 1660 and published it as an anagram in 1676, and later in his celebrated work *'de potentia restitutiva'* in 1678. Hooke [7] remarks

> About two years since I printed this Theory in an Anagram at the end of the book of the Description of Helioscopes, viz, ceiiinosssttuu, idest, Ut tensio sic vis; the power of any Spring is in the same proportion of the tension thereof: that is if one power stretch or bend it one space, two will bend it two, three will bend it three and so forward.

Unfortunately, what one finds in practically every textbook on elasticity, as well as in scholarly articles on the subject, is the erroneous impression that Hooke had a very clear understanding as to which class of materials his constitutive law applied to. Hooke confused issues and got quite a few things wrong. As Moyer [9] remarks:

> AUTHORS OF PHYSICS TEXTBOOKS,[1] as well as scholars writing on the history of science, often portray Robert Hooke as the 'discoverer' of a concise mathematical relationship regarding springs: If we stretch the spring so that its endpoint moves to a position x, the spring will exert a force on the agent doing the stretching given to a good approximation by $F = -kx$, where k is a constant called the force constant of the spring.

And

> While it is a simple and straightforward matter to extract from the early pages of De potentia restitutiva the modern statement of 'Hooke's law' for elastic solids, $F = -kx$, one must first suppress or ignore Hooke's 'trial' involving 'a body of Air.' That is, in modern terms, he inappropriately associated 'Hooke's law'-a direct proportionality between displacement and force-with 'Boyle's law'-an inverse proportionality between the volume of an ideal gas at constant temperature and the pressure to which it is subjected (or, for a gas in a cylinder with a uniform cross-sectional area, between the length of the column of gas and the force on it)."

Interestingly even from a modern viewpoint, an ideal gas is an elastic body, it is an elastic fluid that does not obey Hooke's law. These lecture notes are however dedicated to the discussion of elastic solids, to be more precise a class of approximations to them, and this is precisely why the lecture notes are titled "An Introduction to Linearized Elastic Solids".

Illustrious scientists, the likes of Mariotte, Ricatti, James, John and Daniell Bernoulli, Leibniz, Hugyens, Euler, Lagrange, Coulomb, Young, Navier, Poisson, Girard, Cauchy and others made important contributions to the field of elasticity. A few words concerning Young's research are in order as it is pertinent to material contained in the lecture notes. Young [10] believed in different moduli for compression and extension, and introduced two definitions, the second of which, while not the Young's modulus is related to it (see Todhunter and Pearson [11, 12] for an extensive discussion of the contributions of Young to elasticity):

> 318. Definition: A substance perfectly elastic is initially extended and compressed in equal degrees by equal forces, and proportionally by proportional forces.

---

1 The capitalization is in the original quote by Moyer and not put into place by the author.

> 319. The modulus of elasticity of any substance is, a column of the same substance, capable of producing pressure on its base which is to the weight causing a certain degree of compression as the length of the substance is to its dimunition of its length.

Navier [13] used a molecular approach wherein the basic premise was that the body was comprised of material particles (molecules) and that the force of attraction between two material particles, when the distance between them is slightly increased, is proportional to the product of the distance between them and by which the distance between the particle is increased. This study was the first attempt at a systematic development for the equations governing the response of elastic bodies. Cauchy [14], who was a member of the committee to evaluate the memoir of Navier, introduced the concept of stress as it is currently used and laid out the basic framework for modern nonlinear elasticity. Poisson's [15] work in elasticity was also based on the molecular approach and deserves special mention.

The approximate constitutive relation that describes the linearized response of elastic bodies is obtained in the Cauchy theory of elasticity under the assumption that the maximum of the norm of the displacement gradient at all points belonging to the body, for all time, is sufficiently small that it's square can be ignored with respect to the norm, and it provides an expression for the stress in terms of the linearized strain. However, the linearized strain is often times also expressed in terms of the stress. This indifference to whether the stress should be expressed in terms of the kinematical quantity (linearized strain) or whether it is the kinematical quantity that ought to be expressed in terms of the stress is a fundamental issue that needs to be addressed very carefully, as it concerns the issue of causation, a central idea in Newtonian mechanics. We cannot get into the discussion of this issue here, but merely rest content by observing that Cauchy elasticity is based on expressing the stress in terms of the kinematical quantity, namely the deformation gradient.

In the instance when the stress is expressed in terms of the linearized strain, the material moduli that appear in the constitutive equation are the Lame' constants, while in the case when the linearized strain is expressed in terms of the stress, the material moduli that appear in the constitutive equation are Young's modulus and Poisson's ratio. It is important to recognize that within the context of linearized elasticity, the material moduli have to be constants, they cannot depend on the density, stress or strain. This point cannot be overemphasized as there are numerous experiments which clearly show that the material moduli depend on these quantities in many elastic bodies that undergo small strains.

An important development in elasticity that requires special mention is the seminal contribution made by Green [16]. He observed that if the stress in a Cauchy elastic body is not derivable from a potential (the stored energy), then such a material would be a source of a perpetual motion machine. Green states

-and $\delta\phi$ the exact differential of some function and entirely due to the internal actions of the particles of the medium on each other. Indeed, if $\delta\phi$ were not an exact differential a perpetual motion would be possible, and we have every reason to think, that forces in nature are so disposed as to render this a natural impossibility.

Bodies that have a stored energy associated with them from which the stress can be derived are referred to as Green elastic bodies or Hyperelastic bodies. An interesting question that immediately presents itself is whether a body could be Cauchy elastic but not Green elastic. Carroll [17] recently showed that Cauchy elastic bodies that are not Green elastic exhibit unacceptable physical characteristics. However, it is possible that one could have bodies that are elastic, that are neither Cauchy elastic nor Green elastic (some such elastic bodies are described by implicit constitutive theories and are discussed briefly toward the end of these lecture notes).

As mentioned earlier, Petty in 1674 had characterized an elastic solid as a body that had the ability to retain shape. This seems to be the preferred definition one finds for an elastic body. However, the definition is imprecise for reasons that will become obvious soon. We first provide the various definitions given by some of the leading exponents of the subject and state why they are all vague and imprecise.

Let us start with the definition due to Lord Kelvin. In his article on 'Elasticity' for the Encyclopedia Brittanica, Kelvin [18] introduces an elastic body thusly:

Elasticity of matter is that property in virtue of which a body requires force to change its bulk or shape, and requires a continued application of the force to maintain the change, and springs back when the force is removed, and, if left at rest without the force, does not remain at rest except in its previous bulk and shape. The elasticity is said to be perfect when the body always requires the same force to keep it at rest in the same bulk and shape and same temperature through whatever variations of bulk, shape and temperature it be brought.
A body is said to possess some degree of elasticity if it requires any force to keep it in any particular bulk or shape.

The above explanation is highly unsatisfactory as it introduces hazy notions such as "degree of elasticity", *etc.* In fact, Kelvin [18] even talks about elastic bodies possessing friction and confuses elastic solids with viscoelastic solids (see discussion in [20]). We do see the notion of the ability to retain shape but nothing is clear or precise. We will defer a discussion as to why it is insufficient to merely state a body is capable of recovering shape to the end of this section as several other experts that are quoted below attribute the same characteristic to elastic bodies.

Love [3] provides the following definition of elasticity:

Again solid bodies are not absolutely rigid. By the application of suitable forces they can be made to change both in size and shape. When the induced changes of size and shape are considerable, the body does not, in general, return to its original size and shape after the forces which induced the change have ceased to act. On the other hand when the changes are not too great the recovery may be apparently complete. The property of recovery of original size and shape is the property that is termed elasticity.

Once again, we see the ability to retain shape as the primary characteristic of an elastic body, but no clear mathematical specification as to what this might mean. Also, in the above explanation, we see the bias that elastic bodies are solid bodies.

Robert Southwell [19] defines an elastic body in the following manner:

> It will be observed that our demonstration has involved two distinct postulates: first, that in any circumstance of initial loading a force applied at any point and in any direction will produce a proportionate deflexion of any other point (Hookes's law); secondly, that the body in question will recover its original form after any system of loading has been applied and removed (this may be taken as a definition of elasticity).

Southwell is being too restrictive as elastic bodies need not obey Hooke's law. In fact, many of the early constitutive relations developed to describe the response of elastic bodies were non-linear.

Leibnitz [20] did not believe in Hooke's law and corresponded with James Bernoulli concerning the same. In 1694, James Bernoulli [21] had introduced an exponential relationship

$$\epsilon = a(\sigma)^m,$$

relating the strain ϵ and the stress σ, where a and m are constants, while the relationship

$$\sigma = a(e^{-1/\epsilon}),$$

where a is a constant, was proposed by Ricatti [22], and a parabolic relationship was introduced by Gertsner in 1824 (but he seems to be referring to inelastic response). Wertheim [23] in 1847 proposed the following nonlinear relationship between the strain and the stress to describe the response of tissues,[2] namely

$$\epsilon^2 = a(\sigma)^2 + b(\sigma),$$

where a and b are constants. A student interested in the early nonlinear models for elastic response can find the same in Bell [25] and Truesdell [27]. However, it is far from clear that the response being described in some of the early studies is purely elastic response.

According to Truesdell [27],

> Classical linear elasticity describes the slight deformations of media which are perfectly springy, so that when released from deforming forces they revert to their initial shapes; it is a linear theory, in which an uniformly doubled load necessarily produces a doubled displacement,

2 Tissues are made up of cells which are over 70 % water, and hence it is inappropriate to describe them as elastic bodies, though that is how they are primarily described (see Rajagopal and Rajagopal [24] for a discussion of the relevant issues).

and we see once again the dissatisfying reference to the ability to retain shape. However, later he provides a better definition, which yet does not meet the test for complete clarity:

> For a given body, there exists a fixed natural state, generally regarded as unstressed, such that the body when constrained into a form near to that is its natural state and then released from all external forces returns perfectly to this natural state independently of the manner in which the forces have been applied and removed.

The problem with all the above definitions is that they are made without any recourse to the time scale of observation. Does the body instantaneously recover shape, take a finite amount of time to recover shape, or does it require infinite time to recover shape? There could be bodies that recover their shape on the removal of load taking infinite time to do so, as some viscoelastic solids do. Truesdell does use the terms *"independently of the manner in which the forces have been applied and removed"*, but the notion of time is not used explicitly in the explanation. It does not state how soon the body returns to its original shape. Moreover, as an ideal gas is an elastic body, what does one mean by it being unstressed? To be unstressed, it has to be unconfined and this would mean that the air would just expand to occupy all space. However, in rigorous continuum mechanics propounded by Truesdell, a body by definition is compact.

We conclude this discussion of various definitions of elasticity with that which can be found in a recent mathematical treatment of the subject matter by Sokolnikoff [28] who also asserts that *"A body is called elastic if it possesses the property of recovering its original shape when the forces causing the deformation are removed."*

The proper characterization of an elastic body is that it is a body that is incapable of dissipation while undergoing a purely mechanical process, and this applies to a body, whether it is a solid or a fluid. The above definition requires a thermodynamic setting within which the notion of elasticity has to be discussed. However, since we shall not concern ourselves with thermodynamic issues, we shall not discuss this issue here.

We end this short and cursory discussion of elasticity with some new and novel developments in the field. Recently, Rajagopal [29, 30] has shown that a much larger class of bodies than those envisioned by Cauchy qualify to be referred to as elastic bodies, if by an elastic body we mean a body that is incapable of dissipation, that is converting the time rate of mechanical work (mechanical working) into energy in thermal form (heat). Many such bodies are described by implicit constitutive relations between the stress and the deformation gradient.

An excellent review of the early contributions to elasticity can be found in Todhunter and Pearson [11, 12], Love [3], Timoshenko [32], Truesdell [2], Truesdell and Toupin [33], Truesdell and Noll [34] (see Bell [25] for a discussion of the history of experiments in elasticity). A discussion of implicit constitutive relations to describe the response of elastic bodies can be found in Rajagopal [29, 30] and Bustamante and Rajagopal [31].

We shall end this introductory chapter with some comments concerning what one means by a solid. Contrary to the impression that books in rational mechanics would like to give, the notion of a solid or liquid is not clearly motivated. The very rigorous mathematical definitions that are available are not as convincing as they seem when one scrutinizes them with a healthy dose of skepticism. All these definitions do not address the issue of time, space and force scales which invariably have to come into play in our decision concerning the classification of bodies. Bodies to the naked eye which seem motionless, could in the scale of millions of light years and length scales of nanometers, move, thereby making all definitions which have no recourse to time, rather meaningless.

Goodstein's following statement concerning a fluid can be modified by replacing the word "liquid" by "solid" and it would remain valid. Goldstein [35] remarks *"Precisely what do we mean by the term liquid? Asking what is a liquid is like asking what is life; we usually know it when we see it, but the existence of some doubtful cases make it hard to define precisely."*

The notion of "solid" has not only eluded a proper definition from the point of view of continuum mechanics, it has also escaped the grasp of quantum mechanicians. As Anderson [36] asks *"What is a solid? How does one describe a solid from a really fundamental point of view in which the atomic nuclei as well as electrons are treated truly quantum mechanically? How and why does a solid hold itself together?"* and ends his remarks by stating *"—I have never yet seen a satisfactory fully quantum mechanical description of a solid."*

Though we do not have a firm grasp of what is meant by an elastic body or a solid, we do have some intuitive feel for the same, and we will have to rest content with such an understanding, for this is the nature of any study, as each explanation or answer leads to even more problematic questions.

References for the Introduction

[1] J. Austen. Letters of Jane Austen, easy read edition, Letter XXVI-1880. Accessible Publishing System PTY Limited, 2008.

[2] J. A. Simpson and E. S. C. Weiner, editors. Oxford English Dictionary. Clarendon Press, Oxford, 2nd edition, 2000.

[3] A. E. H. Love. A Treatise on the Mathematical Theory of Elasticity. Dover Publications, New York, 1944.

[4] J. Pecquet. Experimenta Nova Anatomica. Paris, 1651. English translation, as New Anatomical Experiments, Octavian Pulleyn, London (1653).

[5] H. Power. Experimental Philosophy, in Three Books. Printed by T. Roycroft, for John Martin and James Allestry, London, 1664.

[6] R. Boyle. An essay of the great effects of even languid and unheeded whereunto is annexed an experimental discourse of some little observed causes of the insalubrity and salubrity of the air and its effects. Printed by M Flesher for Richard Davis, Bookseller in Oxford, 1685.

[7] R. Hooke, see R. T. Gunther and A. E. Gunther. Early Science in Oxford: the Cutler Lectures of Robert Hooke. Vol. 8 subscribers, 1931.

[8] W. Petty. The discourse made before the Royal Society on the use of duplicate proportion; together with a new hypothesis of springing or elastique motions. Printed for John Martyn, Printer to the Royal Society, at the Bell in St. Paul's churchyard, 1674.
[9] A. B. Moyer. Robert Hooke's ambiguous presentation of 'Hooke's Law'. Isis, 68:275–288, 1975.
[10] T. Young. A Syllabus of a Course of Lectures in Natural and Experimental Philosophy. Press of the Royal Institution, W. Savage, Printer, London, 1802.
[11] I. Todhunter (edited and completed by Pearson K). A History of the Theory of Elasticity and Strength of Materials, Volume II. Printed by C. J. Clay, M. A. and Sons, at the University Press, Cambridge, 1886.
[12] I. Todhunter and K. Pearson (edited and completed by Pearson K). A History of the Theory of Elasticity and Strength of Materials, Volume I: From Galilei to St. Venant. Printed by C. J. Clay, M. A. and Sons, at the University Press, Cambridge, 1886.
[13] C. L. M. Navier. Sur les lois de l'equilibre et du mouvement des corps solides elastiques. In Bulletin des sciences par la Société Philomatique de Paris, pp. 177–181, 1823.
[14] A. Cauchy. Recherches sur l'équilibre et le mouvement intérieur des corps solides ou fluides, élastiques ou non élastiques. 1822.
[15] S. D. Poisson. Mémoire sur l'équilibre et le mouvement des corps élastiques. F. Didot, 1828.
[16] G. Green. On the laws of reflexion and refraction of light at the common surface of two non-crystallized media (1837). Trans Cambr Phil Soc 1839; 7(1839–1842): 1–24. Mathematical Papers of the Late George Green, edited by N. M. Ferris, pp. 245–269. MacMillan and Company, London, 1871.
[17] M. M. Carroll. Must elastic materials be hyperelastic? Math Mech Solid, 14:369–376, 2009.
[18] W. Thomson. On the elasticity and viscosity of metals. Proc Roy Soc London, 14:289–297, 1865.
[19] R. V. Southwell. Elasticity for Engineers and Physicists. Oxford University Press, Oxford, 1941.
[20] G. W. Leibniz. Letter to James Bernoulli, appearing in G. W. Leibniz, Mathematiche Schriften, edited by F. E. Gerhardt, Vol. III, part 1, pp. 13–20. Georg Olms Verlagsbuchhandlung, Heidelsheim, 1855.
[21] J. Bernoulli. Curvatura laminae elasticae. Acta Eruditorum Lipsia, June:262–276, 1694.
[22] J. Ricatti. Reprinted in Sources of Science, Vol. 2, pp. 101–102. Johnson Reprint Corporation, New York and London, 1968.
[23] G. Wertheim. Memoire sur l'elasticite' et la cohesion des principaux tissues du cops humain. Annales de Chimie et de Physique, third series, 21:385–414, 1847 (presented in 1846).
[24] K. R. Rajagopal and K. Rajagopal, Modeling of the aorta: complexities and inadequacies. Aorta, 8:91–97, 2020.
[25] J. F. Bell. Volume I: The Experimental Foundations of Solid Mechanics. Springer, 1973.
[26] C. Truesdell. The Rational Mechanics of flexible or elastic bodies, 1638–1788: Introduction to Leonhardi Euleri Opera Omnia, Vol. X and XI, Seriei Secundae, Zurich, Orel Fussli.
[27] C. Truesdell. Mechanical foundations of elasticity and fluid dynamics. In Continuum Mechanics I. Gordon and Breach, New York, 1966 (Reprinted from J Ration Mech 1952; 1: 125–3000, as corrected in 1953; 2: 595–616 and 1954; 3: 801).
[28] I. S. Sokolnikoff. Mathematical Theory of Elasticity. McGraw Hill, New York, 2nd edition, 1956.
[29] K. R. Rajagopal. Elasticity of elasticity. Zeitschrift fur Angewandte Math Phys, 58:309–417, 2007.
[30] K. R. Rajagopal. Conspectus of concepts of elasticity. Mathematics and Mechanics of Solids, 16:536–562, 2011.
[31] R. Bustamante and K. R. Rajagopal. A review of implicit constitutive theories to describe the response of elastic bodies. In Constitutive Modeling of Solid Continua, 2020.
[32] S. Timoshenko. History of the Strength of Materials. Dover Publications, New York, 1983.
[33] C. Truesdell and R. Toupin. The classical field theories. In Principles of Classical Mechanics and Field Theory/Prinzipien der Klassischen Mechanik und Feldtheorie, pp. 226–858, Springer, 1960.
[34] C. Truesdell and W. Noll. The Non-Linear Field Theories of Mechanics. Handbuch der Physik, III/2. Springer, Berlin, 1965.
[35] D. Goldstein. States of Matter. Dover Publications Inc., New York, 1985.
[36] P. W. Anderson. Concepts in Solids: Lectures on the Theory of Solids, Vol. 58. World Scientific, 1997.

1.2 Preliminary remarks

In the systematic development of a subject, we need to have a clear grasp of the concepts that are assumed to be primitives, namely the concepts for which we have an intuitive understanding (in mathematics what is referred to as the axioms) on which the whole subject is erected. In our current discourse, Space, Time, Mass and Force are physical notions that are primitive notions, concepts which are supposed to be manifest, self-evident. Notions such as displacement, velocity, deformation gradient, stretch, strain, stress, etc., quantities that are obtained in terms of the primitive quantities are amongst the many derived quantities. In addition to the above mentioned primitive concepts on which the edifice of elasticity is raised, concepts such as particles and points are also presumed to be intuitive ideas, but unfortunately, they are far from being intuitive, not that the concepts of space, time, mass and force are any the easier to comprehend. We cannot however get into a discussion of the motivation for these notions but take them for granted as ideas that we intuitively understand.

A central concept in the study of mechanics is the notion of a body. In classical Newtonian particle physics, the body in question is a particle, a particle being a point mass. The problem is that the notion of a point is far from clear in mathematics[3] and hence the notion of a particle is far from clear. When one is referring to a rigid body, one views the body as having finite size, with the distance between any two particles of the rigid body being fixed, the distance incapable of changing. Finally, we come to the way in which the notion of a body is used in continuum mechanics wherein the body is considered as a continuum, requiring us to understand what is meant by a continuum. This notion is not as easy to comprehend as one might think. In fact, for us, continuum is a primitive concept. No real body is a continuum in the sense in which we will use the notion, for in a real body, between any two particles belonging to the body, there might be no other particle touching both of them, that is there are gaps between particle, even if the particles are sub-atomic particles. In fact, there are more such gaps than the so called particles in a body. In reality, a body is a set of disjoint sub-bodies that are called particles. The reason for calling particles as sub-bodies is a consequence that they are really of finite size, albeit small in size. We "homogenize" over the region which we believe the body is occupying so that there are no gaps at the microscopic level, unless we wish to have a macroscopic gap, and call this homogenized body the continuum.

In these lecture notes, we will confine our discussion to purely mechanical issues. We will not concern ourselves with the response of the body to thermal, electrical or magnetic stimuli.

By a body, we refer to a set that has two mathematical structures, a measure, and a topology. The members of the set are particles, and the measure is mass. The topol-

3 Mathematicians and logicians have grappled with this notion and now one finds a definition for the same in terms of lattice theory as a "completely prime filter" (see [2]) which is not the sense in which any person involved in the task of doing mechanics thinks of the concept.

ogy that is endowed on the body is that of open balls in a three-dimensional Euclidean space and in the rigorous development of the subject the body is assumed to be compact. The notion of density is a consequence of the relationship between the volume measure and mass measure. A student interested in details regarding the same can find them in Truesdell [1].

In these lecture notes we will be considering bodies that are deformable, that is, the distance between the particles can change and the angle between material filaments in the body can change. The meaning that we give to the notions of Time and Space is that which has been given by Newton [3]. Newton believed in the concept of absolute time, its duration being measured in terms of hours, days, *etc.*, the notions of “before” and “after” being clearly defined, clearly demarcated on a one-dimensional manifold that is linearly ordered. Time does not speed up or slow down “flowing equably”. To Newton “Space” was immovable, through which bodies move, the movement being detected by our senses, with the notion of “distance” between points in space being clearly defined.

Also, causality is fundamental to our study of elasticity and in view of this we turn our attention to a discussion of this idea now.

References for Preliminaries

[1] C. Truesdell. A First Close in Rational Continuum Mechanics, Volume I. Academic Press, Boston-San Diego-New York, 2nd edition, 1977.
[2] J. Picado and A. Pultr. Frames and Locales: Topology Without Points. Springer Science & Business Media, 2011.
[3] I. Newton. Philosophiae Naturalis Principia Mathematica, Volume I. G. Brookman, 1833.

1.3 Role of causality

Causality and determinism are central tenets of Newtonian mechanics. With regard to continua, one applies forces to bodies and as a consequence the bodies deform. Constitutive relations describe how bodies respond to the application of stimuli, that is they are response relations. Hooke’s law is a constitutive relation, it describes how a certain class of bodies responds when subject to forces, when one restricts oneself to, loosely speaking, small deformations.

Our first acquaintance with the notion of elasticity is a perfectly elastic spring, the response of which is described by

$$F = kx, \tag{1.3.1}$$

where k is the spring constant, F is the applied force and x is the extension of the spring from its unstretched state. Then, we come across one-dimensional Hooke’s Law:

$$\sigma = E\varepsilon, \tag{1.3.2}$$

where σ is stress, ε is the *linearized* strain, and E is Young's modulus. Expressions such as Eq. (1.3.1) and Eq. (1.3.2) are not in keeping with the demands of *causality*. Force is the *cause* and extension is the *effect*. One should express the effect in terms of the cause rather than the cause in terms of the effect. That is, Eq. (1.3.1) ought to be expressed as

$$\forall F, \quad x = \frac{1}{k}F. \tag{1.3.3}$$

Similarly, Eq. (1.3.2) ought to be expressed as

$$\forall \sigma, \quad \varepsilon = (\sigma)/E. \tag{1.3.4}$$

That is, in both these cases, knowing the cause, we can determine the effect. This might seem a trivial observation, but it is not so and it has serious bearing on the development of constitutive relations in elasticity as will become apparent towards the end of the course. For the moment, let us rest satisfied by merely recognizing this issue.

Interestingly, with regard to the description of the response of elastic bodies undergoing small deformations, one finds the response relation being given as the stress being a function of the linearized strain, or the linearized strain being given as a function of the stress. While the former does not comply with the demands of causality, the latter does. The response relation wherein the strain is being expressed as a function of the stress is not a consequence of understanding the requirements that stem from causality, it is just mere happenstance. In the early development of gas laws like Boyle's law, Charles law, etc., it was quite common to express any one of the variables in terms of the other, not as a consequence of any clear philosophical ramifications.

We will see towards the end of the book, the proper attitude that we ought to take with regard to the development of constitutive relations is that the relevant physical quantities that play a role in the response of a body are all associated by a mathematical "relation".

Now, ε in Eq. (1.3.2) is interpreted as strain in the one-dimensional body of length L_0 being extended to the length l, but how exactly is this strain defined? In other words, is the strain ε defined through

$$\varepsilon = \frac{l - L_0}{L_0}? \tag{1.3.5}$$

or is it defined through

$$\varepsilon = \frac{l - L_0}{l}? \tag{1.3.6}$$

When the deformation is very small, by which we mean the numerator is small, the two formulae will both lead essentially to nearly the same answer. For materials like steel for sufficiently small stresses the two ratios are nearly the same. However, for materials like rubber, other elastomers, polymers, and tissues, *etc.*, the ratios can be very different, especially if the applied traction is sufficiently large. A rubber band could be stretched to five times its original length and the two different strains would differ by 320 %. To distinguish between the two expressions, we refer to Eq. (1.3.5) as the *Lagrangian strain* and Eq. (1.3.6) as the *Eulerian strain.*

Note. Hooke's Law is not really a "law", it is a constitutive relation that only holds when the displacement gradients, and hence the strains, are small. Displacement gradients being small imply both the strain and the rotation are small, we shall discuss this in some detail later. Hooke's Law does not describe the response of most elastic bodies when $l - L_0$ becomes large.

We can ask a similar question regarding the stress, σ that appears in Eq. (1.3.2). What does σ stand for in Hooke's Law? Is it

$$\sigma = \frac{F}{A_0}? \tag{1.3.7}$$

or

$$\sigma = \frac{F}{A}? \tag{1.3.8}$$

where A_0 and A are the cross-sectional areas of the cylinder in the stress free reference state, or the current deformed state of the body. These two ratios can be very different. Equation (1.3.7) is the Lagrangian measure of stress while Eq. (1.3.8) is the Eulerian measure of stress. When the extension is small, A_0 and A are nearly the same. Several tensorial stresses can be associated with the Lagrangian measure. The usual one is the Piola-Kirchoff stress. The usual Eulerian tensorial measure of stress is the Cauchy stress. As far as Hooke's law is concerned, it does not matter whether we use the Lagrangean or Eulerian measure for the stress and strains as long as we are consistent in our use of the quantities.

Let us consider a one-dimensional elastic body that has limited extensibility, that is, it cannot be extended beyond a certain limit. Consider the simple idealization considered in Fig. 1.1 of an elastic spring and an inextensible string. When the spring-inextensible string system is extended, energy is stored in the spring. When the spring-inextensible string system goes back to the unextended state all the energy stored in the spring is recovered. The spring-inextensible string system (body) is *elastic*. However, you cannot express the force as a function of the extension, that is,

$$F \neq f(x), \tag{1.3.9}$$

because for one value of extension $x = x_e$, there are infinity of corresponding F. On the other hand, we can express x as a function of F:

$$x = g(F). \tag{1.3.10}$$

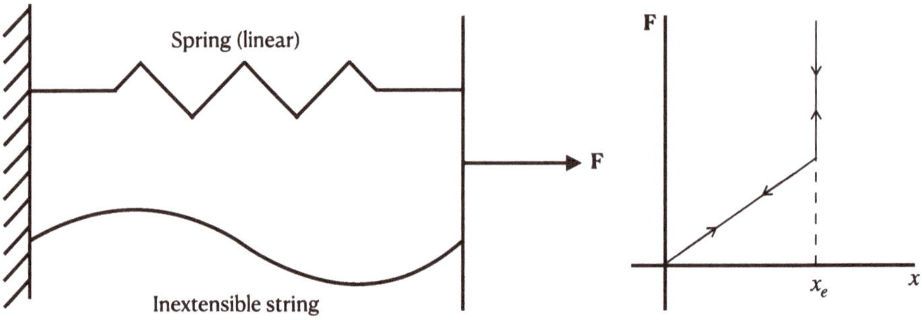

Figure 1.1: Idealization of elastic spring and inextensible string.

For each value of F, there is only one value of x. While the response depicted in the Fig. 1.1 can be expressed in terms of a function of the form

$$x = \begin{cases} \frac{1}{k}F & 0 \leq F \leq F_e \\ \frac{1}{k}F_e & F_e < F < \infty \end{cases} \tag{1.3.11}$$

This form is not invertible and the response cannot be expressed in terms of $F = f(x)$.[4]

Notice that in Eq. (1.3.10), we are expressing the effect in terms of the cause (this also does not always work as sometimes we have to express a relation between cause and effect in the sense of a mathematical relation). In classical non-linear elasticity (i. e., Cauchy/Green elasticity) one expresses the stress in terms of the deformation gradient/the stored energy in terms of the deformation gradient. Causality is turned upside down. As remarked earlier, in classical linearized elasticity one describes the stress in terms of the linearized strain and the linearized strain in terms of the stress as one does not consider the possibility of response of elastic bodies as that depicted in Fig. 1.1. These issues will be discussed toward the end of the lecture notes.

References

[1] K. R. Rajagopal. On implicit constitutive theories. Applications of Mathematics, 28(4):279–319, 2003.

4 See discussion in Rajagopal [1].

2 Kinematics

A detailed treatment of kinematics can be found in Truesdell [1], Malvern [2] and Jaunzemis [3].

2.1 Body

We will consider the abstract body, $\mathcal{B}$, to be composed of particles $P, Q, R, \ldots$ such that $\mathcal{B} := \{\ldots, P, Q, R, \ldots\}$. The idea of an abstract body undergoing deformation is illustrated in Fig. 2.1. In Fig. 2.1, $\kappa_R, \kappa_\tau, \kappa_t, \ldots$ are placers and $\kappa_R(\mathcal{B}), \kappa_\tau(\mathcal{B}), \kappa_t(\mathcal{B}), \ldots$ are configurations of the body in a three dimensional Euclidean space. Suppose that

$$\mathbf{X} = \kappa_R(P), \quad P \in \mathcal{B} \tag{2.1.1a}$$

$$\boldsymbol{\zeta} = \kappa_\tau(P), \quad P \in \mathcal{B} \tag{2.1.1b}$$

$$\mathbf{x} = \kappa_t(P), \quad P \in \mathcal{B}. \tag{2.1.1c}$$

We can assume any configuration as a reference configuration of the body, but one usually chooses the configuration in which the body is stress-free as the reference configuration as this is the configuration that an elastic body tends to when all the forces that are acting on the body are removed. With this in mind, it might be better to use

$$\mathbf{x}_R \quad \text{instead of } \mathbf{X} \tag{2.1.2a}$$

$$\mathbf{x}_\tau \quad \text{instead of } \boldsymbol{\zeta} \tag{2.1.2b}$$

$$\mathbf{x}_t \quad \text{instead of } \mathbf{x}, \tag{2.1.2c}$$

but this makes things unwieldy and cumbersome, so we shall not do so.

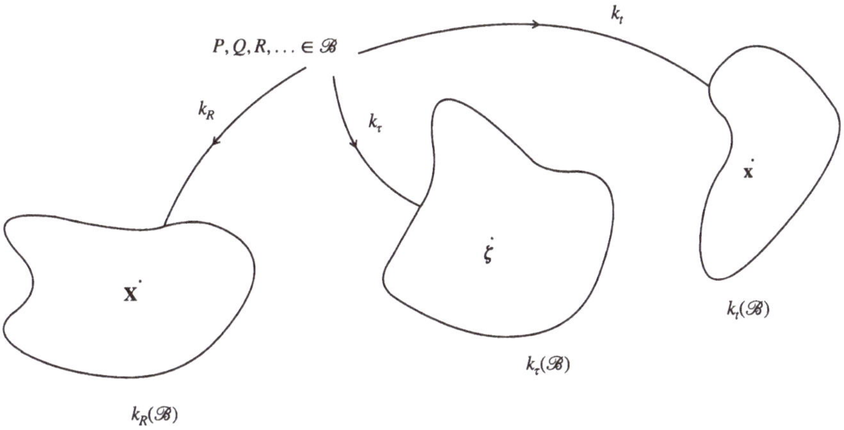

Figure 2.1: An abstract body that is undergoing deformation over time.

https://doi.org/10.1515/9783110789515-002

By motion what one means is a one-parameter family of placers. This immediately implies that we can express the motion as a function χ that assigns to each point $\mathbf{X} \in \kappa_R(\mathcal{B})$, a point $\mathbf{x} \in \kappa_t(\mathcal{B})$. Here, the subscript t is associated with the current time, $\kappa_t(\mathcal{B})$ is considered the current configuration at time t. Therefore, our motion is a mapping

$$\mathbf{x} = \chi(\mathbf{X}, t). \tag{2.1.3}$$

In general, we have to express the motion as

$$\mathbf{x} = \chi_{\kappa_R}(\mathbf{X}, t), \tag{2.1.4}$$

as the form of χ depends on κ_R. That is,

$$\chi_{\kappa_R} : \kappa_R(\mathcal{B}) \times \mathbb{R} \longrightarrow \mathcal{E} \tag{2.1.5}$$

where $\mathbb{R}$ represents the set of real numbers and $\mathcal{E}$ is the Euclidean point space. We shall henceforth drop the subscript κ_R, and assume that the mapping $\chi(\cdot, t)$ is one to one, that is no two particles in the reference configuration get mapped on to the same point $\mathbf{x}$ in the configuration at time t. Thus, we can clearly define

$$\mathbf{X} = \chi^{-1}(\mathbf{x}, t), \tag{2.1.6}$$

where χ^{-1} is the inverse of χ. We shall assume that our motion is sufficiently smooth to render all the derivatives that we take of it to be meaningful.

Next, we observe that

$$\xi = \chi(\mathbf{X}, \tau) = \chi(\chi^{-1}(\mathbf{x}, t), \tau) := \chi_t(\mathbf{x}, \tau). \tag{2.1.7}$$

Here, χ_t is referred to as the Relative Motion. The gradient of χ_t with respect to $\mathbf{x}$ is referred to as the Relative Deformation Gradient.

As ξ and $\mathbf{x}$ are both in configurations that are actually occupied by the body, they can be perceived by observers associated with two different frames. However, if as usually assumed by those following the approach due to Truesdell and Noll, if $\kappa_R(\mathcal{B})$ is not necessarily a configuration actually occupied by the body and merely a surrogate for the abstract body, as far as the two observers are concerned, they perceive members of $\kappa_R(\mathcal{B})$ as being the same, then one finds that the way the deformation gradient $\mathbf{F}$ (defined in section 2.4) transforms is different from how the relative gradient

$$\mathbf{F}_t(\mathbf{x}, \tau) := \frac{\partial \xi}{\partial \mathbf{x}} = \frac{\partial \chi_t(\mathbf{x}, \tau)}{\partial \mathbf{x}} \tag{2.1.8}$$

transforms due to a change of frame. The Cauchy theory of elasticity is formulated using the Cauchy stress to depend on the deformation gradient, however one can develop the theory for the response of elastic bodies using $\mathbf{F}_t(\mathbf{x}, \tau)$ instead of $\mathbf{F}(\mathbf{X}, t)$ (Rajagopal and Tao [4], see also the very general framework for elasticity introduced by Noll [5]).

2.2 Lagrangian and Eulerian descriptions of kinematical quantities

Any quantity defined on $\mathcal{B} \times \mathbb{R}$ can be associated with a quantity defined on $\kappa_R(\mathcal{B}) \times \mathbb{R}$, or $\kappa_t(\mathcal{B}) \times \mathbb{R}$, that is, any φ defined over the abstract body $\mathcal{B}$, at different instants of time, can be expressed as

$$\varphi(\mathcal{P}, t) = \hat{\varphi}(\mathbf{X}, t) = \tilde{\varphi}(\mathbf{x}, t), \tag{2.2.1}$$

where φ could stand for a scalar, vector, or tensor. We can define

$$\nabla_{\mathbf{X}}\varphi = \nabla\varphi = \frac{\partial\hat{\varphi}}{\partial\mathbf{X}}, \quad \frac{\mathrm{d}\varphi}{\mathrm{d}t} = \frac{\partial\hat{\varphi}}{\partial t}, \tag{2.2.2}$$

$$\nabla_{\mathbf{x}}\varphi = \operatorname{grad}\varphi = \frac{\partial\tilde{\varphi}}{\partial\mathbf{x}}, \quad \frac{\partial\varphi}{\partial t} = \frac{\partial\tilde{\varphi}}{\partial t}. \tag{2.2.3}$$

The quantities in (2.2.2) are referred to as Lagrangian expressions, while the quantities in (2.2.3) are referred to as Eulerian expressions. It immediately follows that

$$\frac{\mathrm{d}\varphi}{\mathrm{d}t} = \frac{\partial\varphi}{\partial t} + \nabla_{\mathbf{x}}\varphi \cdot \mathbf{v}, \tag{2.2.4}$$

if φ is a scalar, and

$$\frac{\mathrm{d}\boldsymbol{\varphi}}{\mathrm{d}t} = \frac{\partial\boldsymbol{\varphi}}{\partial t} + [\nabla_{\mathbf{x}}\boldsymbol{\varphi}]\mathbf{v}, \tag{2.2.5}$$

if $\boldsymbol{\varphi}$ is a vector. We can similarly define the derivatives with respect to $\mathbf{X}$ and $\mathrm{d}/\mathrm{d}t$ referred to as the Lagrangian spatial and temporal derivatives, respectively, while derivatives with respect to $\mathbf{x}$ and $\partial/\partial t$ are referred to as Eulerian spatial and temporal derivatives, respectively. The time derivative $\mathrm{d}/\mathrm{d}t$ is also referred to as the total time derivative, material time derivative, or substantial time derivative. The time derivative $\partial/\partial t$ is referred to as the partial time derivative. We could go ahead and define the Lagrangian and Eulerian derivatives of second order tensors and other higher order tensors in a manner similar to that above.

The notion of Lagrangian was first introduced by Euler, and the notion of Eulerian was used before Euler by D'Alembert and Daniel Bernoulli (see Truesdell [7]). In linearized elasticity, we will not make a distinction between these various derivatives. However, it is important to develop the ideas in general and then carry out the approximation rather than start directly with the approximation without understanding the distinction. In view of this, we shall start by considering a fully non-linear approach to elasticity and then carry out the appropriate approximations to arrive at classical linearized elasticity.

The idea of seeding particles in a flowing fluid and tracking those particles is a Lagrangian observation for determining the velocity. Using Laser Doppler Velocimeter for the velocity is an Eulerian measurement. Put loosely, Lagrangian information is historical information about some fixed material point, while Eulerian information is geographical information of the happenings in time at a fixed location in space. That is, path lines are Lagrangian specifications, while stream lines are Eulerian notions. On the other hand, a streak line is neither a Lagrangian nor an Eulerian specification.

2.3 Displacement and displacement gradient

The displacement $\mathbf{u}$ is defined as

$$\mathbf{u} = \mathbf{x} - \mathbf{X} \tag{2.3.1}$$

where $\mathbf{x}$ is the position in the current configuration and $\mathbf{X}$ is the position in the reference configuration. The displacement can be written in terms of the reference configuration or the current configuration by

$$\mathbf{u} = \hat{\mathbf{u}}(\mathbf{X}, t) = \mathbf{x} - \mathbf{X} = \chi(\mathbf{X}, t) - \mathbf{X} \tag{2.3.2}$$

for the Lagrangian (reference) expression or, similarly,

$$\mathbf{u} = \tilde{\mathbf{u}}(\mathbf{x}, t) = \mathbf{x} - \mathbf{X} = \mathbf{x} - \chi^{-1}(\mathbf{x}, t), \tag{2.3.3}$$

for the Eulerian expression.

2.4 Deformation gradient

The deformation gradient $\mathbf{F}$ is defined through

$$\hat{\mathbf{F}}(\mathbf{X}, t) = \frac{\partial \chi}{\partial \mathbf{X}}(\mathbf{X}, t). \tag{2.4.1}$$

The terminology "deformation gradient" is a misnomer as the right hand side of Eq. (2.4.1) is the gradient of the motion, however in view of its usage in the field we will refer to it as deformation gradient. In virtue of Eq. (2.1.6), we can write

$$\hat{\mathbf{F}}(\mathbf{X}, t) = \hat{\mathbf{F}}(\chi^{-1}(\mathbf{x}, t), t) = \tilde{\mathbf{F}}(\mathbf{x}, t) \tag{2.4.2}$$

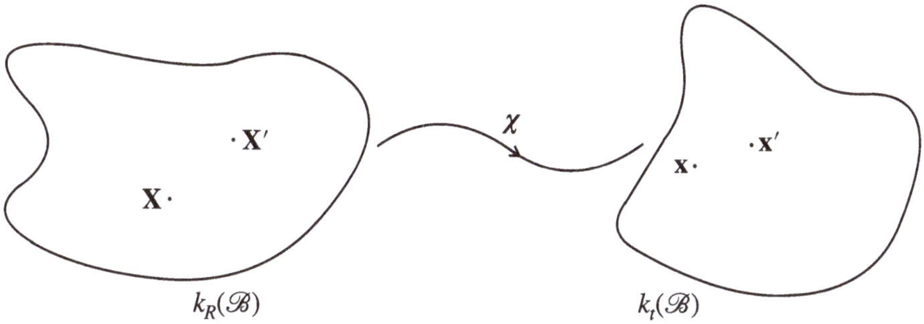

Figure 2.2: Representation of a motion between the reference configuration and the current configuration.

F is a linear transformation.[1] For the following development, we will use the interpretation illustrated in Fig. 2.2. We can write the difference between **x** and **x**′ in terms of the **X** as

$$\mathbf{x}' - \mathbf{x} = \chi(\mathbf{X}', t) - \chi(\mathbf{X}, t) \tag{2.4.3}$$

where $(\mathbf{x}' - \mathbf{x})$ belongs to a three dimensional vector space, $\mathcal{V}$, which is the translation space of $\mathcal{E}$. We can also express Eq. (2.4.3) as

$$\mathbf{x}' - \mathbf{x} = \left[\frac{\partial \chi}{\partial \mathbf{X}}(\mathbf{X}, t)\right][\mathbf{X}' - \mathbf{X}] + \mathcal{O}(|\mathbf{X}' - \mathbf{X}|^2), \tag{2.4.4}$$

where $||\cdot||$ is a norm on the vector space $\mathcal{V}$. That is, $\mathcal{V}$ is a normed vector space. Also, $\mathcal{O}(y)$ stands for something (some function) that is such that

$$\lim_{y \to 0} \frac{\mathcal{O}(y)}{y} \longrightarrow M, \tag{2.4.5}$$

where M is a positive real number. We can now write Eq. (2.4.3) in terms of the deformation gradient, **F**, as

$$\mathbf{x}' - \mathbf{x} = \mathbf{F}(\mathbf{X}' - \mathbf{X}) + \mathcal{O}(|\mathbf{X}' - \mathbf{X}|^2). \tag{2.4.6}$$

Physical meaning of F

If you think of $(\mathbf{X}' - \mathbf{X})$ as an infinitesimal vector $d\mathbf{X}$ (i. e., that is a vector of infinitesimal length), then $d\mathbf{x}$ is the infinitesimal vector such that

$$d\mathbf{x} = \mathbf{F}d\mathbf{X}. \tag{2.4.7}$$

---

1 It transforms vectors that lie in the tangent space at a point at which a particle lies in the reference configuration to vectors that lie in the tangent space at the point corresponding to the same particle in the deformed configuration. We shall not use the notion of tangent space, etc., but rather think in terms of infinitesimal directed line elements (infinitesimal filaments) that have their tails at the point.

That is, $\mathbf{F}$ is a linear transformation that rotates and elongates (or contracts) infinitesimal material filaments in the reference configuration to the infinitesimal material filaments in the current configuration. From just the mathematical standpoint, $\mathbf{F}$ can also provide an inversion in that its determinant is negative (mirror image reflection), but we shall not consider such a possibility in this course.

It is also useful (for future derivations and expressions) to express the gradient of $\mathbf{u}$ in both the reference and current configurations. The Lagrangian gradient or the gradient with respect to the reference configuration takes the form

$$\frac{\partial \mathbf{u}}{\partial \mathbf{X}} = \frac{\partial \hat{\mathbf{u}}}{\partial \mathbf{X}} = \frac{\partial \chi}{\partial \mathbf{X}} - \mathbf{I} = \mathbf{F} - \mathbf{I} \tag{2.4.8}$$

where $\mathbf{F} = \partial\chi/\partial\mathbf{X}$ is the deformation gradient and $\mathbf{I}$ is the identity tensor. We can repeat the process to arrive at the Eulerian expression, and the deformation gradient with respect to the current deformed state takes the form

$$\frac{\partial \mathbf{u}}{\partial \mathbf{x}} = \frac{\partial \tilde{\mathbf{u}}}{\partial \mathbf{x}} = \mathbf{I} - \frac{\partial \mathbf{X}}{\partial \mathbf{x}} = \mathbf{I} - \mathbf{F}^{-1}, \tag{2.4.9}$$

where $\partial\mathbf{X}/\partial\mathbf{x} \equiv \mathbf{F}^{-1}$.

Recall the definition of the determinant of a linear transformation $\mathbf{A}$ that acts on a three dimensional vector space:

$$\det \mathbf{A} = \frac{\mathbf{A}\mathbf{e}_1 \times \mathbf{A}\mathbf{e}_2 \cdot \mathbf{A}\mathbf{e}_3}{\mathbf{e}_1 \times \mathbf{e}_2 \cdot \mathbf{e}_3}, \tag{2.4.10}$$

where $\{\mathbf{e}_1, \mathbf{e}_2, \mathbf{e}_3\}$ is a basis for the vector space under consideration. Also recall that $|\mathbf{e}_1 \times \mathbf{e}_2 \cdot \mathbf{e}_3|$ is the volume of a parallelepiped with edges $\mathbf{e}_1$, $\mathbf{e}_2$, and $\mathbf{e}_3$. So, we can conclude that

$$|\det \mathbf{F}| = \frac{\text{Volume of an infinitesimal parallelepiped after deformation}}{\text{Volume of an infinitesimal parallelepiped before deformation}} \tag{2.4.11}$$

or more precisely, if $d\mathbf{X}_i$, $i = 1, 2, 3$, are infinitesimal filaments that form a parallelepiped, that are mapped to $d\mathbf{x}_i$ by the deformation gradient, then

$$|\det \mathbf{F}| = \frac{|d\mathbf{x}_1 \times d\mathbf{x}_2 \cdot d\mathbf{x}_3|}{|d\mathbf{X}_1 \times d\mathbf{X}_2 \cdot d\mathbf{X}_3|} = \frac{dv}{dV}. \tag{2.4.12}$$

If the $\det \mathbf{F} = 0$, then it means that a $\mathbf{F}$ maps the edges of a parallelepiped whose volume is non-zero to lie on a plane or collapse to a point, so that the volume of the deformed parallelepiped is zero. Since $\det \mathbf{F} = 1$ initially (e. g., if the initial configuration is the reference configuration) and since the $\det \mathbf{F}$ cannot be zero, the, continuity of $\det \mathbf{F}$ implies that $\det \mathbf{F} > 0$. It is important to note that within our framework, we cannot study a volume collapsing to a plane – since we shall assume that $\det \mathbf{F} \neq 0$, $\mathbf{F}^{-1}$ exists.

The definition (2.4.10) for the determinant of $\mathbf{A}$ is given in terms of the basis $\mathbf{e}_i$, $i = 1, 2, 3$. It can be shown that the determinant of a linear transformation is independent of the basis. Also, the definition (2.4.10) refers to a linear transformation that maps vectors from a three-dimensional vector space to another three-dimensional vector space. We recall that the notion of a vector product, and hence a scalar triple product, only holds in a three-dimensional vector space.[2] The definition of the determinant of a linear transformation mapping vectors from a n-dimensional vector space to another n-dimensional vector space can be found in the appendix.

Equation (2.4.12) informs us as to how an infinitesimal volume dV transforms into the infinitesimal volume dv in the deformed configuration at time t. We next determine how an infinitesimal area dA on a surface S_R in the reference configuration transforms to an infinitesimal area da on a surface S_t in the current configuration. Consider the infinitesimal area dA in the reference surface S_R containing the point $\mathbf{X}$, whose edges are infinitesimal filaments $\mathbf{dX}^{(1)}$ and $\mathbf{dX}^{(2)}$, along the directions of unit vectors $\mathbf{N}^{(1)}$ and $\mathbf{N}^{(2)}$ respectively (see Fig. 2.3). After deformation, let the infinitesimal filaments $\mathbf{dX}^{(1)}$ and $\mathbf{dX}^{(2)}$ transform to $\mathbf{dx}^{(1)}$ and $\mathbf{dx}^{(2)}$, respectively in $\kappa_t(\mathcal{B})$ (see Fig. 2.4). Let $\mathbf{dA}$ and $\mathbf{da}$ denote the vector area formed by the infinitesimal filaments $\mathbf{dX}^{(1)}$ and $\mathbf{dX}^{(2)}$, and $\mathbf{dx}^{(1)}$ and $\mathbf{dx}^{(2)}$, respectively. Thus

$$\mathbf{dA} = \mathbf{dX}^{(1)} \times \mathbf{dX}^{(2)}, \quad \mathbf{da} = \mathbf{dx}^{(1)} \times \mathbf{dx}^{(2)}. \tag{2.4.13}$$

We recall that

$$(da)_i = \epsilon_{ijk}(\mathbf{dx}^{(1)})_j(\mathbf{dx}^{(2)})_k, \tag{2.4.14}$$
$$= \epsilon_{ijk}(F_{ji}\mathbf{dX}_l^{(1)})(F_{km}\mathbf{dX}_m^{(2)}), \tag{2.4.15}$$
$$= \epsilon_{ijk}F_{jl}F_{km}\mathbf{dX}_l^{(1)}\mathbf{dX}_m^{(2)}, \tag{2.4.16}$$
$$= \epsilon_{njk}\delta_{ni}F_{jl}F_{km}\mathbf{dX}_l^{(1)}\mathbf{dX}_m^{(2)}, \tag{2.4.17}$$
$$= \epsilon_{njk}F_{np}F_{pi}^{-1}F_{jl}F_{km}\mathbf{dX}_l^{(1)}\mathbf{dX}_m^{(2)}, \tag{2.4.18}$$
$$= \epsilon_{njk}F_{np}F_{jl}F_{km}F_{pi}^{-1}\mathbf{dX}_l^{(1)}\mathbf{dX}_m^{(2)}. \tag{2.4.19}$$

It follows from (2.4.12) that

$$(da)_i = \epsilon_{plm}(\det \mathbf{F})F_{pi}^{-1}\mathbf{dX}_l^{(1)}\mathbf{dX}_m^{(2)}, \tag{2.4.20}$$
$$= (\det \mathbf{F})F_{pi}^{-1}(\epsilon_{plm}\mathbf{dX}_l^{(1)}\mathbf{dX}_m^{(2)}), \tag{2.4.21}$$
$$= (\det \mathbf{F})F_{pi}^{-1}(\mathbf{dX}^{(1)} \times \mathbf{dX}^{(2)})_p. \tag{2.4.22}$$

2 One can define a generalization of vector product, namely the exterior product, in general n-dimensional vector spaces.

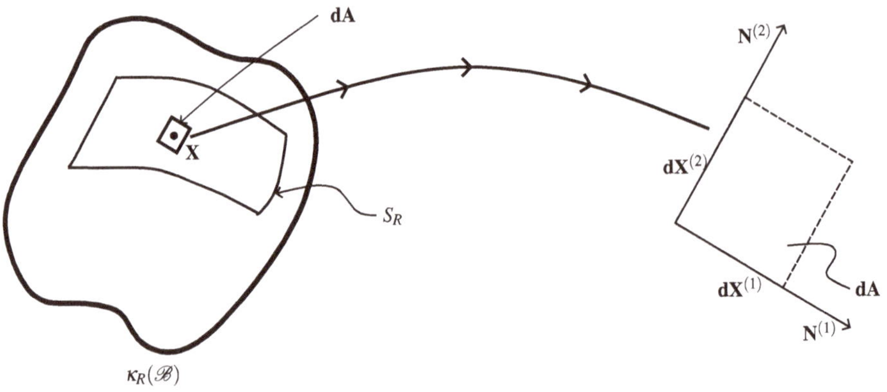

Figure 2.3: Infinitesimal area on a surface in the undeformed (reference) configuration.

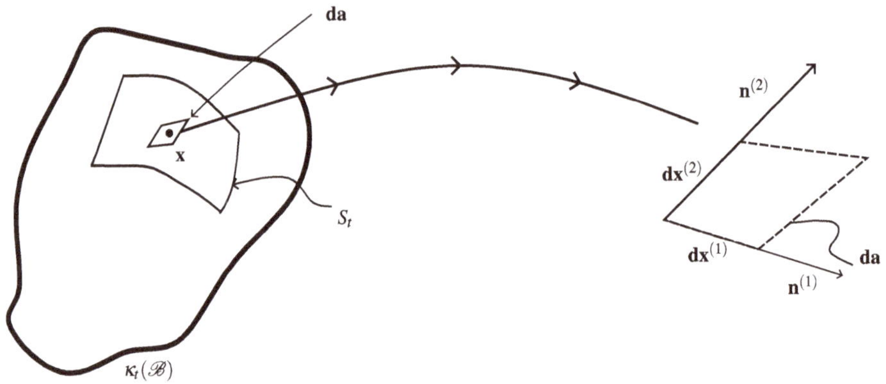

Figure 2.4: Transformation of infinitesimal area due to deformation.

Thus,

$$\mathbf{da} = (\det \mathbf{F})\mathbf{F}^{-T}\mathbf{dA}. \tag{2.4.23}$$

2.5 Velocity

The velocity field is defined through

$$\boldsymbol{\upsilon} = \frac{\partial \chi}{\partial t}(\mathbf{X}, t). \tag{2.5.1}$$

We must note that, like displacement, velocity has both Lagrangian and Eulerian expressions, i. e.,

$$\boldsymbol{\upsilon} = \hat{\boldsymbol{\upsilon}}(\mathbf{X}, t) = \tilde{\boldsymbol{\upsilon}}(\mathbf{x}, t). \tag{2.5.2}$$

2.6 Velocity gradient

The velocity gradient $\mathbf{L}(\mathbf{x}, t)$ is defined through

$$\mathbf{L}(\mathbf{x}, t) = \frac{\partial \boldsymbol{\upsilon}}{\partial \mathbf{x}} = \frac{\partial \tilde{\boldsymbol{\upsilon}}(\mathbf{x}, t)}{\partial \mathbf{x}}. \tag{2.6.1}$$

2.7 Acceleration

The acceleration field can be defined as

$$\mathbf{a} = \frac{\partial^2 \boldsymbol{\chi}}{\partial t^2} = \frac{\partial \hat{\boldsymbol{\upsilon}}}{\partial t} \tag{2.7.1}$$

It is important to note that

$$\mathbf{a} \neq \frac{\partial \tilde{\boldsymbol{\upsilon}}}{\partial t}. \tag{2.7.2}$$

We note that

$$\mathbf{a} = \frac{\partial \hat{\boldsymbol{\upsilon}}}{\partial t} = \frac{d\boldsymbol{\upsilon}}{dt}, \tag{2.7.3}$$

$$= \frac{\partial \tilde{\boldsymbol{\upsilon}}}{\partial t} + \frac{\partial \tilde{\boldsymbol{\upsilon}}}{\partial \mathbf{x}} \frac{d\mathbf{x}}{dt}, \tag{2.7.4}$$

$$= \frac{\partial \tilde{\boldsymbol{\upsilon}}}{\partial t} + \mathbf{L}\tilde{\boldsymbol{\upsilon}}. \tag{2.7.5}$$

The acceleration has both a Lagrangian and Eulerian form:

$$\mathbf{a} = \hat{\mathbf{a}}(\mathbf{X}, t) = \tilde{\mathbf{a}}(\mathbf{x}, t). \tag{2.7.6}$$

2.8 Polar decomposition theorem

We will consider vector spaces defined over the field of $\mathbb{R}$. A linear transformation, $\mathbf{A}$, is said to be positive, semi-definite if

$$(\mathbf{A}\mathbf{x} \cdot \mathbf{x}) \geq 0, \quad \forall \, \mathbf{x} \in \mathcal{V}. \tag{2.8.1}$$

If

$$(\mathbf{A}\mathbf{x} \cdot \mathbf{x}) \geq 0, \quad \forall \, \mathbf{x} \in \mathcal{V}, \tag{2.8.2}$$

and if

$$\mathbf{A}\mathbf{x} \cdot \mathbf{x} = 0 \implies \mathbf{x} = \mathbf{0}, \tag{2.8.3}$$

then **A** is said to be positive definite. If a linear transformation **A** is such that

$$\mathbf{A}\mathbf{A}^{\mathrm{T}} = \mathbf{I}, \tag{2.8.4}$$

then **A** is said to be an orthogonal transformation. If **A** is orthogonal, then we notice that

$$\mathbf{A}^{\mathrm{T}} = \mathbf{A}^{-1}. \tag{2.8.5}$$

Since the motion is invertible and the deformation gradient is invertible, it follows from the Polar Decomposition Theorem (see Halmos [6] for details concerning the same) that the deformation gradient, **F**, can be decomposed into

$$\mathbf{F} = \mathbf{R}\mathbf{U} = \mathbf{V}\mathbf{R}, \tag{2.8.6}$$

where **U**, **V** are positive definite and symmetric and **R** is orthogonal. All three transformations **U**, **V** and **R** are unique. If the deformation gradient is not invertible, then we are not guaranteed the uniqueness of **R**. Here, **U** is referred to as the "right stretch", **V** is referred to as the "left stretch", and **R** is an orthogonal transformation. It is a "rotation" if the det **R** = +1, and it is an inversion if the det **R** = −1.

We observed earlier that since **F** is non-singular and continuous and det **F** = 1 in the reference configuration, it follows that det **F** has to be non-negative. Thus, **R** corresponds to a rotation. Eq. (2.8.6) essentially states that **F** is a

- stretch followed by a rotation (**RU**), or
- rotation followed by a stretch (**VR**).

That is,

$$\mathbf{F}d\mathbf{X} = \mathbf{R}(\mathbf{U}d\mathbf{X}) = \mathbf{V}(\mathbf{R}d\mathbf{X}) \tag{2.8.7}$$

where $\mathbf{R}(\mathbf{U}d\mathbf{X})$ is stretch of $d\mathbf{X}$ followed by a rotation **R**, that is, $\mathbf{R}(\mathbf{U}d\mathbf{X})$, and $\mathbf{V}(\mathbf{R}d\mathbf{X})$ is a rotation of $d\mathbf{X}$ followed by a stretch.

Suppose that v_i are the eigen-values and $\mathbf{u}_i$ are the eigen-vectors of **U**, $i = 1, 2, 3$. Now, since **RU** = **VR**, we can write

$$\mathbf{V}\mathbf{R}\mathbf{u}_i = \mathbf{R}\mathbf{U}\mathbf{u}_i = \mathbf{R}(v_i\mathbf{u}_i) \quad \text{no sum on } i, i = 1, 2, 3 \tag{2.8.8}$$

So

$$\mathbf{V}(\mathbf{R}\mathbf{u}_i) = v_i(\mathbf{R}\mathbf{u}_i) \quad \text{no sum on } i, i = 1, 2, 3. \tag{2.8.9}$$

This means v_i are the eigen-values of **V** and $(\mathbf{R}\mathbf{u}_i)$ are the eigen-vectors of **V**. Thus, the deformation gradient takes vectors that are in the eigen-vector direction of **U**, into the directions of the eigen-vector of **V**.

Let us consider the material fibers satisfying $|\mathbf{dX}| = \text{constant}$. This defines the surface of a sphere. We are interested in how the sphere deforms. In this sphere, let us consider three infinitesimal filaments along the directions of the eigen-vectors of **U**. That is, we choose $\mathbf{dX}^{(i)} = \mathbf{u}_i|\mathbf{dX}^{(i)}| = \mathbf{u}_i|\mathbf{dX}|$, no sum on i, $i = 1, 2, 3$. Next,

$$\mathbf{dx}^{(i)} = \mathbf{RU}(\mathbf{u}_i|\mathbf{dX}|), \quad i = 1, 2, 3 \tag{2.8.10a}$$

$$= \mathbf{R}|\mathbf{dX}|\mathbf{Uu}_i \tag{2.8.10b}$$

$$= \mathbf{R}|\mathbf{dX}|\nu_i\mathbf{u}_i \quad \text{no sum on } i,\ i = 1, 2, 3 \tag{2.8.10c}$$

$$= \nu_i|\mathbf{dX}|\mathbf{Ru}_i, \quad \text{no sum on } i,\ i = 1, 2, 3. \tag{2.8.10d}$$

Let $\mathbf{v}_i$ be the eigen-vectors of **V**. Then,

$$\mathbf{dx}^{(i)} = \nu_i\mathbf{v}_i|\mathbf{dX}| \quad \text{no sum on } i, i = 1, 2, 3. \tag{2.8.11}$$

Therefore, the infinitesimal filaments which before deformation were along the eigen-vectors of **U** are now along the eigen-vectors of **V**. The sphere has now deformed into and ellipsoid whose principal axes are along the direction of the eigen-vectors of **V**. Since we are considering the sphere $|\mathbf{dX}| = \text{constant}$, it follows that

$$|\mathbf{dX}| = (\mathbf{dX} \cdot \mathbf{dX})^{1/2} \tag{2.8.12a}$$

$$= (\mathbf{F}^{-1}\mathbf{dx} \cdot \mathbf{F}^{-1}\mathbf{dx})^{1/2} \tag{2.8.12b}$$

$$= (\mathbf{F}^{-T}\mathbf{F}^{-1}\mathbf{dx} \cdot \mathbf{dx})^{1/2} \tag{2.8.12c}$$

where, we have used the definition of the transpose and the notation, $\mathbf{F}^{-T} = (\mathbf{F}^{-1})^T$.

Now, let us define

$$\mathbf{B} = \mathbf{FF}^T, \tag{2.8.13}$$

where **B** is called the *left Cauchy-Green tensor.*[3] Then, we can rewrite Eq. (2.8.12c) as

$$|\mathbf{dX}| = (\mathbf{B}^{-1}\mathbf{dx} \cdot \mathbf{dx})^{1/2}. \tag{2.8.14}$$

In like fashion, we can define

$$\mathbf{C} = \mathbf{F}^T\mathbf{F}, \tag{2.8.15}$$

where **C** is referred to as the right Cauchy-Green tensor. We notice that

---

3 The reason it is called the left Cauchy-Green tensor is because **B** is the square of the Left stretch tensor. Some authors refer to it as the Cauchy-Green stretch tensor, but this is not correct as it is the square of the stretch. A similar remark applies to **C**, it is not a stretch but the square of the stretch.

$$\mathbf{B} = \mathbf{V}^2 \tag{2.8.16}$$

and

$$\mathbf{C} = \mathbf{U}^2. \tag{2.8.17}$$

We now derive how $\mathbf{C}$ is related to $\partial\hat{\mathbf{u}}/\partial\mathbf{X}$ and how $\mathbf{B}$ is related to $\partial\tilde{\mathbf{u}}/\partial\mathbf{x}$. Using Eq. (2.4.8) and Eq. (2.8.15), we can write $\mathbf{C}$ in terms of the displacement gradient:

$$\mathbf{C} = \left(\mathbf{I} + (\nabla\mathbf{u})^{\mathrm{T}}\right)(\mathbf{I} + \nabla\mathbf{u}) \tag{2.8.18a}$$

$$= \mathbf{I} + (\nabla\mathbf{u}) + (\nabla\mathbf{u})^{\mathrm{T}} + (\nabla\mathbf{u})^{\mathrm{T}}(\nabla\mathbf{u}) \tag{2.8.18b}$$

Notice that $\mathbf{C}$ is a nonlinear function of the displacement gradient. Similarly, we can use Eq. (2.4.9) and Eq. (2.8.13) to express $\mathbf{B}$ in terms of the Eulerian displacement gradient:

$$\mathbf{B}^{-1} = \left[\mathbf{F}\mathbf{F}^{\mathrm{T}}\right]^{-1} \tag{2.8.19a}$$

$$= \mathbf{F}^{-\mathrm{T}}\mathbf{F}^{-1} \tag{2.8.19b}$$

$$= (\mathbf{I} - \operatorname{grad}\mathbf{u})^{\mathrm{T}}(\mathbf{I} - \operatorname{grad}\mathbf{u}) \tag{2.8.19c}$$

$$= \mathbf{I} - \operatorname{grad}\mathbf{u} - (\operatorname{grad}\mathbf{u})^{\mathrm{T}} + (\operatorname{grad}\mathbf{u})^{\mathrm{T}}(\operatorname{grad}\mathbf{u}). \tag{2.8.19d}$$

2.9 Green-St. Venant strain

The Cauchy-Green tensors, $\mathbf{B}$ and $\mathbf{C}$, are not measures of strain. When there is no deformation, we expect the strain to be zero, and this is not the case with regard to $\mathbf{C}$ and $\mathbf{B}$. When there is no deformation, the deformation gradient is the identity tensor and thus both the Cauchy-Green tensors are also identity tensors.

Next, let us consider the ratio

$$\frac{|d\mathbf{x}|^2 - |d\mathbf{X}|^2}{|d\mathbf{X}|^2}, \tag{2.9.1}$$

where $d\mathbf{X} = \mathbf{N}|d\mathbf{X}|$, where $\mathbf{N}$ is a unit vector associated with the reference configuration (see Figure 2.5). We now seek to express the numerator in terms of the reference configuration:

$$|d\mathbf{x}|^2 - |d\mathbf{X}|^2 = \left[(\mathbf{F}d\mathbf{X}\cdot\mathbf{F}d\mathbf{X})^{1/2}\right]^2 - |d\mathbf{X}|^2 \tag{2.9.2a}$$

$$= \left(\mathbf{F}^{\mathrm{T}}\mathbf{F}d\mathbf{X}\cdot d\mathbf{X}\right) - |d\mathbf{X}|^2 \tag{2.9.2b}$$

$$= (\mathbf{C}d\mathbf{X}\cdot d\mathbf{X}) - |d\mathbf{X}|^2 \tag{2.9.2c}$$

$$= |d\mathbf{X}|^2(\mathbf{C}\mathbf{N}\cdot\mathbf{N} - \mathbf{I}) \tag{2.9.2d}$$

$$= |d\mathbf{X}|^2(\mathbf{C} - \mathbf{I})\mathbf{N}\cdot\mathbf{N}. \tag{2.9.2e}$$

So, we can now say that

$$\frac{|\mathbf{dx}|^2 - |\mathbf{dX}|^2}{|\mathbf{dX}|^2} = (\mathbf{C} - \mathbf{I})\mathbf{N} \cdot \mathbf{N}. \tag{2.9.3}$$

Let us define

$$\mathbf{E} = \frac{1}{2}(\mathbf{C} - \mathbf{I}), \tag{2.9.4}$$

where $\mathbf{E}$ is called the Green-St. Venant strain. Notice that when there is no deformation, $\mathbf{F} = \mathbf{I}$, $\mathbf{C} = \mathbf{I}$ and $\mathbf{E} = 0$. Also, notice $\mathbf{E}$ is symmetric. It follows from Eq. (2.8.18b) and Eq. (2.9.4) that

$$\mathbf{E} = \frac{1}{2}[(\nabla\mathbf{u}) + (\nabla\mathbf{u})^{\mathrm{T}} + (\nabla\mathbf{u})^{\mathrm{T}}(\nabla\mathbf{u})], \tag{2.9.5}$$

or, in indicial notation,

$$E_{ij} = \frac{1}{2}\left[\frac{\partial u_i}{\partial X_j} + \frac{\partial u_j}{\partial X_i} + \frac{\partial u_k}{\partial X_i}\frac{\partial u_k}{\partial X_j}\right]. \tag{2.9.6}$$

2.10 Almansi-Hamel strain

Next, let us consider the ratio

$$\frac{|\mathbf{dx}|^2 - |\mathbf{dX}|^2}{|\mathbf{dx}|^2}, \tag{2.10.1}$$

but now let us consider the case when $\mathbf{dx} = \mathbf{n}|\mathbf{dx}|$, where $\mathbf{n}$ associated with the deformed configuration at time t (see Figure 2.5). That is

$$\frac{|\mathbf{dx}|^2 - |\mathbf{dX}|^2}{|\mathbf{dx}|^2} = \frac{|\mathbf{dx}|^2 - (\mathbf{F}^{-1}\mathbf{dx} \cdot \mathbf{F}^{-1}\mathbf{dx})}{|\mathbf{dx}|^2} \tag{2.10.2a}$$

$$= \frac{|\mathbf{dx}|^2(\mathbf{I} - \mathbf{B}^{-1}\mathbf{n} \cdot \mathbf{n})}{|\mathbf{dx}|^2} \tag{2.10.2b}$$

$$= (\mathbf{I} - \mathbf{B}^{-1})\mathbf{n} \cdot \mathbf{n}. \tag{2.10.2c}$$

Again, when there is no deformation $\mathbf{F} = \mathbf{I}$, $\mathbf{B} = \mathbf{I}$ and $(\mathbf{I} - \mathbf{B}^{-1}) = \mathbf{0}$. Let us define the Almansi-Hamel strain $\boldsymbol{e}$ as

$$\boldsymbol{e} = \frac{1}{2}(\mathbf{I} - \mathbf{B}^{-1}). \tag{2.10.3}$$

When there is no deformation, $\boldsymbol{e} = 0$. Also, note that $\boldsymbol{e}$ is symmetric. It follows from Eq. (2.8.19d) and Eq. (2.10.3) that

$$\boldsymbol{e} = \frac{1}{2}[(\operatorname{grad}\mathbf{u}) + (\operatorname{grad}\mathbf{u})^{\mathrm{T}} - (\operatorname{grad}\mathbf{u})^{\mathrm{T}}(\operatorname{grad}\mathbf{u})] \tag{2.10.4}$$

that is

$$e_{ij} = \frac{1}{2}\left[\frac{\partial u_i}{\partial x_j} + \frac{\partial u_j}{\partial x_i} - \left(\frac{\partial u_k}{\partial x_i}\right)\left(\frac{\partial u_k}{\partial x_j}\right)\right]. \tag{2.10.5}$$

2.11 Problems

2.1 Show that $\mathbf{B}^{-1}\mathrm{d}\mathbf{x}\cdot\mathrm{d}\mathbf{x} =$ constant defines an ellipsoid.

2.2 If $\mathbf{B}$ is positive definite, prove that $\mathbf{B}^{-1}$ is positive definite. Show that the eigen values of $\mathbf{B}$ and $\mathbf{B}^{-1}$ are v_i^2 and $1/v_i^2$, respectively. What are the eigen-vectors of $\mathbf{B}$ and $\mathbf{B}^{-1}$?

2.3 Obtain expressions for the $\operatorname{tr}\mathbf{B}$, $1/2[(\operatorname{tr}\mathbf{B})^2 - \operatorname{tr}\mathbf{B}^2]$ and $\det\mathbf{B}$, in terms of the eigen values of $\mathbf{V}$.

2.12 Stretch along a particular direction

We will now study the stretch undergone by an infinitesimal filament, which after deformation is along the direction $\mathbf{n}$ in the current configuration at time t. Let $\lambda_{\mathbf{n}}$ denote the ratio of the current length of an infinitesimal filament that is currently along the direction $\mathbf{n}$, to its original length, at the point $\mathbf{x}$ in the current configuration, that is

$$\lambda_{\mathbf{n}} = \frac{|\mathrm{d}\mathbf{x}|}{|\mathrm{d}\mathbf{X}|}, \tag{2.12.1}$$

where $\mathbf{n}$ is a unit vector, $\mathrm{d}\mathbf{x} = \mathbf{n}|\mathrm{d}\mathbf{x}|$, $\mathrm{d}\mathbf{x}$ being the infinitesimal filament along the direction $\mathbf{n}$ shown in Fig. 2.5. $\lambda_{\mathbf{n}}$ is referred to as the stretch at a point $\mathbf{x}$ of an infinitesimal filament along the direction $\mathbf{n}$. It follows from Eq. (2.12.1) and the definition of the mag-

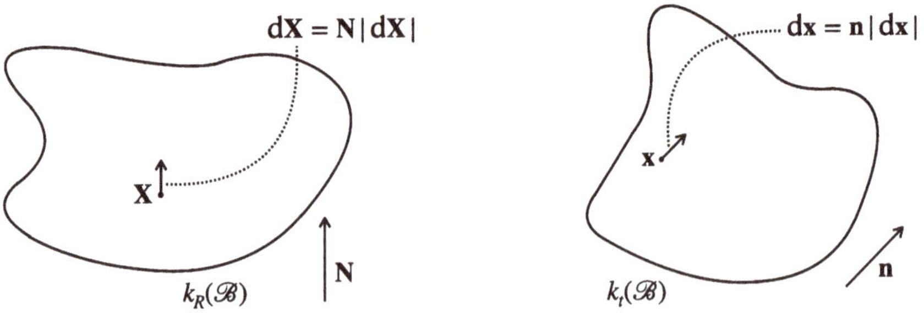

Figure 2.5: Infinitesimal filament shown in both reference and current configurations.

nitude of a vector, we can express the stretch along the direction $\mathbf{n}$ as

$$\lambda_{\mathbf{n}} = \frac{|\mathrm{d}\mathbf{x}|}{|\mathrm{d}\mathbf{X}|} = \frac{|\mathrm{d}\mathbf{x}|}{(\mathbf{B}^{-1}\mathrm{d}\mathbf{x}\cdot\mathrm{d}\mathbf{x})^{1/2}} \tag{2.12.2a}$$

$$= \frac{|\mathrm{d}\mathbf{x}|}{(\mathbf{B}^{-1}\mathbf{n}|\mathrm{d}\mathbf{x}|\cdot\mathbf{n}|\mathrm{d}\mathbf{x}|)^{1/2}} \tag{2.12.2b}$$

$$= \frac{1}{(\mathbf{B}^{-1}\mathbf{n}\cdot\mathbf{n})^{1/2}}. \tag{2.12.2c}$$

Next, let us determine the stretch at the material point $\mathbf{X}$ in the reference configuration on an infinitesimal filament which is along the direction $\mathbf{N}$, when the body is subject to deformation:

$$\lambda_{\mathbf{N}} = \frac{|\mathrm{d}\mathbf{x}|}{|\mathrm{d}\mathbf{X}|} = \frac{(\mathbf{F}\mathrm{d}\mathbf{X}\cdot\mathbf{F}\mathrm{d}\mathbf{X})^{1/2}}{|\mathrm{d}\mathbf{X}|} \tag{2.12.3a}$$

$$= \frac{(\mathbf{F}^{\mathrm{T}}\mathbf{F}\mathrm{d}\mathbf{X}\cdot\mathrm{d}\mathbf{X})^{1/2}}{|\mathrm{d}\mathbf{X}|} \tag{2.12.3b}$$

$$= \frac{(\mathbf{C}\mathbf{N}|\mathrm{d}\mathbf{X}|\cdot\mathbf{N}|\mathrm{d}\mathbf{X}|)^{1/2}}{|\mathrm{d}\mathbf{X}|} \tag{2.12.3c}$$

$$= (\mathbf{C}\mathbf{N}\cdot\mathbf{N})^{1/2}. \tag{2.12.3d}$$

2.13 Shear

Let $\mathbf{N}_I$ and $\mathbf{N}_J$ be two unit vectors perpendicular to one another, and let us consider two infinitesimal fibers in the reference configuration $\kappa_R(\mathcal{B})$ along the direction of $\mathbf{N}_I$ and $\mathbf{N}_J$, $I \neq J$, respectively. We are interested in determining how the angle between the two infinitesimal fibers change due to deformation (see Fig. 2.6).

Let the infinitesimal fibers which were along the directions $\mathbf{N}_I$ and $\mathbf{N}_J$ in the reference configuration, after deformation, be along the directions $\mathbf{n}_i$ and $\mathbf{n}_j$, in the deformed

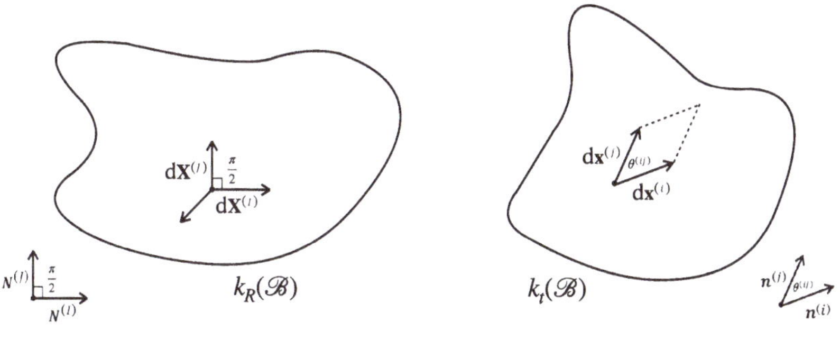

Figure 2.6: Shear which two infinitesimal material filaments currently making the angle $\theta^{(ij)}$ have undergone, when in the undeformed configuration the corresponding infinitesimal material filaments were along the direction of the orthonormal vectors $\mathbf{N}^{(I)}$ and $\mathbf{N}^{(J)}$.

configuration. Let us then define

$$\gamma_{\mathbf{N}_I\mathbf{N}_J} := \gamma^{(IJ)} = \frac{\pi}{2} - \theta^{(ij)} \begin{cases} > 0 & \text{if angle between } \mathbf{N}_I \text{ and } \mathbf{N}_J \text{ closes} \\ < 0 & \text{if angle between } \mathbf{N}_I \text{ and } \mathbf{N}_J \text{ opens} \end{cases} \tag{2.13.1}$$

Since $d\mathbf{X}^{(I)} = \mathbf{N}^{(I)}|d\mathbf{X}^{(I)}|$, and $d\mathbf{X}^{(J)} = \mathbf{N}^{(J)}|d\mathbf{X}^{(J)}|$, no sum of I or J, we find that

$$\cos\theta^{(ij)} = \frac{\mathbf{C}\mathbf{N}^{(I)}\cdot\mathbf{N}^{(J)}}{\lambda_{\mathbf{N}^{(I)}}\lambda_{\mathbf{N}^{(J)}}}, \quad I \neq J. \tag{2.13.2}$$

Since $\mathbf{N}^{(I)}$ and $\mathbf{N}^{(J)}$ are orthonormal, in virtue of Eq. (2.9.4),

$$\cos\theta^{(ij)} = \frac{2\mathbf{E}\mathbf{N}^{(I)}\cdot\mathbf{N}^{(J)}}{\lambda_{\mathbf{N}^{(I)}}\lambda_{\mathbf{N}^{(J)}}}. \tag{2.13.3}$$

In indicial notation, if $\mathbf{e}_I$, $I = 1,2,3$ is a basis,

$$\cos\theta^{(ij)} = \frac{2E_{IJ}}{(1+2E_{II})^{(1/2)}(1+2E_{JJ})^{(1/2)}}, \quad I,J = 1,2,3,\ I \neq J, \tag{2.13.4}$$

where $\mathbf{E} = E_{IJ}\mathbf{e}_I \otimes \mathbf{e}_J$. Thus, by Eq. (2.13.1),

$$\sin\gamma_{\mathbf{N}_I\mathbf{N}_J} = \cos\theta^{(ij)} = \frac{2E_{IJ}}{(1+2E_{II})^{(1/2)}(1+2E_{JJ})^{(1/2)}}, \quad I,J = 1,2,3,\ I \neq J. \tag{2.13.5}$$

If $\gamma_{\mathbf{N}_I\mathbf{N}_J}$ is small, then

$$\gamma_{\mathbf{N}_I\mathbf{N}_J} = \sin\gamma_{\mathbf{N}_I\mathbf{N}_J} = \frac{2E_{IJ}}{(1+2E_{II})^{(1/2)}(1+2E_{JJ})^{(1/2)}}, \quad I,J = 1,2,3,\ I \neq J. \tag{2.13.6}$$

Next, consider two infinitesimal fibers that are along the directions $\mathbf{n}^{(i)}$ and $\mathbf{n}^{(j)}$, where $\mathbf{n}^{(i)}$ and $\mathbf{n}^{(j)}$ are orthonormal vectors (see Fig. 2.7).

Let $\Theta^{(IJ)}$ denote the angle between these two infinitesimal fibers make in the undeformed reference configuration. Let us define

$$\gamma_{\mathbf{n}^{(i)}\mathbf{n}^{(j)}} = \Theta^{(IJ)} - \frac{\pi}{2}. \tag{2.13.7}$$

Then

$$\cos\Theta^{(IJ)} = \frac{d\mathbf{X}^{(I)}\cdot d\mathbf{X}^{(J)}}{|d\mathbf{X}^{(I)}||d\mathbf{X}^{(J)}|}, \quad I,J = 1,2,3,\ I \neq J, \tag{2.13.8}$$

$$= \frac{2\mathbf{e}\mathbf{n}^{(i)}\cdot\mathbf{n}^{(j)}}{\lambda_{\mathbf{n}^{(i)}}\lambda_{\mathbf{n}^{(j)}}} \tag{2.13.9}$$

$$= \frac{2\mathbf{e}\mathbf{n}^{(i)}\cdot\mathbf{n}^{(j)}}{(1-2\mathbf{e}\mathbf{n}^{(i)}\cdot\mathbf{n}^{(i)})^{(1/2)}(1-2\mathbf{e}\mathbf{n}^{(j)}\cdot\mathbf{n}^{(j)})^{(1/2)}}, \quad i,j = 1,2,3,\ i \neq j, \tag{2.13.10}$$

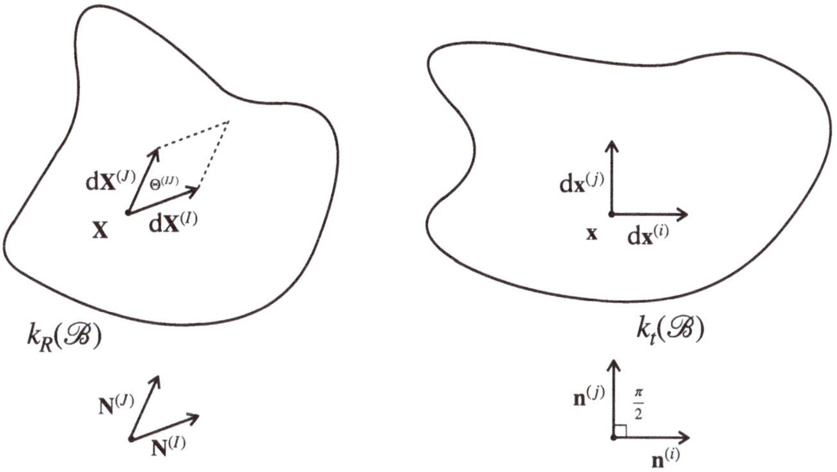

Figure 2.7: Shear which two infinitesimal material filaments currently along the orthonormal vectors $\mathbf{n}^{(i)}$ and $\mathbf{n}^{(j)}$ have undergone, when in the undeformed configuration the corresponding infinitesimal material filament made the angle $\Theta^{(IJ)}$.

which in indicial notation reads

$$\cos\Theta^{(IJ)} = \frac{2e_{ij}}{(1-2e_{ii})^{(1/2)}(1-2e_{jj})^{(1/2)}}, \quad i,j = 1,2,3, \ i \neq j. \tag{2.13.11}$$

2.14 Simple shear

Consider the transformation representing simple shear in Fig. 2.8. The motion that takes $(X_1, X_2, X_3) \to (x_1, x_2, x_3)$ is such that

$$x_1 = X_1 + \frac{\delta}{h}X_2, \quad x_2 = X_2, \quad x_3 = X_3 \tag{2.14.1}$$

where $(X_1, X_2, X_3) \in \kappa_R(\mathcal{B})$ and $(x_1, x_2, x_3) \in \kappa_t(\mathcal{B})$. This means that we can determine $\mathbf{F}$, whose matrix with respect to a rectangular coordinate system is given by

$$[\mathbf{F}] = \begin{bmatrix} 1 & \delta/h & 0 \\ 0 & 1 & 0 \\ 0 & 0 & 1 \end{bmatrix}. \tag{2.14.2}$$

From the deformation gradient, we can construct the gradient of the displacement whose matrix with respect to the rectangular Cartesian coordinate system is given by

$$[\nabla\mathbf{u}] = [\mathbf{F} - \mathbf{I}] = \begin{bmatrix} 0 & \delta/h & 0 \\ 0 & 0 & 0 \\ 0 & 0 & 0 \end{bmatrix}. \tag{2.14.3}$$

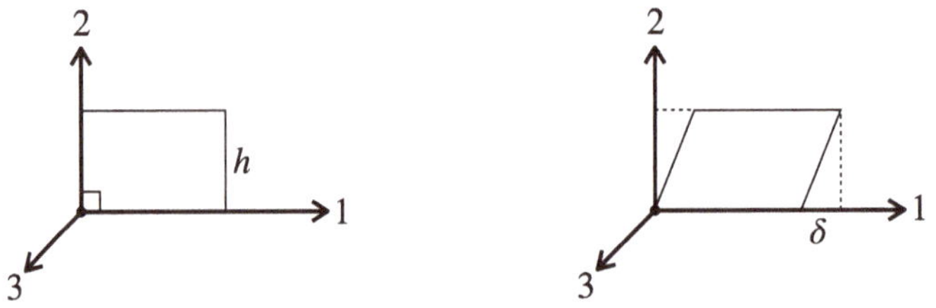

$$(X_1, X_2, X_3) \rightarrow (x_1, x_2, x_3)$$

Figure 2.8: Shear deformation illustrated in the body under consideration.

So it follows from Eq. (2.9.5) that

$$[\mathbf{E}] = \begin{bmatrix} 0 & 1/2(\delta/h) & 0 \\ 1/2(\delta/h) & 1/2(\delta/h)^2 & 0 \\ 0 & 0 & 0 \end{bmatrix}. \tag{2.14.4}$$

Thus, by Eq. (2.13.3) and Eq. (2.14.4),

$$\cos\theta^{(12)} = \frac{\delta/h}{(1 + (\delta/h)^2)^{1/2}}. \tag{2.14.5}$$

In virtue of Eq. (2.13.5) and Eq. (2.14.5), we have

$$\sin\gamma^{(12)} = \frac{\delta/h}{(1 + (\delta/h)^2)^{1/2}}. \tag{2.14.6}$$

Thus, if $(\delta/h) \ll 1$, then

$$\tan\gamma^{(12)} = \frac{\delta}{h}, \tag{2.14.7}$$

and

$$\tan\gamma^{(12)} = \gamma^{(12)} = \frac{\delta}{h}. \tag{2.14.8}$$

2.15 Axial strain

Next, let $\mathrm{d}\mathbf{X}^{(1)}$, $\mathrm{d}\mathbf{X}^{(2)}$, and $\mathrm{d}\mathbf{X}^{(3)}$ denote three mutually perpendicular infinitesimal filaments at $\mathbf{X}$. The *axial strain* or normal strain in the ith direction is defined as

$$\gamma^{(i)} = \frac{|\mathrm{d}\mathbf{x}^{(i)}| - |\mathrm{d}\mathbf{X}^{(i)}|}{|\mathrm{d}\mathbf{X}^{(i)}|}. \tag{2.15.1}$$

This is the definition one initially encounters in a first course in the strength of materials, namely *increase in length/original length*. At times, the axial strain is also referred to as Extensional strain. This however might give the incorrect impression that the infinitesimal filament can only extend. It can also be compressed. If $\gamma^{(i)}$ is positive, the axial strain is tensile and if it is negative, the axial strain is compressive.

We next determine the relationship between E_{ii} (no sum on i) and $\gamma^{(i)}$. Recall from (Eq. (2.9.3) and (2.9.4)) that

$$|\mathbf{dx}|^2 - |\mathbf{dX}|^2 = 2\mathbf{E}\mathbf{dX} \cdot \mathbf{dX}. \tag{2.15.2}$$

It follows that

$$|\mathbf{dx}^{(1)}|^2 - |\mathbf{dX}^{(1)}|^2 = 2\mathbf{E}\mathbf{dX}^{(1)} \cdot \mathbf{dX}^{(1)}. \tag{2.15.3}$$

Let $\mathbf{e}_1$ denote a unit vector along the 1-direction. Recall that any tensor $\mathbf{A}$ can be expressed as

$$\mathbf{A} = A_{ij}\mathbf{e}_i \otimes \mathbf{e}_j, \tag{2.15.4}$$

where $\{\mathbf{e}_i, i = 1, 2, 3\}$ is a basis for $\mathcal{V}$. Then,

$$\mathbf{E} = E_{ij}\mathbf{e}_i \otimes \mathbf{e}_j. \tag{2.15.5}$$

Let us now compute $\mathbf{E}\mathbf{dX}^{(1)} \cdot \mathbf{dX}^{(1)}$, when $\mathbf{dX}^{(i)}$ are along $\mathbf{e}_i$:

$$\begin{aligned}
\mathbf{E}\mathbf{dX}^{(1)} &= |\mathbf{dX}^{(1)}|\mathbf{E}\mathbf{e}_1 && \text{(2.15.6a)}\\
&= |\mathbf{dX}^{(1)}|E_{pq}(\mathbf{e}_p \otimes \mathbf{e}_q)\mathbf{e}_1 && \text{(2.15.6b)}\\
&= |\mathbf{dX}^{(1)}|E_{pq}\mathbf{e}_p\delta_{q1} && \text{(2.15.6c)}\\
&= |\mathbf{dX}^{(1)}|E_{p1}\mathbf{e}_p. && \text{(2.15.6d)}
\end{aligned}$$

So,

$$\begin{aligned}
\mathbf{E}\mathbf{dX}^{(1)} \cdot \mathbf{dX}^{(1)} &= |\mathbf{dX}^{(1)}|^2 E_{p1}\mathbf{e}_p \cdot \mathbf{e}_1 && \text{(2.15.7a)}\\
&= |\mathbf{dX}^{(1)}|^2 E_{p1}\delta_{p1} && \text{(2.15.7b)}\\
&= |\mathbf{dX}^{(1)}|^2 E_{11}. && \text{(2.15.7c)}
\end{aligned}$$

Since in virtue of (2.15.1)

$$\gamma^{(1)} = \frac{|\mathbf{dx}^{(1)}| - |\mathbf{dX}^{(1)}|}{|\mathbf{dX}^{(1)}|}, \tag{2.15.8}$$

it follows that

$$|\mathbf{dx}^{(1)}| = (\gamma^{(1)} + 1)|\mathbf{dX}^{(1)}|, \tag{2.15.9}$$

which leads to

$$|\mathbf{dx}^{(1)}|^2 = (\gamma^{(1)} + 1)^2 |\mathbf{dX}^{(1)}|^2. \tag{2.15.10}$$

So,

$$|\mathbf{dx}^{(1)}|^2 - |\mathbf{dX}^{(1)}|^2 = [(\gamma^{(1)} + 1)^2 - 1] |\mathbf{dX}^{(1)}|^2. \tag{2.15.11}$$

It follows from Eq. (2.15.3), Eq. (2.15.7c), and Eq. (2.15.11) that

$$2E_{11} = (\gamma^{(1)} + 1)^2 - 1, \tag{2.15.12}$$

or

$$\gamma^{(1)} = [(1 + 2E_{11})^{\frac{1}{2}} - 1]. \tag{2.15.13}$$

Thus, we can finally state

$$\gamma^{(i)} = [(1 + 2E_{ii})^{\frac{1}{2}} - 1], \quad i = 1, 2, 3, \text{ no sum on } i. \tag{2.15.14}$$

When E_{ii} is small (in a sense to be defined precisely later),

$$\gamma^{(i)} \approx E_{ii}. \tag{2.15.15}$$

Thus, what we learn as axial strain, namely $\gamma^{(i)}$, is approximately equal to the normal component of the Green-St. Venant strain $\mathbf{E}$, only when E_{ii} is small in some sense to be defined later.

2.16 Linearization of the non-linear strain

Recall a scalar product between two linear transformations $\mathbf{A}$ and $\mathbf{B}$ can be defined through

$$\mathbf{A} \cdot \mathbf{B} = \mathrm{tr}(\mathbf{A}\mathbf{B}^{\mathrm{T}}) = \mathrm{tr}(\mathbf{A}^{\mathrm{T}}\mathbf{B}), \tag{2.16.1}$$

and the Frobenius norm induced through the above scalar product is defined through

$$||\mathbf{A}||^2 = \mathrm{tr}(\mathbf{A}\mathbf{A}^{\mathrm{T}}) = \mathbf{A} \cdot \mathbf{A}. \tag{2.16.2}$$

In indicial notation

$$||\mathbf{A}|| = \sqrt{A_{ij}A_{ij}}. \tag{2.16.3}$$

Thus, if $\mathbf{u}$ is the displacement, then

$$||\nabla \mathbf{u}|| = \left\{ \left(\frac{\partial u_1}{\partial X_1} \right)^2 + \left(\frac{\partial u_1}{\partial X_2} \right)^2 + \cdots + \left(\frac{\partial u_3}{\partial X_3} \right)^2 \right\}^{1/2} \tag{2.16.4}$$

where $\mathbf{u} = u_1\mathbf{e}_1 + u_2\mathbf{e}_2 + u_3\mathbf{e}_3$, $\{\mathbf{e}_i, i = 1, 2, 3\}$ being the basis. Suppose that for all time t belonging to the set of real numbers and for all $\mathbf{X}$ belonging to $\kappa_R(\mathcal{B})$, the $\max||\nabla \mathbf{u}|| = \mathcal{O}(\delta)$, $\delta \ll 1$. Recall from equation Eq. (2.4.5) that a function f is $\mathcal{O}(\delta)$ if

$$\lim_{\delta \to 0} \frac{\mathcal{O}(\delta)}{\delta} = M, \tag{2.16.5}$$

where M is a positive real constant. Since

$$\mathbf{F} = \mathbf{I} + \nabla \mathbf{u}, \quad \mathbf{F} = \mathbf{I} + \mathcal{O}(\delta). \tag{2.16.6}$$

Also,

$$\mathbf{E} = \frac{1}{2}[(\nabla \mathbf{u}) + (\nabla \mathbf{u})^{\mathrm{T}} + (\nabla \mathbf{u})^{\mathrm{T}}(\nabla \mathbf{u})] \tag{2.16.7a}$$

$$= \frac{1}{2}[(\nabla \mathbf{u}) + (\nabla \mathbf{u})^{\mathrm{T}}] + \mathcal{O}(\delta^2). \tag{2.16.7b}$$

Let us define

$$\varepsilon = \frac{1}{2}[(\nabla \mathbf{u}) + (\nabla \mathbf{u})^{\mathrm{T}}], \tag{2.16.8}$$

where ε is the linearized strain. It is this quantity that we first come across as the three dimensional strain tensor in a first course in strength of materials.

In the matrix for the linearized strain tensor, the diagonal terms represent the axial strains along the directions 1, 2, and 3, and the off-diagonal terms represent the linearized shear strains:

$$[\varepsilon] = \begin{bmatrix} \varepsilon_{11} & \varepsilon_{12} & \varepsilon_{13} \\ \varepsilon_{21} & \varepsilon_{22} & \varepsilon_{23} \\ \varepsilon_{31} & \varepsilon_{32} & \varepsilon_{33} \end{bmatrix}, \tag{2.16.9}$$

that is, ε_{ii}, $i = 1, 2, 3$, no sum on i, are the extensional strains and $\varepsilon_{ij} = (1/2)\gamma_{ij}$, $i, j = 1, 2, 3$, are the shear strains.

Recall that

$$\mathbf{F} = \mathbf{R}\mathbf{U} \implies \mathbf{R} = \mathbf{F}\mathbf{U}^{-1}. \tag{2.16.10}$$

Since $\mathbf{U}^2 = \mathbf{C} = \mathbf{I} + 2\mathbf{E}$,

$$\mathbf{U} = (\mathbf{I} + 2\mathbf{E})^{\frac{1}{2}}. \tag{2.16.11}$$

When $\max||\nabla \mathbf{u}|| = \mathcal{O}(\delta)$, $\delta \ll 1$ holds, $\mathbf{E} \approx \varepsilon$, and so Eq. (2.16.11) implies that

$$\mathbf{U} \approx (\mathbf{I} + \mathbf{E}) \approx (\mathbf{I} + \varepsilon). \tag{2.16.12}$$

Next, in virtue of Eq. (2.4.8)

$$\mathbf{F} = \mathbf{I} + \nabla\mathbf{u} = \mathbf{I} + \mathcal{O}(\delta), \tag{2.16.13}$$

and by Eq. (2.16.10) and Eq. (2.16.13)

$$\mathbf{R} = (\mathbf{I} + \nabla\mathbf{u})(\mathbf{I} - \varepsilon) + \mathcal{O}(\delta^2) \tag{2.16.14a}$$

$$\approx (\mathbf{I} + \nabla\mathbf{u})\left[\mathbf{I} - \frac{1}{2}(\nabla\mathbf{u} + (\nabla\mathbf{u})^{\mathrm{T}})\right] \tag{2.16.14b}$$

$$\approx \mathbf{I} + \frac{1}{2}[(\nabla\mathbf{u}) - (\nabla\mathbf{u})^{\mathrm{T}}]. \tag{2.16.14c}$$

Let us define the linearized rotation tensor

$$\omega = \frac{1}{2}[(\nabla\mathbf{u}) - (\nabla\mathbf{u})^{\mathrm{T}}]. \tag{2.16.15}$$

Notice that since the gradient of the displacement is small, the rotation is also small. Then, by Eq. (2.16.8) and Eq. (2.16.15),

$$\nabla\mathbf{u} = \varepsilon + \omega, \tag{2.16.16}$$

where ε is the linearized strain (i. e., symmetric part of $\nabla\mathbf{u}$) and ω is the linearized rotation (i. e., skew part of $\nabla\mathbf{u}$).

2.17 Rigid motion

A rigid motion is a mapping $\chi(\mathbf{X}, t)$ such that

$$\mathbf{x} = \chi(\mathbf{X}, t) = \mathbf{c}(t) + \mathbf{Q}(t)[\mathbf{X} - \mathbf{X}_0], \tag{2.17.1}$$

where $\mathbf{c}(t)$ is a time dependent vector, $\mathbf{X}_0$ is a fixed point not necessarily in $\kappa_R(\mathcal{B})$, and $\mathbf{Q}(t)$ is a proper orthogonal transformation. $\mathbf{c}(t)$ corresponds to a translation and $\mathbf{Q}(t)[\mathbf{X}-\mathbf{X}_0]$ is a rotation about the fixed point $\mathbf{X}_0$. In fact, if by a rigid motion, one means that the motion is such that the distance between any two points remain the same, then one can prove that such a motion has to be defined through (2.17.1).

2.18 Problems

2.1 Show that

$$\boldsymbol{e} = \varepsilon + \mathcal{O}(\delta^2). \tag{2.18.1}$$

2.2 Recall that

$$\gamma^{(i)} = (1 + 2E_{ii})^{\frac{1}{2}} - 1, \tag{2.18.2}$$

with no sum on i. Use binomial theorem and neglect terms of $\mathcal{O}(\delta^2)$ to show that

$$\gamma^{(i)} \approx E_{ii} \approx e_{ii} \approx \varepsilon_{ii}, \qquad \text{no sum on } i, \tag{2.18.3a}$$

$$\gamma^{(ij)} \approx 2E_{ij} \approx 2e_{ij} \approx 2\varepsilon_{ij}, \quad i \neq j, \tag{2.18.3b}$$

References

[1] C. A. Truesdell. First Course in Rational Continuum Mechanics V1. Academic Press, 1992.

[2] L. E. Malvern. Introduction to the Mechanics of a Continuous Medium, 1967.

[3] W. Jaunzemis. Continuum Mechanics. Macmillan, 1967.

[4] K. R. Rajagopal and L. Tao. On the response of non-dissipative solids. Communications in Nonlinear Science and Numerical Simulation, 13(6):1089–1100, 2008.

[5] W. Noll. A general framework for problems in the-statics of finite elasticity. In North-Holland Mathematics Studies, Vol. 30, pp. 363–387. Elsevier, 1978.

[6] P. R. Halmos. Finite Dimensional Vector Spaces. Van Nostrand, Princeton, 2nd edition, 1958.

[7] C. Truesdell. Notes on the history of the General Equations of Hydrodynamics. The American Mathematical Monthly, 60:445–458, 1953.

3 Compatibility of the linearized strain

Given the motion that a body undergoes, one can determine the displacement of the particles belonging to the body, and thereby determine the linearized strain (as well as the nonlinear Green-St. Venant and the Almansi-Hamel strains). In many problems, one guesses the possible state of the linearized strain in the body. It is possible that our guess work might be incorrect in that the state of strain assumed is not possible in the sense that it cannot be derived from a displacement field. We investigate the necessary and sufficient conditions that will guarantee that our assumption for the linearized strain field is permissible.

Henceforth, let us assume that the configuration of the body, $\kappa_R(\mathcal{B})$, under consideration is simply connected and star-shaped (see [1] for a definition of the same). We would like to answer the question: under what conditions could we, given $\hat{\varepsilon}$ in $\kappa_R(\mathcal{B})$, find $\mathbf{u}$? In order to find the displacement field $\mathbf{u}$, given

$$\frac{1}{2}\left[\left(\frac{\partial \mathbf{u}}{\partial \mathbf{X}}\right)+\left(\frac{\partial \mathbf{u}}{\partial \mathbf{X}}\right)^{\mathrm{T}}\right]=\hat{\varepsilon}, \tag{3.0.1}$$

or

$$\frac{1}{2}\left[\left(\frac{\partial u_i}{\partial X_j}\right)+\left(\frac{\partial u_j}{\partial X_i}\right)\right]=\hat{\varepsilon}_{ij}, \tag{3.0.2}$$

a mixed first order partial differential equation for the displacement field $\mathbf{u}$ needs to be solved. In this section, we seek to answer the following questions:

1. Does a $\mathbf{u}$ solving Eq. (3.0.1) exist?
2. If $\mathbf{u}$ exists, is it unique?

Since we know that two displacements that differ by a rigid body motion will lead to the same linearized strain, the displacement corresponding to a given linearized strain is not unique. This leads us to inquire if the displacement corresponding to a given linearized strain is unique to within a rigid body motion.

3.1 Uniqueness of the displacement field to within rigid body motion for a given linearized strain

The answer to our questions in the previous section is given by the following

Theorem 3.1.1 (Uniqueness). *Let* $\mathbf{u}_1(\mathbf{X}, t)$ *and* $\mathbf{u}_2(\mathbf{X}, t) \in \mathcal{C}^2(\kappa_R(\mathcal{B}))$ *and satisfy Eq.* (3.0.1) *where* $\hat{\varepsilon}$ *is given. Let us consider the partial differential equation (Eq.* (3.0.1)*) at a fixed instant of time. Then*

$$\mathbf{u}_1=\mathbf{u}_2+\mathbf{u}^0, \tag{3.1.1}$$

https://doi.org/10.1515/9783110789515-003

where $\mathbf{u}^0$ *is a rigid displacement. That is,*

$$u_i^0 = c_i + W_{ij}X_j, \tag{3.1.2}$$

$$i.e., \mathbf{u}^0 = \mathbf{c} + \mathbf{W}\mathbf{X}, \tag{3.1.3}$$

where $\mathbf{c}$ *and* $\mathbf{W}$ *are functions of time,* $\mathbf{W}$ *being a skew-symmetric tensor.*

Proof. Let us define $\bar{\mathbf{u}} = \mathbf{u}_1 - \mathbf{u}_2$ where $\mathbf{u}_1$ and $\mathbf{u}_2$ both meet (3.0.1). Substituting $\bar{\mathbf{u}}$ into Eq. (3.0.1), we see that

$$\frac{1}{2}\left[\left(\frac{\partial \bar{\mathbf{u}}}{\partial \mathbf{X}}\right) + \left(\frac{\partial \bar{\mathbf{u}}}{\partial \mathbf{X}}\right)^{T}\right] = \mathbf{0}. \tag{3.1.4}$$

Equation (3.1.4) implies that

$$\bar{u}_{i,j} = -\bar{u}_{j,i}. \tag{3.1.5}$$

Now, taking a spatial derivative of the term on the left hand side a second time yields

$$\bar{u}_{i,jk} = \bar{u}_{i,kj}, \tag{3.1.6}$$

since we assume that the function under consideration is sufficiently smooth, the order of derivatives can be interchanged. In virtue of Eq. (3.1.5), we see that

$$\bar{u}_{i,kj} = -\bar{u}_{k,ij} = -\bar{u}_{k,ji} = \bar{u}_{j,ki} = \bar{u}_{j,ik} = -\bar{u}_{i,jk} = -\bar{u}_{i,kj} = 0. \tag{3.1.7}$$

Now, $\bar{u}_{i,jk} = 0$ implies that

$$\bar{u}_{i,j} = W_{ij}, \tag{3.1.8}$$

where W_{ij} can be at most a function of time, which in turn on integration implies that

$$\bar{u}_i = c_i + W_{ij}X_j, \tag{3.1.9}$$

where c_i can at most be a function of time. In virtue of (3.1.5) and (3.1.8) it follows that W_{ij} is skew-symmetric. □

3.2 Necessary and sufficient conditions for compatibility

We next turn to the question whether a body could be deformed in a manner wherein the components of the linearized strain tensor could be arbitrary or whether the components have to satisfy certain compatibility conditions. The next theorem provides a necessary condition that the linearized strain field has to meet.

Theorem 3.2.1 (Necessary condition). *Let* $\hat{\varepsilon}(\mathbf{X}, t) \in C^2(\kappa_R(\mathcal{B}))$. *Then a necessary condition that there exists* $\mathbf{u} \in C^3(\kappa_{R(B)})$ *and which meets*

$$\hat{\varepsilon} = \frac{1}{2}(\nabla\mathbf{u} + (\nabla\mathbf{u})^T) \tag{3.2.1}$$

is that

$$\boldsymbol{\Omega} = \text{curl}\,\text{curl}\,\hat{\varepsilon} = \mathbf{0}, \tag{3.2.2}$$

that is

$$\Omega_{ij} \equiv \epsilon_{irs}\epsilon_{jkl}\hat{\varepsilon}_{rk,sl} = 0 \quad \textit{in } \Omega, \tag{3.2.3}$$

where $\boldsymbol{\Omega}$ *is symmetric.*

Proof. Substituting Eq. (3.2.1) into Eq. (3.2.3), we see that

$$\Omega_{ij} = \frac{1}{2}\epsilon_{irs}\epsilon_{jkl}(u_{r,k} + u_{k,r})_{,sl}, \tag{3.2.4a}$$

$$= \frac{1}{2}\epsilon_{irs}\epsilon_{jkl}(u_{r,ksl} + u_{k,rsl}). \tag{3.2.4b}$$

Since the components of the alternator (a third order tensor) satisfy $\epsilon_{irs} = -\epsilon_{isr}$, $u_{r,ksl} = u_{r,lsk}$, and $u_{k,rsl} = u_{k,srl}$ implies that

$$\Omega_{ij} = 0. \tag{3.2.5}$$

Next, we state and prove a proposition that we will use to obtain an expression for the displacement $\hat{\mathbf{u}}$ that corresponds to a given $\hat{\varepsilon}$.

Proposition 3.2.1. Let $\Gamma_{mnpq} = \epsilon_{imp}\epsilon_{jnq}\Omega_{ij}$. Then,

$$\Gamma_{mnpq} = 0, \quad \text{if and only if } \Omega_{ij} = 0. \tag{3.2.6}$$

Proof. Now, $\Gamma_{mnpq} = \epsilon_{imp}\epsilon_{jnq}\Omega_{ij}$ implies that

$$\Gamma_{mnpp} = \epsilon_{imp}\epsilon_{jnp}\Omega_{ij} \tag{3.2.7a}$$

$$= \epsilon_{pim}\epsilon_{pjn}\Omega_{ij} \tag{3.2.7b}$$

$$= (\delta_{ij}\delta_{mn} - \delta_{in}\delta_{mj})\Omega_{ij} \tag{3.2.7c}$$

$$= \delta_{ij}\delta_{mn}\Omega_{ij} - \Omega_{nm}. \tag{3.2.7d}$$

We have used the fact that $\epsilon_{pim}\epsilon_{pjn} = (\delta_{ij})(\delta_{mn}) - (\delta_{in})(\delta_{mj})$. We notice that if $\Omega_{ij} = 0$, then $\Gamma_{mnpq} = 0$. Now, the right hand side of Eq. (3.2.7d) can be re-written as $\delta_{mn}\Omega_{ii} - \Omega_{mn}$, and so if $\Gamma_{mnpp} = 0$, then

$$\delta_{ij}\delta_{mn}\Omega_{ij} - \Omega_{nm} = 0. \tag{3.2.8}$$

Suppose that $m \neq n$. Then Eq. (3.2.8) implies that

$$\Omega_{nm} = 0. \tag{3.2.9}$$

If $m = n$, then we get

$$\Omega_{11} + \Omega_{22} = 0, \tag{3.2.10a}$$
$$\Omega_{22} + \Omega_{33} = 0, \tag{3.2.10b}$$
$$\Omega_{33} + \Omega_{11} = 0, \tag{3.2.10c}$$

and it immediately follows that

$$\Omega_{11} = \Omega_{22} = \Omega_{33} = 0, \tag{3.2.11}$$

and thus

$$\Gamma_{mnpp} = 0 \implies \Omega_{ij} = 0. \tag{3.2.12}$$

Thus, we have proved the proposition. □

Cesaro representation

Now we shall show that $\Omega_{ij} = 0$ is sufficient for the existence $\mathbf{u}$ of such that

$$\frac{1}{2}\left(\nabla\mathbf{u} + (\nabla\mathbf{u})^{\mathrm{T}}\right) = \hat{\varepsilon}, \tag{3.2.13}$$

where $\hat{\varepsilon}$ is given. Now, in indicial notation, the above becomes

$$\frac{1}{2}(u_{i,j} + u_{j,i}) = \hat{\varepsilon}_{ij}. \tag{3.2.14}$$

It follows from Eq. (3.2.14) that

$$\frac{1}{2}(u_{i,jk} + u_{j,ik}) = \hat{\varepsilon}_{ij,k}, \tag{3.2.15a}$$
$$\frac{1}{2}(u_{j,ki} + u_{k,ji}) = \hat{\varepsilon}_{jk,i}, \tag{3.2.15b}$$
$$\frac{1}{2}(u_{k,ij} + u_{i,jk}) = \hat{\varepsilon}_{ki,j}. \tag{3.2.15c}$$

Adding Eq. (3.2.15a) and Eq. (3.2.15c) and subtracting Eq. (3.2.15b) yields

$$u_{i,jk} = \hat{\varepsilon}_{ij,k} + \hat{\varepsilon}_{ki,j} - \hat{\varepsilon}_{jk,i}. \tag{3.2.16}$$

Integrating Eq. (3.2.16) along a curve C from $X(0)$ to $X(\gamma)$ gives

$$\int_0^\gamma u_{i,jk}(X(\lambda))\frac{dX_k(\lambda)}{d\lambda}\,d\lambda = u_{i,j}(\gamma) - u_{i,j}(0) \tag{3.2.17a}$$

$$= \hat{\varepsilon}_{ij}(\gamma) - \hat{\varepsilon}_{ij}(0) + \int_0^\gamma [\hat{\varepsilon}_{ki,j}(\lambda) - \hat{\varepsilon}_{jk,i}(\lambda)]\frac{dX_k}{d\lambda}\,d\lambda. \tag{3.2.17b}$$

Next, integrating Eq. (3.2.17b) from 0 to λ yields

$$\begin{aligned}
&\int_0^\lambda u_{i,j}(\gamma)\frac{dX_j(\gamma)}{d\gamma}\,d\gamma - u_{i,j}(0)[X_j(\lambda) - X_j(0)] \\
&= \int_0^\lambda \hat{\varepsilon}_{ij}(\gamma)\frac{dX_j(\gamma)}{d\gamma}\,d\gamma - \hat{\varepsilon}_{ij}(0)[X_j - X_j(0)] \\
&\quad + \int_0^\lambda \left\{\int_0^\gamma [\hat{\varepsilon}_{ki,j}(\lambda) - \hat{\varepsilon}_{jk,i}(\lambda)]\frac{dX_k(\lambda)}{d\lambda}\,d\lambda\right\}\frac{dX_j(\gamma)}{d\gamma}\,d\gamma,
\end{aligned} \tag{3.2.18}$$

or

$$\begin{aligned}
&u_i(\lambda) - u_i(0) - u_{i,j}(0)[X_j(\lambda) - X_j(0)] \\
&= \int_0^\lambda \hat{\varepsilon}_{ij}(\gamma)\frac{dX_j(\gamma)}{d\gamma}\,d\gamma - \hat{\varepsilon}_{ij}(0)[X_j(\lambda) - X_j(0)] \\
&\quad + \int_0^\lambda \left\{\int_0^\gamma [\hat{\varepsilon}_{ki,j}(\lambda) - \hat{\varepsilon}_{jk,i}(\lambda)]\frac{dX_k(\lambda)}{d\lambda}\,d\lambda\right\}\frac{dX_j(\gamma)}{d\gamma}\,d\gamma.
\end{aligned} \tag{3.2.19}$$

Let us integrate the last term in Eq. (3.2.19) by parts to find

$$\begin{aligned}
&\int_0^\lambda \left\{\int_0^\gamma [\hat{\varepsilon}_{ki,j}(\lambda) - \hat{\varepsilon}_{jk,i}(\lambda)]\frac{dX_k(\lambda)}{d\lambda}\,d\lambda\right\}\frac{dX_j(\gamma)}{d\gamma}\,d\gamma \\
&= \int_0^\lambda \frac{d}{d\gamma}\left[X_j(\gamma)\int_0^\gamma (\hat{\varepsilon}_{ki,j}(\lambda) - \hat{\varepsilon}_{jk,i}(\lambda))\frac{dX_k(\lambda)}{d\lambda}\,d\lambda\right]d\gamma \\
&\quad - \int_0^\lambda X_j(\gamma)\frac{d}{d\gamma}\left[\int_0^\gamma (\hat{\varepsilon}_{ki,j}(\lambda) - \hat{\varepsilon}_{jk,i}(\lambda))\frac{dX_k(\lambda)}{d\lambda}\,d\lambda\right]d\gamma
\end{aligned}$$

$$= X_j(\lambda)\int_0^\lambda [\hat{\varepsilon}_{ki,j}(\gamma) - \hat{\varepsilon}_{jk,i}(\gamma)]\frac{dX_k(\gamma)}{d\gamma}\,d\gamma$$
$$-\int_0^\lambda X_j(\gamma)[\hat{\varepsilon}_{ki,j}(\gamma) - \hat{\varepsilon}_{jk,i}(\gamma)]\frac{dX_k(\gamma)}{d\gamma}\,d\gamma. \tag{3.2.20}$$

Thus, the displacement can be represented as

$$u_i(\lambda) = u_i(0) + [u_{i,j}(0) - \hat{\varepsilon}_{ij}(0)](X_j(\lambda) - X_j(0)) + \int_0^\lambda U_{ik}(\gamma, X)\frac{dX_k(\lambda)}{d\gamma}\,d\gamma, \tag{3.2.21}$$

where

$$U_{ik}(\gamma, X) \equiv \hat{\varepsilon}_{ik}(\gamma) + (X_j(\lambda) - X_j(\gamma))[\hat{\varepsilon}_{ki,j}(\gamma) - \hat{\varepsilon}_{jk,i}(\gamma)]. \tag{3.2.22}$$

The above representation for the displacement field is known as the Cesaro Representation. We will now verify that the above representation for the displacement leads to

$$\varepsilon_{ij} = \frac{1}{2}(u_{i,j} + u_{j,i}). \tag{3.2.23}$$

We have

$$u_i(\lambda) = u_i(0) + [u_{i,j}(0) - \hat{\varepsilon}_{ij}(0)][X_j(\lambda) - X_j(0)] + \int_0^\lambda U_{ik}(\tau, X)\frac{dX_k(\tau)}{d\tau}\,d\tau \tag{3.2.24}$$

The above equation is the same as the one given below

$$u_i(X(\lambda)) = u_i(X(0)) + [u_{i,j}(X(0)) - \hat{\varepsilon}_{ij}(X(0))][X_j(\lambda) - X_j(0)]$$
$$+ \int_0^\lambda U_{ik}[X(\lambda), X(\tau)]dX_k(\tau). \tag{3.2.25}$$

Now,

$$u_{i,l}(X(\lambda)) = [u_{i,j}(X(0)) - \hat{\varepsilon}_{ij}(X(0))]\delta_{jl} + \frac{d}{dX_l(\lambda)}\int_0^\lambda U_{ik}dX_k(\tau) \tag{3.2.26}$$

$$= [u_{i,l}(X(0)) - \hat{\varepsilon}_{il}(X(0))] + \int_0^\lambda [\hat{\varepsilon}_{ki,j} - \hat{\varepsilon}_{jk,i}]\delta_{jl}dX_k(\tau) + \hat{\varepsilon}_{il}(X(S)) \tag{3.2.27}$$

$$= [u_{i,l}(X(0)) - \hat{\varepsilon}_{il}(X(0))] + \int_0^\lambda [\hat{\varepsilon}_{ki,l}(X(\tau)) - \hat{\varepsilon}_{lk,i}(X(\tau))]dX_k(\tau) + \hat{\varepsilon}_{il}(X(\lambda)) \tag{3.2.28}$$

and

$$u_{l,i}(X(\lambda)) = [u_{l,i}(X(0)) - \hat{\varepsilon}_{li}(X(0))] + \int_0^\lambda [\hat{\varepsilon}_{kl,i}(X(\tau)) - \hat{\varepsilon}_{ik,l}(X(\tau))]dX_k(\tau) + \hat{\varepsilon}_{li}(X(\lambda)). \tag{3.2.29}$$

Hence,

$$\begin{aligned} u_{i,l}(X(\lambda)) + u_{l,i}(X(\lambda)) = {} & \hat{\varepsilon}_{il}(X(\lambda)) + \hat{\varepsilon}_{li}(X(\lambda)) \\ & + [u_{i,l}(X(0)) - \hat{\varepsilon}_{il}(X(0)) + u_{l,i}(X(0)) - \hat{\varepsilon}_{li}(X(0))] \\ & + \int_0^\lambda [\hat{\varepsilon}_{ki,l}(X(\tau)) - \hat{\varepsilon}_{lk,i}(X(\tau)) + \hat{\varepsilon}_{kl,i}(X(\tau)) - \hat{\varepsilon}_{ik,l}(X(\tau))]dX_k(\tau), \end{aligned} \tag{3.2.30}$$

but

$$\hat{\varepsilon}_{il} = \hat{\varepsilon}_{li}, \tag{3.2.31}$$

and

$$\hat{\varepsilon}_{ki,l} = \hat{\varepsilon}_{ik,l}; \quad \hat{\varepsilon}_{lk,i} = \hat{\varepsilon}_{kl,i}. \tag{3.2.32}$$

Hence, the terms within the integral vanish. Therefore,

$$u_{i,l}(X(\lambda)) + u_{l,i}(X(\lambda)) = 2\hat{\varepsilon}_{il}(X(\lambda)) + u_{i,l}(X(0)) + u_{l,i}(X(0)) - 2\hat{\varepsilon}_{il}(X(0)) \tag{3.2.33}$$

Since

$$u_{i,l}(X(0)) + u_{l,i}(X(0)) = 2\hat{\varepsilon}_{il}(X(0)), \tag{3.2.34}$$

thus

$$\hat{\varepsilon}_{il} = \frac{1}{2}[u_{i,l} + u_{l,i}]. \tag{3.2.35}$$

Let us now examine the integral

$$\int_0^\lambda U_{ik}(\gamma, X)\frac{dX_k(\gamma)}{d\gamma}\, d\gamma, \tag{3.2.36}$$

for fixed i. Now, Eq. (3.2.36) looks like

$$\int_0^\lambda \mathbf{f} \cdot \hat{\gamma}\, d\gamma. \tag{3.2.37}$$

So, the solution is single-valued if and only if

$$\oint_0^\lambda U_{ik}(\gamma, X) \frac{dX_k}{d\gamma}\, d\gamma = 0 \quad \forall \text{ closed paths in } \kappa_R(\mathcal{B}). \tag{3.2.38}$$

It follows from Stokes' theorem that if $\boldsymbol{\Omega}$ is simply connected this integral will have the above property if and only if

$$\hat{\epsilon}_{plk} U_{ik,l}(\gamma) = 0 \quad \text{in } \Omega \tag{3.2.39}$$

and thus

$$U_{ik,l} = U_{il,k} \quad \text{in } \Omega. \tag{3.2.40}$$

Next,

$$U_{ik,l}(\gamma) = \hat{\varepsilon}_{ik,l} + (X_j(0) - X_j(\gamma))(\hat{\varepsilon}_{ki,jl} - \hat{\varepsilon}_{jk,il}) - (\hat{\varepsilon}_{ki,l} - \hat{\varepsilon}_{lk,i}), \tag{3.2.41}$$

and

$$U_{il,k}(\gamma) = \hat{\varepsilon}_{il,k} + (X_j(0) - X_j(\gamma))(\hat{\varepsilon}_{li,jk} - \hat{\varepsilon}_{jl,ik}) - (\hat{\varepsilon}_{li,k} - \hat{\varepsilon}_{kl,i}). \tag{3.2.42}$$

Hence,

$$(X_j(0) - X_j(\gamma))(\hat{\varepsilon}_{ki,jl} - \hat{\varepsilon}_{jk,il} - \hat{\varepsilon}_{li,jk} + \hat{\varepsilon}_{jl,ik}) = 0, \tag{3.2.43}$$

but $X_j(0)$ and $X_j(\gamma)$ are arbitrary. Thus,

$$\hat{\varepsilon}_{ki,jl} - \hat{\varepsilon}_{jk,il} - \hat{\varepsilon}_{li,jk} + \hat{\varepsilon}_{jl,ik} = 0. \tag{3.2.44}$$

So, we have shown that

$$\Gamma_{ijkl} = 0. \tag{3.2.45}$$

□

We have established the following theorem concerning the necessary and sufficient conditions for the existence of a single valued displacement field, given a linearized strain field.

Theorem 3.2.2 (Necessary and sufficient condition for the existence of a displacement field corresponding to a prescribed strain field). *Let $\hat{\varepsilon} \in \mathcal{C}^2(\kappa_R(\mathcal{B}))$. Then, if $\kappa_R(\mathcal{B})$ is simply connected there exists a single valued $\mathbf{u} \in \mathcal{C}^3(\kappa_R(\mathcal{B}))$ such that*

$$\hat{\varepsilon} = \frac{1}{2}(\nabla\mathbf{u} + (\nabla\mathbf{u})^T) \quad in\ \kappa_R(\mathcal{B}) \tag{3.2.46}$$

if and only if

$$\Omega_{ij} = \epsilon_{irs}\epsilon_{jkl}\hat{\varepsilon}_{rk,sl} = 0 \quad in\ \kappa_R(\mathcal{B}). \tag{3.2.47}$$

Moreover, $\mathbf{u}$ *is given by*

$$u_i(\lambda) = rigid + \int_0^\lambda U_{ik}(\gamma, X)\frac{dX_k(\gamma)}{d\gamma}\,d\gamma, \tag{3.2.48}$$

where

$$U_{ik} = \hat{\varepsilon}_{ik}(\gamma) + [X_j(\gamma) - X_j(\gamma)](\hat{\varepsilon}_{ki,j}(\gamma) - \hat{\varepsilon}_{jk,i}(\gamma)), \tag{3.2.49}$$

independent of the path chosen. Now, $\hat{\varepsilon}_{ik}(\gamma) = \hat{\varepsilon}_{ik}(X(\gamma))$. The displacement field is not allowed to be multivalued.[1]

Recall that

$$\int_{\mathbf{X}_1}^{\mathbf{X}_2} \mathbf{f}(\mathbf{X}) \cdot \hat{\gamma}(\mathbf{X})\,d\lambda, \tag{3.2.50}$$

is independent of path if and only if curl $\mathbf{f} = \mathbf{0}$ in $\kappa_R(\mathcal{B})$. Moreover, if Eq. (3.2.50) holds in a simply connected domain, $\kappa_R(\mathcal{B})$, then

$$\phi(\mathbf{y}) \equiv \int_{\mathbf{X}_1}^{\mathbf{y}} \mathbf{f}(\mathbf{X}) \cdot \hat{\gamma}(\mathbf{X})\,d\lambda, \tag{3.2.51}$$

is single-valued, $\phi \in \mathcal{C}^2(\kappa_R(\mathcal{B}))$, and $\nabla\phi = \mathbf{f}$ in $\kappa_R(\mathcal{B})$.

This theorem is not true if $\kappa_R(\mathcal{B})$ is not simply connected. If $\kappa_R(\mathcal{B})$ is not simply connected, ϕ is Eq. (3.2.51) will still meet $\nabla\phi = \mathbf{f}$, but ϕ need not be single valued. For *uniqueness*, given $\mathbf{f}$, assume the existence of ϕ. Then,

1 Ordinary functions that one uses in Real Analysis are by definition single valued. However, one comes across set valued functions where a function assigns a set to a single element belonging to its domain. One also comes across the notion of multivalued functions in complex analysis.

$$\mathbf{f} = \nabla\phi_1 = \nabla\phi_2 \implies \nabla(\phi_1 - \phi_2) = 0, \tag{3.2.52}$$

or

$$\phi_1 = \phi_2 + C \quad \text{in } \kappa_R(\mathcal{B}) \tag{3.2.53}$$

Let ε be smooth enough in a simply connected domain. Then, there exist a $\mathbf{u}$ such that $\frac{1}{2}[(\nabla\mathbf{u}) + (\nabla\mathbf{u})^{\mathrm{T}}] = \varepsilon$ in the domain if and only if

$$\Omega_{ij} = \epsilon_{irs}\epsilon_{jkl}\varepsilon_{rk,sl} = 0 \tag{3.2.54}$$

in the domain. Moreover $\mathbf{u}$ is given by

$$u_i(s) = \text{rigid} + \int_0^s U_{ik}(\tau, \mathbf{x})\frac{dx_k}{d\tau}\,d\tau, \tag{3.2.55}$$

where

$$U_{ik} = \varepsilon_{ik} + [x_j(s) + x_j(\tau)](\varepsilon_{ki,j} - \varepsilon_{jk,i}), \tag{3.2.56}$$

and $\mathbf{u}$ is path independent.

The compatibility equations, namely the condition that $\Omega_{ij} = 0$, can be expressed in the Cartesian coordinate system by a system of six equations, due to the symmetry of $\boldsymbol{\Omega}$, as

$$\frac{\partial^2\varepsilon_{xx}}{\partial y\partial z} = \frac{\partial}{\partial x}\left[-\frac{\partial\varepsilon_{yz}}{\partial x} + \frac{\partial\varepsilon_{zx}}{\partial y} + \frac{\partial\varepsilon_{xy}}{\partial z}\right] \tag{3.2.57a}$$

$$\frac{\partial^2\varepsilon_{yy}}{\partial z\partial x} = \frac{\partial}{\partial y}\left[-\frac{\partial\varepsilon_{zx}}{\partial y} + \frac{\partial\varepsilon_{xy}}{\partial z} + \frac{\partial\varepsilon_{yz}}{\partial x}\right] \tag{3.2.57b}$$

$$\frac{\partial^2\varepsilon_{zz}}{\partial x\partial y} = \frac{\partial}{\partial z}\left[-\frac{\partial\varepsilon_{xy}}{\partial z} + \frac{\partial\varepsilon_{yz}}{\partial x} + \frac{\partial\varepsilon_{zx}}{\partial y}\right] \tag{3.2.57c}$$

$$2\frac{\partial^2\varepsilon_{xy}}{\partial x\partial y} = \frac{\partial^2\varepsilon_{xx}}{\partial y^2} + \frac{\partial^2\varepsilon_{yy}}{\partial x^2} \tag{3.2.57d}$$

$$2\frac{\partial^2\varepsilon_{yz}}{\partial y\partial z} = \frac{\partial^2\varepsilon_{yy}}{\partial z^2} + \frac{\partial^2\varepsilon_{zz}}{\partial y^2} \tag{3.2.57e}$$

$$2\frac{\partial^2\varepsilon_{zx}}{\partial z\partial x} = \frac{\partial^2\varepsilon_{zz}}{\partial x^2} + \frac{\partial^2\varepsilon_{xx}}{\partial z^2}. \tag{3.2.57f}$$

In two dimensions, namely the x-y plane, the compatibility equations reduce to (in Cartesian coordinates)

$$2\frac{\partial^2\varepsilon_{xy}}{\partial x\partial y} = \frac{\partial^2\varepsilon_{xx}}{\partial y^2} + \frac{\partial^2\varepsilon_{yy}}{\partial x^2} \tag{3.2.58}$$

For a more detailed discussion of the notion of compatibility, see [2].

Example 3.2.1. Consider the following strain field:

$$\varepsilon_{xx} = 3x + \sin y, \quad \varepsilon_{yy} = x^3, \quad \varepsilon_{xy} = \frac{x}{2}\cos y + \frac{a}{2}x^2 y.$$

Find the displacement by picking an appropriate value for a.

Solution. Using the compatibility conditions, we find that

$$2\frac{\partial^2 \varepsilon_{xy}}{\partial x \partial y} = \frac{\partial^2 \varepsilon_{xx}}{\partial y^2} + \frac{\partial^2 \varepsilon_{yy}}{\partial x^2} \implies a = 3. \tag{3.2.59}$$

Thus,

$$u = \frac{3x^2}{2} + x\sin y + C(t)y + D(t) \tag{3.2.60a}$$

$$v = x^3 y - C(t)x + E(t). \tag{3.2.60b}$$

Example 3.2.2. Consider the following strain field:

$$\varepsilon_{xx} = m[\cos mx \cos ny]e^{-at}$$
$$\varepsilon_{yy} = n[\cos mx \cos ny]e^{-at}$$
$$\varepsilon_{xy} = \frac{-(m+n)}{\beta}[\sin mx \sin ny]e^{-at}.$$

Compatibility equations are satisfied if $\beta = 2$. So we have

$$\frac{\partial^2 \varepsilon_{xx}}{\partial y^2} = -mn^2(\cos mx \cos ny)e^{-at}, \tag{3.2.61a}$$

$$\frac{\partial^2 \varepsilon_{yy}}{\partial x^2} = -nm^2(\cos mx \cos ny)e^{-at}, \tag{3.2.61b}$$

$$2\frac{\partial^2 \varepsilon_{xy}}{\partial x \partial y} = -2\frac{(nm^2 + mn^2)}{\beta}(\cos mx \cos ny)e^{-at}. \tag{3.2.61c}$$

Then it follows that

$$u = (\sin mx \cos ny)e^{-at} + f(t), \tag{3.2.62a}$$

$$v = (\cos mx \sin ny)e^{-at} + g(t). \tag{3.2.62b}$$

Example 3.2.3. The beam shown in Fig. 3.1 is in pure bending. Then it follows that (let us assume the strains for now and assume that Young's modulus and Poisson's that have not been defined are some given constants)

$$\varepsilon_{11} = \varepsilon_{22} = \frac{\nu M}{EI}x_1 \tag{3.2.63a}$$

$$\varepsilon_{33} = -\frac{M}{EI}x_1 \tag{3.2.63b}$$

$$\varepsilon_{ij} = 0, \quad i \neq j \tag{3.2.63c}$$

$$\frac{M}{EI} \equiv k, \tag{3.2.63d}$$

where ν is Poisson's ratio, E is Young's modulus, and I is the moment of inertia.

Solution. Step (I) – check compatibility (it is satisfied). Note that the domain is simply connected. Step (II) – use the Cesaro representation. Let us find u_1 (i. e., the component of **u** along the 1-direction). We need to find U_{11}, U_{12}, and U_{13}.

$$U_{11} = \varepsilon_{11} + \left[x_1(s) + x_1(\tau)\right](\varepsilon_{11,1} - \varepsilon_{11,1}) \tag{3.2.64a}$$

$$= \varepsilon_{11} = \nu k x_1 \tag{3.2.64b}$$

$$U_{12} = -\left[x_2(s) - x_2(\tau)\right]\varepsilon_{22,1} \tag{3.2.64c}$$

$$= -\nu k\left[x_2(s) - x_2(\tau)\right] \tag{3.2.64d}$$

$$U_{13} = -\left[x_3(s) - x_3(\tau)\right]\varepsilon_{33,1} \tag{3.2.64e}$$

$$= k\left[x_3(s) - x_3(\tau)\right]. \tag{3.2.64f}$$

Then we have

$$u_1(s) = \text{rigid} + \cdots$$
$$\int_0^s \left\{\nu k x_1(\tau)\frac{dx_1(\tau)}{d\tau} - \nu k\left[x_2(s) - x_2(\tau)\right]\frac{dx_2(\tau)}{d\tau} + k\left[x_3(s) - x_3(\tau)\right]\frac{dx_3(\tau)}{d\tau}\right\}d\tau. \tag{3.2.65}$$

Note that $x_1(0) = 0$. Thus we have

$$u_1(s) = \text{rigid} + \frac{\nu k}{2}x_1^2(s) - \frac{\nu k}{2}x_2^2(s) + \frac{k}{2}x_3^2(s), \tag{3.2.66}$$

where $\varepsilon_{11} = \partial u_1/\partial x_1$.

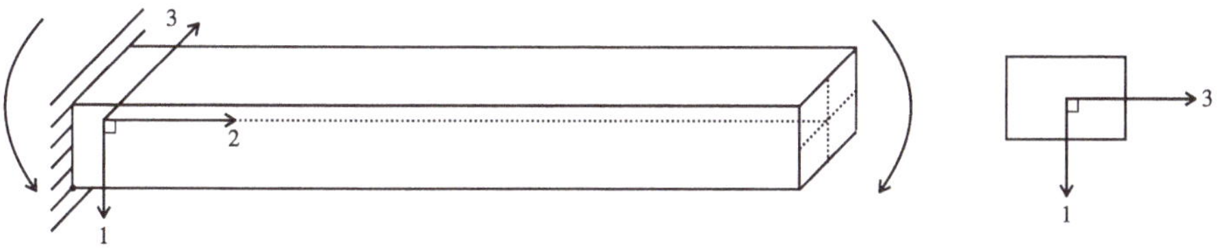

Figure 3.1: Beam in pure bending.

Example 3.2.4. For a dilational state of strain we have

$$\boldsymbol{\varepsilon} = -e\mathbf{I}, \quad \varepsilon_{ij} = -e\delta_{ij}. \tag{3.2.67}$$

Check when $\Omega_{ij} = 0$.

Solution. By definition,

$$\Omega_{ij} = \epsilon_{irs}\epsilon_{jkl}\varepsilon_{rk,sl}, \tag{3.2.68}$$

where by substituting in the given strain state yields

$$\Omega_{ij} = \epsilon_{irs}\epsilon_{jkl}(e\delta_{rk})_{,sl} \tag{3.2.69a}$$
$$= \epsilon_{irs}\epsilon_{jkl}(e_{,sl})\delta_{rk}. \tag{3.2.69b}$$

Now we set $r = k$, giving us

$$\Omega_{ij} = \epsilon_{iks}\epsilon_{jkl}(e_{,sl}) \tag{3.2.70a}$$
$$= \epsilon_{ksi}\epsilon_{klj}(e_{,sl}) \tag{3.2.70b}$$
$$= (\delta_{sl}\delta_{ij} - \delta_{sj}\delta_{il})e_{,sl} \tag{3.2.70c}$$
$$= (\delta_{ij}e_{,ss}) - e_{,ji}. \tag{3.2.70d}$$

We need

$$\delta_{ij}e_{,ss} - e_{,ij} = 0, \tag{3.2.71}$$

which implies that we need

$$e_{,ss} = 0, \quad e_{,ij} = 0. \tag{3.2.72}$$

References

[1] O. D. Kellogg. Foundations of Potential Theory. Dover Publications, New York, 2010.
[2] I. S. Sokolnikoff. Mathematical Theory of Elasticity, Vol. 83. McGraw-Hill, New York, 1956.

4 Frames, changes of frames, frame indifference

There is considerable debate concerning what one means by "Frame Indifference". We shall not get into an extended or even limited discussion of the philosophical issues here, suffice it is to say that the frame indifference of physical quantities plays a critical role in the development of constitutive relations. Truesdell [1] and like minded researchers make assumptions that imply that the linearized strain is not frame-indifferent, while there are others that take the point of view that the linearized strain *is* frame-indifferent (see the paper by Steigmann [2] concerning this issue).[1]

Here, we will adopt the point of view advocated by Truesdell. A detailed discussion of events, frames, and changes of frame can be found in Truesdell [1]. Here, we are interested in merely giving some motivation to the relevant issues as these ideas play a very important role in the development of constitutive theories. We would however be remiss if we do not emphasize that the requirement of frame indifference to place restrictions on constitutive relations is a matter of much controversy. Since frame indifference, and the weaker requirement of Galilean invariance, have had a fair amount of success in the development of constitutive theories, we shall discuss the same, albeit briefly, here.

Crudely speaking, since an "event", e, takes place at a particular location in space, $\mathbf{x}$, and at a particular instant of time, t, we can think of a frame as an assignment of a particular location, $\mathbf{x}$, and a particular time, t, to an event, e. The notion of an event, in such a point of view, is primitive and that which identifies the ordered pair $\{\mathbf{x}, t\}$ to the event e is called the frame. The set of all events is called the "event-world". Thus, a frame can be thought of as an observer, who can assign spatial locations where events occur, and assign the instant of time when the event occurred. For two such observers to be able to discuss the various events that comprise the event world, or put differently, for two frames to be meaningfully correlated, first both the observers have to be able to see the event taking place, and they have to agree on the notion of distance between locations in space, time intervals, and the sense of time. A mapping of the event-world onto itself, that is an assignment of a $\{\mathbf{x}, t\}$ to a $\{\mathbf{x}^*, t^*\}$ that preserves distance, interval of time, and sense of time, takes the form

1 The philosophy of the author is quite different from both these approaches. The author is of the opinion that one ought to develop the theory appealing only to configurations actually taken by the body. In the case of the relative deformation gradients, we only deal with real configurations that a body takes in a three dimensional Euclidean space. The notion of deformation gradient which uses the reference configuration as a surrogate for the abstract body and then presumes that particles in this reference configuration are not accessible to the frames, which in turn implies that the deformation gradient transforms in a certain manner with regard to frame indifference, is philosophically flawed as it presumes that particles in the reference configuration are the same with regard to all observers. If the reference configuration is a configuration taken by the body at a specific time t, then that particular configuration is not available to an observer at some other time, unless the body has not moved. If the reference configuration is not one taken by the body, then it has no relevance to observers, as observers have to be able to behold the body. We cannot get into a detailed discussion of this issue here.

https://doi.org/10.1515/9783110789515-004

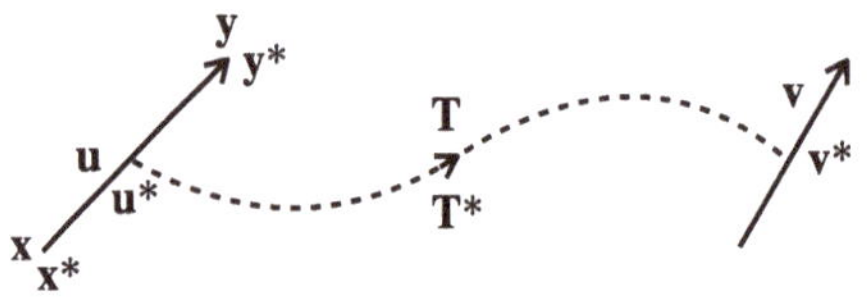

Figure 4.1: Transformation of an arbitrary directed infinitesimal filament (vector) seen by two observers.

$$\mathbf{x}^* = \mathbf{c}(t) + \mathbf{Q}(t)[\mathbf{x} - \mathbf{x}^0], \tag{4.0.1a}$$

$$t^* = t - a, \tag{4.0.1b}$$

where $\mathbf{Q}(t)$ is an orthogonal transformation, $\mathbf{x}^0$ is a fixed point in a three-dimensional Euclidean space, a is a scalar constant, and $\mathbf{c}(t)$ is a vector. The vector $\mathbf{c}(t)$ can be viewed as a translation, and the term $\mathbf{Q}(t)[\mathbf{x} - \mathbf{x}^0]$ is a rotation or reflection about the point $\mathbf{x}^0$. Since $\mathbf{Q}(t)$ is orthogonal, $\det \mathbf{Q}(t) = \pm 1$. When $\det \mathbf{Q}(t) = +1$, we have a rotation, and when $\det \mathbf{Q}(t) = -1$ we have a reflection.

A quantity is said to be a frame-indifferent if the quantity is invariant under all changes of frame given by Eq. (4.0.1a) and Eq. (4.0.1b). A scalar ϕ is said to be frame indifferent if $\phi^* = \phi$. A vector, $\boldsymbol{\phi}$ is said to be frame indifferent if

$$\boldsymbol{\phi}^* = [\mathbf{Q}(t)]\boldsymbol{\phi}. \tag{4.0.2}$$

It is imperative to mention that there is much debate concerning whether one ought to require that relations such as (4.0.2) and (4.0.3) hold for all orthogonal transformation or restrict it to only proper orthogonal transformation.

A second-order tensor, $\boldsymbol{\Phi}$, is said to be frame indifferent if

$$\boldsymbol{\Phi}^* = [\mathbf{Q}(t)]\boldsymbol{\Phi}[\mathbf{Q}(t)]^{\mathrm{T}}. \tag{4.0.3}$$

Let us consider two locations in space, which according to one frame are $\mathbf{x}$ and $\mathbf{y}$, while according to another frame are $\mathbf{x}^*$ and $\mathbf{y}^*$. Then, by (4.0.1a)

$$(\mathbf{y}^* - \mathbf{x}^*) = [\mathbf{Q}(t)](\mathbf{y} - \mathbf{x}). \tag{4.0.4}$$

Finite dimensional vectors are isomorphic to the set of “directed line segments,” and vectors transforming according to (4.0.4), that is physical quantities, mathematically represented as vectors, that transform like directed line segments, are said to be frame-indifferent.

Let $\mathbf{u} = \mathbf{y} - \mathbf{x}$ and $\mathbf{u}^* = \mathbf{y}^* - \mathbf{x}^*$. That is, the same directed line segment is viewed by the two observers as $\mathbf{u}$ and $\mathbf{u}^*$, respectively (i. e., in the two frames, the same directed line segment is identified as $\mathbf{u}$ and $\mathbf{u}^*$, respectively). Suppose a linear transformation $\mathbf{T}$ assigns the vector $\boldsymbol{\upsilon}$ to the vector $\mathbf{u}$ in the unstarred frame, i. e., $\boldsymbol{\upsilon} = \mathbf{T}\mathbf{u}$. In the starred frame, it would be viewed as a linear transformation $\mathbf{T}^*$, assigning to the vector $\mathbf{u}^*$ the vector $\boldsymbol{\upsilon}^*$. Now, since

$$\mathbf{\upsilon}^* = \mathbf{T}^*\mathbf{u}^*, \tag{4.0.5}$$

in virtue of Eq. (4.0.4),

$$\mathbf{\upsilon}^* = \mathbf{Q}\mathbf{\upsilon} = \mathbf{T}^*\mathbf{Q}\mathbf{u} \implies \mathbf{\upsilon} = [\mathbf{Q}(t)]^{\mathrm{T}}\mathbf{T}^*[\mathbf{Q}(t)]\mathbf{u}, \tag{4.0.6}$$

and thus,

$$\mathbf{T}^* = [\mathbf{Q}(t)]\mathbf{T}[\mathbf{Q}(t)]^{\mathrm{T}}. \tag{4.0.7}$$

We have used the fact that $\mathbf{Q}(t)$ is orthogonal. Thus, second-order tensors, namely linear transformations that map a vector from one vector space to another vector space, as they would directed line segments, are called frame indifferent tensors.

We now consider how some physical quantities transform due to a change of frame.

(i) Velocity: in the starred frame, let $\mathbf{\upsilon}^*$ denote the velocity. That is

$$\mathbf{\upsilon}^* = \frac{\mathrm{d}\mathbf{x}^*}{\mathrm{d}t^*} = \frac{\mathrm{d}\mathbf{x}^*}{\mathrm{d}t}\frac{\mathrm{d}t}{\mathrm{d}t^*} = \frac{\mathrm{d}\mathbf{c}}{\mathrm{d}t} + [\mathbf{Q}(t)]\frac{\mathrm{d}\mathbf{x}}{\mathrm{d}t} + [\dot{\mathbf{Q}}(t)][\mathbf{x} - \mathbf{x}^0] \tag{4.0.8}$$

$$= \frac{\mathrm{d}\mathbf{c}}{\mathrm{d}t} + [\mathbf{Q}(t)]\mathbf{\upsilon} + [\dot{\mathbf{Q}}(t)][\mathbf{x} - \mathbf{x}^0]. \tag{4.0.9}$$

Thus, velocity is not frame indifferent. It follows from Eq. (4.0.1a) that

$$[\mathbf{x} - \mathbf{x}^0] = [\mathbf{Q}^{\mathrm{T}}(t)][\mathbf{x}^* - \mathbf{c}(t)]. \tag{4.0.10}$$

Thus, on using Eq. (4.0.10), Eq. (4.0.9) can be expressed as

$$\mathbf{\upsilon}^* = \frac{\mathrm{d}\mathbf{c}}{\mathrm{d}t} + \mathbf{Q}(t)\mathbf{\upsilon} + \dot{\mathbf{Q}}(t)\mathbf{Q}^{\mathrm{T}}(t)(\mathbf{x}^* - \mathbf{c}(t)). \tag{4.0.11}$$

Let

$$\mathbf{A} \equiv \dot{\mathbf{Q}}\mathbf{Q}^{\mathrm{T}}, \tag{4.0.12}$$

so that

$$\mathbf{\upsilon}^* = \frac{\mathrm{d}\mathbf{c}}{\mathrm{d}t} + \mathbf{Q}(t)\mathbf{\upsilon} + \mathbf{A}(\mathbf{x}^* - \mathbf{c}(t)). \tag{4.0.13}$$

(ii) Acceleration: Now,

$$\mathbf{a}^* = \frac{\mathrm{d}\mathbf{\upsilon}^*}{\mathrm{d}t^*} = \frac{\mathrm{d}\mathbf{\upsilon}^*}{\mathrm{d}t}\frac{\mathrm{d}t}{\mathrm{d}t^*} = \frac{\mathrm{d}\mathbf{\upsilon}^*}{\mathrm{d}t} \tag{4.0.14}$$

Thus,

$$\mathbf{a}^* = \frac{\mathrm{d}^2\mathbf{c}}{\mathrm{d}t^2} + [\ddot{\mathbf{Q}}(t)](\mathbf{x} - \mathbf{x}^0) + 2\dot{\mathbf{Q}}(t)\mathbf{\upsilon} + \mathbf{Q}\mathbf{a}. \tag{4.0.15}$$

Thus, in general, acceleration is not frame indifferent. Eq. (4.0.15) can be expressed as

$$\mathbf{a}^* = [\mathbf{Q}(t)]\mathbf{a} + \frac{d^2\mathbf{c}}{dt^2} + 2\mathbf{A}\left[\dot{\mathbf{x}}^* - \frac{d\mathbf{c}}{dt}\right] + [\dot{\mathbf{A}} - \mathbf{A}^2](\mathbf{x}^* - \mathbf{c}(t)). \tag{4.0.16}$$

Two frames that are related by a change of frame, wherein

$$\frac{d^2\mathbf{c}}{dt^2} = \mathbf{0} \quad \text{and} \quad \mathbf{Q} = \text{constant} \tag{4.0.17}$$

are said to be related by a Galilean transformation. That is, the set of transformations under which acceleration becomes frame-indifferent is the basis for a Galilean transformation.

We note that if

$$\frac{d\mathbf{c}}{dt} = \mathbf{0} \quad \text{and} \quad \mathbf{A} = \mathbf{0}, \tag{4.0.18}$$

the velocity becomes frame indifferent relative to these two frames.

(iii) Deformation Gradient: Let $\mathbf{F}^*$ denote the deformation gradient in the starred frame. Then

$$\mathbf{F}^* = \frac{\partial \mathbf{x}^*}{\partial \mathbf{X}} = \frac{\partial \mathbf{x}^*}{\partial \mathbf{x}}\frac{\partial \mathbf{x}}{\partial \mathbf{X}}. \tag{4.0.19}$$

Thus,

$$\mathbf{F}^* = \mathbf{QF}. \tag{4.0.20}$$

Hence, the deformation gradient is not frame indifferent.

In the derivation above, the point $\mathbf{X}$ belonging to the body in the reference configuration is assumed to be perceived as the same, by observers associated with both the frames. That is, the reference configuration is treated as a surrogate for the abstract body which is not placed in a configuration in a three dimensional Euclidean Space accessible to the observers at times t and t^*. However, in the real world we only perceive a body in terms of its configuration in real space at some instant of time. This being so, the quantity which is physically meaningful and useful is the Relative Deformation Gradient, and we determine how this quantity changes due to a change of frame in what follows.

(iv) Relative Deformation Gradient: Let $\mathbf{F}^*_{t^*}(\tau^*)$ denote the relative deformation gradient in the starred frame. Then,

$$\mathbf{F}^*_{t^*}(\tau^*) = \frac{\partial \boldsymbol{\xi}^*}{\partial \mathbf{x}^*} = \frac{\partial \boldsymbol{\xi}^*}{\partial \boldsymbol{\xi}}\frac{\partial \boldsymbol{\xi}}{\partial \mathbf{x}}\frac{\partial \mathbf{x}}{\partial \mathbf{x}^*}. \tag{4.0.21}$$

Thus,

$$\mathbf{F}^*_{t^*}(\tau^*) = \mathbf{Q}(\tau)\mathbf{F}_t(\tau)\mathbf{Q}^{\mathrm{T}}(t). \tag{4.0.22}$$

Thus, the relative deformation gradient is not frame indifferent, as $\mathbf{Q}(\tau)$ can be different from $\mathbf{Q}(t)$. If, however, we restrict our changes of frame to only include $\mathbf{Q}$s that are independent of time, the quantity will transform like a frame independent quantity.

We shall not consider the frame indifference of the numerous quantities that one comes across in continuum mechanics. We merely provide how a few of the quantities that are relevant to a discussion of elasticity, transform.

(v) Cauchy-Green Tensors: It is trivial to show that

$$\mathbf{C}^* = \mathbf{C} \tag{4.0.23}$$

and

$$\mathbf{B}^* = \mathbf{Q}(t)\mathbf{B}\mathbf{Q}^{\mathrm{T}}(t). \tag{4.0.24}$$

Thus, $\mathbf{C}$ is not frame indifferent, while $\mathbf{B}$ is frame indifferent.

It is straightforward to consider the frame indifference of right and left Relative Cauchy Green Tensors, but we shall not do so here.

(vi) Velocity Gradient: It is straightforward to show that

$$\mathbf{L}^* = \dot{\mathbf{Q}}(t)\mathbf{Q}^{\mathrm{T}}(t) + [\mathbf{Q}(t)]\mathbf{L}[\mathbf{Q}^{\mathrm{T}}(t)]. \tag{4.0.25}$$

It immediately follows that

$$\mathbf{D}^* = \frac{1}{2}(\mathbf{L}^* + (\mathbf{L}^*)^{\mathrm{T}}) = [\mathbf{Q}(t)]\mathbf{D}[\mathbf{Q}^{\mathrm{T}}(t)], \tag{4.0.26}$$

and

$$\mathbf{W}^* = \frac{1}{2}[\dot{\mathbf{Q}}(t)\mathbf{Q}^{\mathrm{T}}(t) - \mathbf{Q}(t)\dot{\mathbf{Q}}^{\mathrm{T}}(t)] + \mathbf{Q}(t)\mathbf{W}\mathbf{Q}^{\mathrm{T}}(t). \tag{4.0.27}$$

Thus, while $\mathbf{D}$ is frame indifferent, $\mathbf{L}$ and $\mathbf{W}$ are not frame indifferent.

References

[1] C. A. Truesdell. First Course in Rational Continuum Mechanics V1. Academic Press, 1992.

[2] D. Steigmann. On frame invariance of linear elasticity theory. Zeitschrift für Angewandte Mathematik und Physik, 58:121–136, 2007.

5 Elastic bodies

As mentioned in the introduction, the class of elastic bodies is far broader than those that are usually regarded as elastic bodies. In this course, as we are primarily concerned with the classical linearized elastic solid and we shall rest content with discussing only the class of Cauchy elastic bodies and its special subclass, Green elastic bodies. Later, we will discuss a more general class of elastic bodies, which when linearized under the assumption that the displacement gradient is small leads to constitutive relations that possess a nonlinear relationship between the linearized strain and the stress.

5.1 Cauchy elastic body

A body is said to be Cauchy elastic (see [1]) if the stress $\mathbf{T}$ in the body depends on the deformation $\mathbf{F}$ at the point $\mathbf{X}$, the density ρ, and the point $\mathbf{X}$ that is

$$\mathbf{T} = \mathbf{g}(\rho, \mathbf{F}, \mathbf{X}). \tag{5.1.1}$$

Since the density ρ is related to the reference density ρ_{R} through the determinant of $\mathbf{F}$, in virtue of the balance of mass, we do not need to express the dependence on the density. However, in inhomogeneous elastic bodies, we ought to include the dependence on ρ_R. In the case of a homogeneous elastic body, we will not have any dependence on ρ_R. Since we shall be primarily interested in homogeneous bodies we shall suppress the dependence of the reference density, unless we specifically consider inhomogeneous bodies. Since we shall not consider internal couples acting on the elastic bodies under consideration, it follows from the balance of angular momentum that the Cauchy stress is symmetric.

Frame-indifference places restrictions on how the function $\mathbf{f}$ depends on the deformation gradient $\mathbf{F}$. The following theorem makes this clear.

Theorem. *Let* $\mathbf{U}$ *be the unique positive definite symmetric tensor in the polar decomposition* $\mathbf{F} = \mathbf{RU}$. *Then,*

$$\mathbf{T} = \mathbf{Rg}(\mathbf{U}, \mathbf{X})\mathbf{R}^T. \tag{5.1.2}$$

Proof. Recall that frame-indifference of the stress implies that

$$\mathbf{T}^* = \mathbf{QTQ}^{\mathrm{T}} = \mathbf{Qg}(\mathbf{F}, \mathbf{X})\mathbf{Q}^{\mathrm{T}} \tag{5.1.3}$$

$$= \mathbf{Qg}(\mathbf{RU}, \mathbf{X})\mathbf{Q}^{\mathrm{T}}. \tag{5.1.4}$$

However, frame-indifference implies that

$$\mathbf{T}^* = \mathbf{g}(\mathbf{F}^*, \mathbf{X}) \tag{5.1.5}$$

$$= \mathbf{g}(\mathbf{QF}, \mathbf{X}) \tag{5.1.6}$$

$$= \mathbf{g}(\mathbf{QRU}, \mathbf{X}) \tag{5.1.7}$$

https://doi.org/10.1515/9783110789515-005

It follows from Eq. (5.1.4) and (5.1.7) that

$$\mathbf{g}(\mathbf{QRU},\mathbf{X}) = \mathbf{Q}\mathbf{g}(\mathbf{RU},\mathbf{X})\mathbf{Q}^{\mathrm{T}}. \tag{5.1.8}$$

However, the above should be true for all orthogonal $\mathbf{Q}$. Pick $\mathbf{Q} = \mathbf{R}^{\mathrm{T}}$. Then, it follows that

$$\mathbf{g}(\mathbf{U},\mathbf{X}) = \mathbf{R}^{\mathrm{T}}\mathbf{g}(\mathbf{RU},\mathbf{X})\mathbf{R} \tag{5.1.9}$$

$$\mathbf{R}\mathbf{g}(\mathbf{U},\mathbf{X})\mathbf{R}^{\mathrm{T}} = \mathbf{g}(\mathbf{RU},\mathbf{X}) \tag{5.1.10}$$

$$= \mathbf{g}(\mathbf{F},\mathbf{X}) = \mathbf{T} \tag{5.1.11}$$

Thus,

$$\mathbf{T} = \mathbf{R}\mathbf{g}(\mathbf{U},\mathbf{X})\mathbf{R}^{\mathrm{T}}. \tag{5.1.12}$$

Recall $\mathbf{U}^2 = \mathbf{C}$. Since, $\mathbf{U}$ is positive definite symmetric, $\mathbf{C}$ has a unique square root which is associated with $\mathbf{U}$. We can thus define

$$\hat{\hat{\mathbf{g}}}(\mathbf{C},\mathbf{X}) = \mathbf{g}(\mathbf{U},\mathbf{X}). \tag{5.1.13}$$

Recall $\frac{1}{2}(\mathbf{C}-\mathbf{I}) \equiv \mathbf{E}$. Then,

$$\mathbf{g}(\mathbf{U},\mathbf{X}) = \boldsymbol{\beta}(\mathbf{E},\mathbf{X}). \tag{5.1.14}$$

Thus,

$$\mathbf{T} = \mathbf{R}\boldsymbol{\beta}(\mathbf{E},\mathbf{X})\mathbf{R}^{\mathrm{T}}. \tag{5.1.15}$$

□

Linearization

Next, we shall consider the approximation of (5.1.15) when we restrict ourselves to deformations wherein the displacement gradient is small. We shall quantify this smallness by requiring that the Frobenius norm of the displacement gradient is small (note that we are not merely requiring that the strain is small. We are requiring that the norm of the displacement gradient is small. This implies both the linearized strain and the linearized rotation are small).

Thus we assume that $\max||\nabla\mathbf{u}|| = \mathcal{O}(\delta)$, $\delta \ll 1$, $\forall\, \mathbf{X} \in \mathcal{B}_t$, $\forall\, t \in \mathbb{R}$. Recall

$$\mathbf{E} \approx \boldsymbol{\varepsilon} \approx \mathcal{O}(\delta). \tag{5.1.16}$$

Now, let us fix $\mathbf{X}$. Note $\boldsymbol{\beta}(\mathbf{E},\mathbf{X})$ is a symmetric tensor. Drop $\mathbf{X}$ for convenience and suppose $\boldsymbol{\beta}(\mathbf{0},\mathbf{X}) = \mathbf{0}$. Then:

$$\beta_{ij}(E_{mp}) = \cancelto{0}{\beta_{ij}(0)} + \left.\frac{\partial\beta_{ij}}{\partial E_{mp}}\right|_{\mathbf{E}=\mathbf{0}}(E_{mp}) + \mathcal{O}(\delta^2) \tag{5.1.17}$$

$$\beta_{ij}(E_{mp}) = \frac{\partial \beta_{ij}}{\partial E_{mp}}\bigg|_{\mathbf{E}=\mathbf{0}} (\varepsilon_{mp}) + \mathcal{O}(\delta^2). \tag{5.1.18}$$

Next, define

$$\frac{\partial \beta_{ij}}{\partial E_{mp}}\bigg|_{\mathbf{E}=\mathbf{0}} = \mathcal{C}_{ijmp}. \tag{5.1.19}$$

$\mathcal{C}$, is a fourth order tensor (a linear transformation that transforms elements belonging to the vector space that is the set of all second order tensors to the same vector space of the set of all second order tensors, that is it assigns to one second order tensor another second order tensor), and $\mathcal{C}_{ijkl}$ are its components with respect to the Cartesian base vectors.

$$\beta_{ij} = \mathcal{C}_{ijmp}\varepsilon_{mp} + \mathcal{O}(\delta^2), \tag{5.1.20}$$

where ε_{mp} are the components of the linearized strain tensor. Next, recall

$$\mathbf{R} \approx \mathbf{I} + \boldsymbol{\omega}, \tag{5.1.21}$$

where $\boldsymbol{\omega}$ is skew and of $\mathcal{O}(\delta)$. So,

$$\mathbf{R} = \mathbf{I} + \mathcal{O}(\delta). \tag{5.1.22}$$

Finally, (5.1.15), (5.1.17) and (5.1.21) imply that

$$T_{ij} = \mathcal{C}_{ijmp}\varepsilon_{mp} + \mathcal{O}(\delta^2), \tag{5.1.23}$$

where ε_{mp} is the linearized strain tensor. On neglecting terms of order $\mathcal{O}(\delta^2)$, we obtain

$$T_{ij} = \mathcal{C}_{ijmp}\varepsilon_{mp}, \tag{5.1.24}$$

or in bold-faced notation

$$\mathbf{T} = \boldsymbol{\mathcal{C}}\boldsymbol{\varepsilon}. \tag{5.1.25}$$

$\boldsymbol{\mathcal{C}}$ is referred to as the elasticity tensor. We notice that the matrix associated with the tensor has 81 components. While the fourth order elasticity tensor has 81 entries in its matrix representation, not all of them are independent. Since the stress and the linearized strain are symmetric, it immediately follows that

$$\mathcal{C}_{ijkl} = \mathcal{C}_{ijlk} = \mathcal{C}_{jikl}. \tag{5.1.26}$$

Equation (5.1.26) immediately implies that $\boldsymbol{\mathcal{C}}$ has only 36 independent entries. In the case of a Green elastic solid wherein we have stored energy associated with the elastic body, we can show that there are only 21 independent entries in the matrix for the elasticity tensor.

The number of independent components becomes fewer in number depending on the material symmetry the body possesses, that is invariance of the constitutive relation to certain group of transformations, and we discuss this in the next Chapter.

The starting point for Cauchy elasticity is the assumption that the Cauchy stress in the body depends on density and the deformation gradient. As mentioned in the introduction, it is possible that bodies described by implicit constitutive relations of the form

$$\mathbf{f}(\rho, \mathbf{T}, \mathbf{F}, \mathbf{X}) = \mathbf{0}, \tag{5.1.27}$$

can also exhibit elastic response (see Rajagopal [3, 4]). We discuss the mechanics of such bodies at the end of the book. The above development of the constitutive theory for the representation of the stress for elastic bodies does not appeal to the elastic body possessing the means to store energy due to the deformation, the energy being recoverable when the body returns to the undeformed state. This characteristic, which is usually associated with the notion of an elastic body, was first articulated by Green [2]. Such elastic bodies, as mentioned earlier, are called Green elastic bodies or hyperelastic bodies. The class of Green elastic bodies also has their counterparts in the more general class of bodies defined through implicit constitutive relations (see Rajagopal and Srinivasa [5]).

5.2 Green elastic body

In Cauchy elastic bodies that were defined earlier, the stress is assumed to be a function of the deformation gradient. No notion of energy that is stored in the elastic body that can be recovered completely is appealed to in the development of the constitutive expression. Green recognized that if one did not associate the ability of an elastic body to store energy, from which the notion of stress can be derived, one could have bodies from which infinite energy can be recovered, that is, one could use such bodies to develop perpetual motion machines. In this section, we turn our attention to a discussion of Green elastic bodies.

5.2.1 Strain energy

In order to introduce the strain energy due to the deformation of the elastic body, we need to introduce some basic quantities such as the power expended in deforming a part of the body, the kinetic energy associated with a part of the body, etc.

5.2.2 Mechanical working (time rate at which work is done)

Let $\mathcal{W}(\mathcal{D}_t)$ denote the mechanical working on a part $\mathcal{D}_t \subseteq \kappa_t(\mathcal{B})$ (see Figure 5.1). Then

$$\mathcal{W}(\mathcal{D}_t) \equiv \int_{\partial\mathcal{D}_t} \mathbf{t_n} \cdot \mathbf{v}\, da + \int_{\mathcal{D}_t} \rho\mathbf{b} \cdot \mathbf{v}\, dv \quad \forall\, \mathcal{D}_t \subseteq \kappa_t(\mathcal{B}). \tag{5.2.1}$$

The kinetic energy $\mathcal{K}(\mathcal{D}_t)$ of the sub-part $\mathcal{D}_t \subseteq \kappa_t(\mathcal{B})$ is given by

$$\mathcal{K}(\mathcal{D}_t) \equiv \int_{\mathcal{D}_t} \frac{1}{2}\rho|\mathbf{v}|^2\, dv \quad \forall\, \mathcal{D}_t \subseteq \kappa_t(\mathcal{B}). \tag{5.2.2}$$

Let $p(\mathcal{D}_t)$ denote the power of deformation, i. e.,

$$p(\mathcal{D}_t) = \mathcal{W}(\mathcal{D}_t) - \dot{\mathcal{K}}(\mathcal{D}_t) \quad \forall\, \mathcal{D}_t \subseteq \kappa_t(\mathcal{B}). \tag{5.2.3}$$

In the above equation, the superscript "dot" denotes derivative with respect to time.

5.2.3 Power theorem

$$\mathcal{W}(\mathcal{D}_t) = \frac{d\mathcal{K}(\mathcal{D}_t)_+}{dt} \int_{\mathcal{D}_t} (\mathbf{T}^{\mathrm{T}} \cdot \operatorname{grad} \mathbf{v})\, dv \quad \forall\, \mathcal{D}_t \subseteq \kappa_t(\mathcal{B}). \tag{5.2.4}$$

The above theorem states the time rate at which external forces are doing work on a sub-part $\mathcal{D}_t$ of $\kappa_t(\mathcal{B})$ equals the time rate of change of the kinetic energy of the sub-part $\mathcal{D}_t$ plus the rate at which the internal forces are doing work on the sub-part $\mathcal{D}_t$.

Recall the balance of linear momentum

$$\rho\frac{d\mathbf{v}}{dt} = \operatorname{div} \mathbf{T}^T + \rho\mathbf{b}. \tag{5.2.5}$$

On forming the scalar product of the above equation with the velocity $\mathbf{v}$ and integrating over the sub-part $\mathcal{D}_t$ leads to

$$\int_{\mathcal{D}_t} \left[(\operatorname{div} \mathbf{T}^T) \cdot \mathbf{v} + \rho\mathbf{b} \cdot \mathbf{v} - \rho\frac{d\mathbf{v}}{dt} \cdot \mathbf{v}\right] dv = 0 \quad \forall\, \mathcal{D}_t \subseteq \kappa_t(\mathcal{B}). \tag{5.2.6}$$

We note that

$$\int_{\mathcal{D}_t} (\operatorname{div} \mathbf{T}^T) \cdot \mathbf{v} dv = \int_{\partial\mathcal{D}_t} \mathbf{t_n} \cdot \mathbf{v} da - \int_{\mathcal{D}_t} \mathbf{T}^T \cdot \operatorname{grad} \mathbf{v}\, dv \quad \forall\, \mathcal{D}_t \subseteq \kappa_t(\mathcal{B}). \tag{5.2.7}$$

In obtaining the above identity we have appealed to the divergence theorem to express

$$\int_{\mathcal{D}_t} \operatorname{div}(\mathbf{T}^T \mathbf{\upsilon})\, dv = \int_{\partial \mathcal{D}_t} \mathbf{T}^T \mathbf{n} \cdot \mathbf{\upsilon}\, da \quad \forall\, \mathcal{D}_t \subseteq \kappa_t(\mathcal{B}) \tag{5.2.8}$$

and

$$(\operatorname{div} \mathbf{T}^T) \cdot \mathbf{\upsilon} = \operatorname{div}(\mathbf{T}^T \mathbf{\upsilon}) - \mathbf{T}^T \cdot \operatorname{grad} \mathbf{\upsilon}. \tag{5.2.9}$$

The power of deformation can also be expressed as

$$\mathcal{W}(\mathcal{D}_t) \equiv \frac{d\mathcal{K}(\mathcal{D}_t)}{dt} + \int_{\mathcal{D}_t} \frac{1}{(\det \mathbf{F})} \mathbf{S}^{\mathrm{T}} \cdot \dot{\mathbf{F}}\, dv \quad \forall\, \mathcal{D}_t \subseteq \kappa_t(\mathcal{B}) \tag{5.2.10}$$

$$\mathcal{W}(\mathcal{D}_t) \equiv \frac{d\mathcal{K}(\mathcal{D}_t)}{dt} + \int_{\mathcal{D}_R} \mathbf{S}^{\mathrm{T}} \cdot \dot{\mathbf{F}}\, dV \quad \forall\, \mathcal{D}_R \subseteq \kappa_R(\mathcal{B}) \tag{5.2.11}$$

where $\mathcal{D}_R$ is the pre-image of $\mathcal{D}_t$ and $\mathbf{S} = (\det \mathbf{F})\mathbf{F}^{-1}\mathbf{T}$. The quantity $\mathbf{S}$ is referred to as the Piola-Kirchhoff tensor.

Hence, it follows from elementary calculation that

$$p(\mathcal{D}_t) = \int_{\mathcal{D}_t} \mathbf{T} \cdot \operatorname{grad} \mathbf{\upsilon}\, dv \quad \forall\, \mathcal{D}_t \subseteq \kappa_t(\mathcal{B}), \tag{5.2.12}$$

$$= \int_{\mathcal{D}_0} \mathbf{S}^T \cdot \mathbf{F}\, dV \quad \forall\, \mathcal{D}_0 \subseteq \kappa_t(\mathcal{B}). \tag{5.2.13}$$

5.2.4 Work of deformation

Consider a given time interval (t_1, t_2), and define

$$\mathscr{W}(\mathcal{D}_{t_1}, \mathcal{D}_{t_2}) = \int_{t_1}^{t_2} p(\mathcal{D}_t)\, dt. \tag{5.2.14}$$

Now, $\mathscr{W}(\mathcal{D}_{t_1}, \mathcal{D}_{t_2}) > 0$ implies that work is expended during the interval of time while $\mathscr{W}(\mathcal{D}_{t_1}, \mathcal{D}_{t_2}) < 0$ implies deformation work is extracted during the interval of time.

Work postulate

Let κ_0 denote the natural configuration of the elastic body (i. e., zero stress and undistorted). Consider a motion $\kappa_{t_0} \longrightarrow \kappa_{t_f}$, $t \in [t_0, t_f]$, then for any $\mathcal{D}_{t_f} \subseteq \kappa_t(\mathcal{B})$ which is distorted we must have

$$\mathscr{W}(\mathcal{D}_{t_0}, \mathcal{D}_{t_f}) > 0. \tag{5.2.15}$$

Here, $\mathcal{D}_{t_f}$ and $\mathcal{D}_{t_0}$ are not related by a rigid transformation, i. e., $\mathbf{F}(t_f) \neq \mathbf{Q}$ (everywhere) where $\mathbf{Q} \in \mathscr{O}^+$. Any motion which meets the above postulate is called work *admissible*.

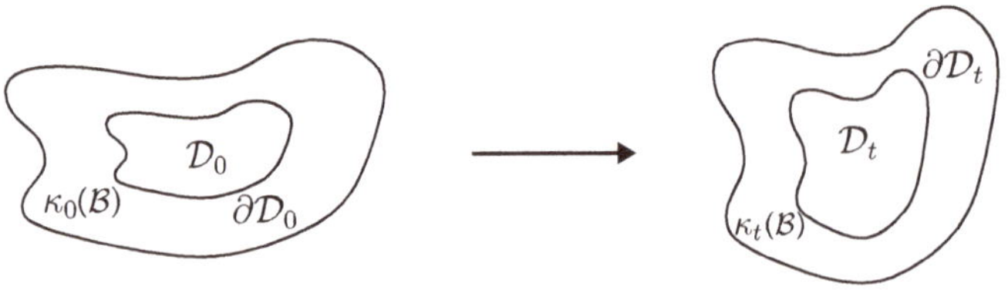

Figure 5.1: Sub-domains in the reference and current configurations and their boundaries.

Definition 5.2.1 (Closed deformation). A motion for κ_0 is said to be a *closed deformation* for the time interval (t_1, t_2) if $\mathcal{D}_{t_1} = \mathcal{D}_{t_2}\ \forall\ \mathcal{D}_t \subseteq \Omega_t$.

Theorem 5.2.1 (Elastic solids). *Assume all motions are work admissible. Then, for any motion that is a "closed deformation" over the interval* $[t_1, t_2]$, *the work of deformation meets*

$$\mathscr{W}(\mathcal{D}_{t_1}, \mathcal{D}_{t_2}) = 0 \quad \forall\ \mathcal{D}_t \subset \Omega_t \tag{5.2.16}$$

provided $\mathcal{D}_{t_1} = \mathcal{D}_{t_2}\ \forall\ \mathcal{D}_t \subset \kappa_t$.

Proof of the theorem depends on the following lemma.

Lemma 5.2.2. *Assume all motions are work admissible. Let* $\kappa_0(\mathcal{B})$ *be a natural configuration. Then every closed deformation during* $(0, t_f)$ *has zero work of deformation.*

Remark. This lemma is the same as the theorem except that $\kappa_0(\mathcal{B})$ is natural (stress free), where $\kappa_{t_1}(\mathcal{B})$ and $\kappa_{t_2}(\mathcal{B})$ are not necessarily natural.

Remark. The lemma proves Theorem 5.2.1.

Suppose we have a closed deformation in $[t_1, t_2]$.

Proof. Assume we are given an arbitrary but possible motion with closed deformation $\kappa_0(\mathcal{B}) \longrightarrow \kappa_{t_f}(\mathcal{B}) \equiv \kappa_0(\mathcal{B})$. The deformation gradient $\mathbf{F}(t)$ is given for $t \in [0, t_f]$, $\mathbf{F}(0) = \mathbf{F}(t_f) = \mathbf{I}$ (see Figure 5.2).

Consider the following motion $\kappa_0(\mathcal{B}) \longrightarrow \kappa_{t_f^+}(\mathcal{B})$ where $t_f^+ > t_f$. Then

$$\mathbf{F}^+(t) \equiv \begin{cases} \mathbf{F}(t) & t \in [0, t_f] \\ \mathbf{I} + (t - t_f)\mathbf{A} & t \in (t_f, t_f^+], \end{cases} \tag{5.2.17}$$

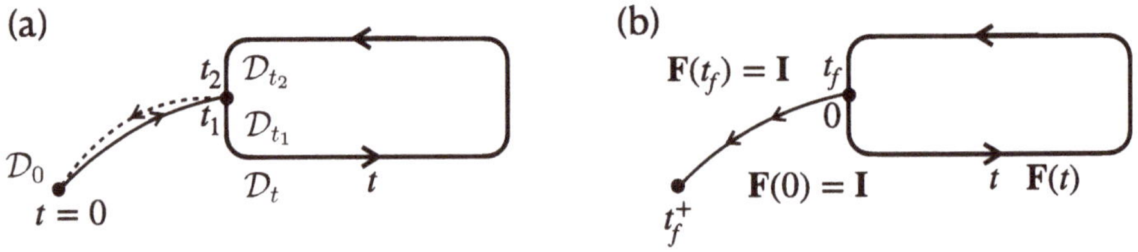

Figure 5.2: History of deformation.

where $\mathbf{A}$ is an arbitrary constant positive-definite, symmetric tensor, and $t_f^+ > t_f$ is defined such that

$$\det(\mathbf{I} + (t - t_f)\mathbf{A}) > 0 \quad \forall\, t \in (t_f, t_f^+]. \tag{5.2.18}$$

Remark. There exists $t_f^+ > t_f$ such that this is met, since $\det(\mathbf{I} + (t - t_f)\mathbf{A})$ is a continuous function of time and has a value equal to $+1$ at $t = t_f$.

Remark. Every $\mathcal{D}_t \subseteq \Omega_t$ is distorted during the interval $(t_f, t_f^+]$.

Remark. $\mathbf{I} + (t - t_f)\mathbf{A} \notin \mathcal{O}$. Suppose $\mathbf{I} + (t - t_f)\mathbf{A} \in \mathcal{O}$, then

$$(\mathbf{I} + (t - t_f)\mathbf{A})^{\mathrm{T}}(\mathbf{I} + (t - t_f)\mathbf{A}) = \mathbf{I} \quad t \in (t_f, t_f^+), \tag{5.2.19}$$

which implies

$$(t - t_f)\mathbf{A}[2\mathbf{I} + (t - t_f)\mathbf{A}] = \mathbf{0} \quad \forall\, \mathbf{A} \in \mathcal{J}_S, \tag{5.2.20}$$

where $\mathcal{J}_S$ is the set of all symmetric linear transformations.

Note, $\mathbf{A}$ is also positive-definite. Therefore,

$$(t - t_f)[2\mathbf{I} + (t - t_f)\mathbf{A}] = \mathbf{0}. \tag{5.2.21}$$

Thus, since t is not equal to t_f,

$$2\mathbf{I} + (t - t_f)\mathbf{A} = \mathbf{0}. \tag{5.2.22}$$

But $\mathbf{A}$ is positive-definite, and since $(t - t_f) > 0$, Eq. (5.2.22) cannot be met.

The work postulate implies that

$$P^+(\mathcal{D}_0, \mathcal{D}_\tau) \equiv \int_0^\tau p^+(\mathcal{D}_t)\, dt > 0 \quad \forall\, \tau \in (t_f, t_f^+], \tag{5.2.23}$$

where

$$p^+(\mathcal{D}_t) = \int_{\mathcal{D}_0} \operatorname{tr}(\mathbf{S}^+ \dot{\mathbf{F}}^+)\, dV. \tag{5.2.24}$$

Here, $\mathbf{S}^+ = \mathbf{S}(\mathbf{F}^+, \mathbf{X})$ is the given response function. Now,

$$\lim_{\tau \to t_f} P^+(\mathcal{D}_0, \mathcal{D}_\tau) = P(\mathcal{D}_0, \mathcal{D}_{t_f}) \geq 0, \tag{5.2.25}$$

where

$$P(\mathcal{D}_0, \mathcal{D}_{t_f}) = \int_0^{t_f} p(\mathcal{D}_t)\, dt \tag{5.2.26}$$

and

$$p(\mathcal{D}_t) = \int_{\mathcal{D}_t} \operatorname{tr}(\mathbf{S}\dot{\mathbf{F}})\, dV. \tag{5.2.27}$$

Now consider $\kappa_0(\mathcal{B}) \longrightarrow \kappa_{t_f}(\mathcal{B}) \longrightarrow \kappa_{t_f^+}(\mathcal{B})$ where $\kappa_0(\mathcal{B}) \longrightarrow \kappa_{t_f}(\mathcal{B})$ is reversed. Then,

$$^{-}\mathbf{F} = \begin{cases} \mathbf{F}(t - t_f) & t \in [0, t_f] \\ \mathbf{I} + (t - t_f)\mathbf{A} & t \in [t_f, t_f^+]. \end{cases} \tag{5.2.28}$$

The work postulate implies that

$$^{-}P(\mathcal{D}_0, \mathcal{D}_\tau) \equiv \int_0^{\tau} {}^{-}p(\mathcal{D}_t)\, dt > 0 \tag{5.2.29}$$

provided that $\tau \in (t_f, t_f^+)$. Here,

$$^{-}p(\mathcal{D}_t) = \int_{\mathcal{D}_t} \operatorname{tr}({}^{-}\mathbf{S}\,{}^{-}\dot{\mathbf{F}})\, dV \tag{5.2.30}$$

and

$$^{-}\mathbf{S} = \mathbf{S}({}^{-}\mathbf{F}, \mathbf{X}). \tag{5.2.31}$$

Now,

$$\lim_{\tau \to t_f} {}^{-}P(\mathcal{D}_0, \mathcal{D}_\tau) = {}^{-}P(\mathcal{D}_0, \mathcal{D}_{t_f}) \geq 0 \quad \forall\, \mathcal{D}_t \subset \Omega_t, \tag{5.2.32}$$

where

$$P(\mathcal{D}_0, \mathcal{D}_{t_f}) = \int_0^{t_f} {}^{-}p(\mathcal{D}_t)\, dt \tag{5.2.33}$$

and

$$^{-}\mathbf{S} = \mathbf{S}({}^{-}\mathbf{F}, \mathbf{X}) = \mathbf{S}(\mathbf{F}(t - t_f), \mathbf{X}) \tag{5.2.34}$$

and

$$ {}^{-}\dot{\mathbf{F}} = \frac{d\mathbf{F}}{dt}(t_f - t) = \frac{d}{d\lambda}F(\lambda)\bigg|_{\lambda=t_f-t}. \tag{5.2.35}$$

Here, we change t to $\lambda = t_f - t$. Then

$$ {}^{-}P(\mathcal{D}_0, \mathcal{D}_{t_f}) = -\int_{t_f}^{0} {}^{-}p(\mathcal{D}_{t_f-\lambda})\, d\lambda = \int_{0}^{t_f} {}^{-}p(\mathcal{D}_{t_f-\lambda})\, d\lambda, \tag{5.2.36}$$

where

$$ {}^{-}p(\mathbf{D}_{t_f-\lambda}) = -\int_{\mathcal{D}_0} \operatorname{tr}\left[\mathbf{S}(\mathbf{F}(\lambda), \mathbf{X})\, \frac{d\mathbf{F}(\lambda)}{d\lambda}\right] dV. \tag{5.2.37}$$

Therefore,

$$ {}^{-}P(\mathcal{D}_0, \mathcal{D}_{t_f}) = -P(\mathcal{D}_0, \mathcal{D}_{t_f}). \tag{5.2.38}$$

Therefore,

$$ P(\mathcal{D}_0, \mathcal{D}_{t_f}) = 0 \quad \forall\, \mathcal{D}_t \subset \Omega_t. \tag{5.2.39}$$

Hence, Lemma 5.2.2 is proved. This then proves Theorem 5.2.1. □

Corollary 5.2.2.1. *We have*

$$ \oint_{t_1}^{t_2} \operatorname{tr}(\mathbf{S}\dot{\mathbf{F}})\, dt = 0 \quad \forall\, \mathbf{X} \in \Omega_0, \tag{5.2.40}$$

for every closed deformation in (t_1, t_2).

Proof. Theorem 5.2.1 implies Eq. (5.2.40). □

5.2.5 Work postulate (and its linearization)

We recall that within the context of linearization, the Cauchy stress and Piola-Kirchhoff stress both reduce to the stress in a linearized elastic body, and this is given within order ϵ^2 by

$$ \mathbf{S} \doteq \mathbf{T} \doteq \boldsymbol{\tau} = \mathcal{C}[\varepsilon], \tag{5.2.41}$$

where $\mathbf{S}$, $\mathbf{T}$, $\boldsymbol{\tau}$ are the Piola-Kirchhoff stress, Cauchy stress, and the stress in a linearized elastic body, respectively, and $\mathcal{C}$ is the elasticity tensor and ε is the linearized strain tensor. The deformation gradient and the time rate of deformation gradient are given by

$$\mathbf{F} = \mathbf{I} + \nabla\mathbf{u} \implies \dot{\mathbf{F}} = \dot{\overline{\nabla\mathbf{u}}} = \nabla\dot{\mathbf{u}}. \tag{5.2.42}$$

Now,

$$\operatorname{tr}(\mathbf{S}\dot{\mathbf{F}}) = \mathbf{S}^{\mathrm{T}} \cdot \dot{\mathbf{F}} = \operatorname{tr}(\boldsymbol{\tau}\nabla\dot{\mathbf{u}}) + \mathcal{O}(\epsilon^2). \tag{5.2.43}$$

Recall that

$$\boldsymbol{\tau} = \boldsymbol{\tau}^{\mathrm{T}}, \quad \boldsymbol{\varepsilon} = \frac{1}{2}(\nabla\mathbf{u} + (\nabla\mathbf{u})^{\mathrm{T}}). \tag{5.2.44}$$

Thus,

$$\operatorname{tr}(\mathbf{S}\dot{\mathbf{F}}) \doteq \operatorname{tr}(\boldsymbol{\tau}\dot{\boldsymbol{\varepsilon}}). \tag{5.2.45}$$

Corollary 5.2.2.1 implies that for all closed deformations on the interval (t_1, t_2), we have

$$\oint_{t_1}^{t_2} \operatorname{tr}(\boldsymbol{\tau}\dot{\boldsymbol{\varepsilon}})\, dt = 0 \quad \forall\, \mathbf{X} \in \Omega_0 \tag{5.2.46}$$

Theorem 5.2.3. *A necessary and sufficient condition for Eq.* (5.2.46) *to hold is*

$$\mathcal{C}_{ijkl} = \mathcal{C}_{klij}. \tag{5.2.47}$$

Proof. Consider

$$u_i = \epsilon\{(\cos t - 1)A_{ij}X_j + \sin tB_{ij}X_j\}, \tag{5.2.48}$$

where $\epsilon < 1$ and $\mathbf{A} = \mathbf{A}^{\mathrm{T}}$ and $\mathbf{B} = \mathbf{B}^{\mathrm{T}}$. Using this displacement, we can compute

$$\int_0^{\pi} \operatorname{tr}(\boldsymbol{\tau}\dot{\boldsymbol{\varepsilon}})\, dt = -\pi\epsilon^2(\mathcal{C}_{ijkl} - \mathcal{C}_{klij})A_{ij}B_{kl}. \tag{5.2.49}$$

Now, Eq. (5.2.46) implies that

$$(\mathcal{C}_{ijkl} - \mathcal{C}_{klij})A_{ij}B_{kl} = 0 \quad \forall\, \mathbf{A}, \mathbf{B} \in \mathcal{T}_s. \tag{5.2.50}$$

Therefore,

$$\mathcal{C}_{klij} = \mathcal{C}_{ijkl}, \tag{5.2.51}$$

and sufficiency is proved. To prove the necessary condition, consider

$$\tau_{ij}\dot{\varepsilon}_{ij} = \mathcal{C}_{ijkl}\varepsilon_{kl}\dot{\varepsilon}_{ij} = \mathcal{C}_{ijkl}\dot{\varepsilon}_{ij}\varepsilon_{kl} = \mathcal{C}_{klij}\dot{\varepsilon}_{ij}\varepsilon_{kl} = \dot{\tau}_{kl}\varepsilon_{kl}. \tag{5.2.52}$$

Thus,

$$\overline{\tau_{ij}\varepsilon_{ij}}^{\cdot} = \dot{\tau}_{ij}\varepsilon_{ij} + \tau_{ij}\dot{\varepsilon}_{ij} = 2\tau_{ij}\dot{\varepsilon}_{ij}, \tag{5.2.53}$$

and

$$\oint_{t_1}^{t_2} \operatorname{tr}(\boldsymbol{\tau}\dot{\boldsymbol{\varepsilon}})\, dt = \frac{1}{2}\operatorname{tr}(\boldsymbol{\tau}\boldsymbol{\varepsilon})\Big|_{t_1}^{t_2} = 0 \tag{5.2.54}$$

for all motions with closed deformation on the interval (t_1, t_2). Therefore, Eq. (5.2.46) is satisfied and the necessary condition is proved. □

Definition 5.2.2 (Strain energy density for a linearized elastic body). The "linearized" strain energy density per unit reference volume is given by

$$\mathbb{E}(\boldsymbol{\varepsilon}, \mathbf{X}) \equiv \frac{1}{2}C_{ijkl}(\mathbf{X})\varepsilon_{ij}\varepsilon_{kl}, \tag{5.2.55}$$

or

$$\mathbb{E} = \frac{1}{2}\operatorname{tr}(\boldsymbol{\tau}\boldsymbol{\varepsilon}). \tag{5.2.56}$$

If $\boldsymbol{C}$ is invertible, we can define Γ (to imply complementary energy density)

$$\Gamma = \tilde{\Gamma}(\boldsymbol{\tau}, \mathbf{X}) = K_{ijkl}\tau_{ij}\tau_{kl}, \tag{5.2.57}$$

where $\mathbf{K}$ denotes the compliance tensor and

$$\tau_{ij} = \frac{\partial \mathbb{E}(\boldsymbol{\varepsilon}, \mathbf{X})}{\partial \varepsilon_{ij}}. \tag{5.2.58}$$

Let us now return to the work postulate. We have

$$P(\mathcal{D}_0, \mathcal{D}_{t_f}) = \int_0^{t_f} p(\mathcal{D}_t)\, dt > 0, \tag{5.2.59}$$

where $\mathcal{D}_{t_f} \subset \kappa_{t_f}$ is distorted for all $\mathcal{D}_0$. Then,

$$p(\mathcal{D}_t) = \int_{\mathcal{D}_0} \operatorname{tr}(\mathbf{S}\dot{\mathbf{F}})\, dv \doteq \int_{\mathcal{D}_0} \operatorname{tr}(\boldsymbol{\tau}\dot{\boldsymbol{\varepsilon}})\, dv. \tag{5.2.60}$$

Therefore,

$$\int_0^{t_f} \operatorname{tr}(\boldsymbol{\tau}\dot{\boldsymbol{\varepsilon}})\, dt \geq 0, \tag{5.2.61}$$

whenever $\varepsilon(t_f) \neq \varepsilon(0) = \mathbf{0}$ ($\varepsilon(0) = \mathbf{0}$ since the natural state is unstrained). But $\mathrm{tr}(\boldsymbol{\tau}\dot{\varepsilon}) = \dot{\mathbb{E}}(\varepsilon, \mathbf{X})$, so

$$\int_0^t \dot{\mathbb{E}}\, dt \geq 0, \tag{5.2.62}$$

implying that $\mathbb{E}(t_f) - \mathbb{E}(0) \geq 0$. Since $\varepsilon(t_f) \neq \varepsilon(0) = \mathbf{0}$, $\mathbb{E}(t_f) \geq 0$. Therefore,

$$C_{ijkl}\varepsilon_{ij}\varepsilon_{kl} \geq 0 \quad \forall\, \varepsilon \neq \mathbf{0},\ \forall\, \mathbf{X} \in \Omega_0. \tag{5.2.63}$$

Let us assume that

$$C_{ijkl}\varepsilon_{ij}\varepsilon_{kl} \geq 0 \quad \forall\, \varepsilon \neq \mathbf{0}, \tag{5.2.64}$$

with the equality holding if and only if $\varepsilon \equiv \mathbf{0}$, i. e., the strain energy is a positive definite quadratic form. Then,

Theorem 5.2.4. *A necessary and sufficient condition that $\mathbb{E}$ be positive definite is*

$$\det \boldsymbol{C} > 0. \tag{5.2.65}$$

Consider the strain energy of the body that is defined through

$$\mathcal{U} \equiv \int_{\Omega_0} \mathbb{E}(\varepsilon, \mathbf{X})\, dV. \tag{5.2.66}$$

This gives

$$\dot{\mathcal{U}} + \dot{\mathcal{K}} = \int_{\partial\Omega_0} \mathbf{t} \cdot \dot{\mathbf{u}}\, da + \int_{\Omega_0} \rho_0 \mathbf{b} \cdot \dot{\mathbf{u}}\, dV, \tag{5.2.67}$$

where

$$\dot{\mathcal{U}} = \int_{\Omega_0} \dot{\mathbb{E}}\, dV = \int_{\Omega_0} \tau_{ij}\dot{\varepsilon}_{ij}\, dV = \int_{\Omega_0} \tau_{ij}\dot{u}_{i,j}\, dV \tag{5.2.68}$$

$$= \int_{\partial\Omega_0} \tau_{ij}\dot{u}_i n_j\, da - \int_{\Omega_0} \tau_{ij,j}\dot{u}_i\, dV \tag{5.2.69}$$

$$= \int_{\partial\Omega_0} \mathbf{t} \cdot \dot{\mathbf{u}}\, da + \int_{\Omega_0} \rho_0 \mathbf{b} \cdot \dot{\mathbf{u}}\, dV - \int_{\Omega_0} \rho_0 \ddot{\mathbf{u}} \cdot \dot{\mathbf{u}}\, dV. \tag{5.2.70}$$

Now, we consider the strain energy for isotropic materials (see next chapter for discussion of material symmetry). That is, bodies whose constitutive relations are given by

$$\boldsymbol{\tau} = \lambda(\mathrm{tr}\,\varepsilon)\mathbf{I} + 2\mu\varepsilon, \tag{5.2.71}$$

$$\mathbb{E}(\varepsilon, \mathbf{X}) = \frac{\lambda}{2}(\mathrm{tr}\,\varepsilon)^2 + \mu\,\mathrm{tr}\,\varepsilon^2, \tag{5.2.72}$$

where λ and μ are Lame's first and second constant, respectively. Equation (5.2.72) implies that the strain energy density can be written as

$$\mathbb{E} = \frac{E}{2(1+\nu)}\left\{\frac{\nu}{1-2\nu}(\mathrm{tr}\,\varepsilon)^2 + \mathrm{tr}\,\varepsilon^2\right\} \tag{5.2.73}$$

$$= \frac{1}{2E}\{(1+\nu)\,\mathrm{tr}\,\boldsymbol{\tau}^2 - \nu(\mathrm{tr}\,\boldsymbol{\tau})^2\}, \tag{5.2.74}$$

where E is Young's modulus and ν is Poisson's ratio where

$$E = \frac{\mu(2\lambda + 3\mu)}{\lambda + \mu}$$

$$\nu = \frac{\lambda}{2(\lambda + \mu)}.$$

Theorem 5.2.5. *A necessary and sufficient condition for $\mathbb{E}(\varepsilon, \mathbf{X})$ to be positive definite (isotropic media) is either $\mu > 0$, $(3\lambda + 2\mu) > 0$ or $E > 0$, $-1 < \nu < 1/2$.*

Proof. Recall that we can decompose the linearized strain tensor such that

$$\varepsilon = \varepsilon\mathbf{I} + \bar{\varepsilon}, \quad \mathrm{tr}\,\bar{\varepsilon} = 0. \tag{5.2.75}$$

Now,

$$(\mathrm{tr}\,\varepsilon)^2 = 9\varepsilon^2 \tag{5.2.76}$$

$$\mathrm{tr}\,\varepsilon^2 = \mathrm{tr}\{(\varepsilon\mathbf{I} + \bar{\varepsilon})(\varepsilon\mathbf{I} + \bar{\varepsilon})\} = 3\varepsilon^2 + \mathrm{tr}\,\bar{\varepsilon}^2. \tag{5.2.77}$$

Thus,

$$\mathbb{E} = \frac{3}{2}(3\lambda + 2\mu)\varepsilon^2 + \mu\,\mathrm{tr}\,\bar{\varepsilon}^2, \tag{5.2.78}$$

where ε and $\bar{\varepsilon}$ are independent. Therefore, $\mathbb{E}$ for all $\varepsilon \neq 0$ gives
1. $(3\lambda + 2\mu)\varepsilon^2 \quad \forall\,\varepsilon \neq 0.$
2. $\mu\,\mathrm{tr}\,\bar{\varepsilon}^2 > 0 \quad \forall\,\bar{\varepsilon}.$ □

Example 5.2.1. Decompose $\boldsymbol{\tau}$ into isotropic and pure shear states such that

$$\boldsymbol{\tau} = -p\mathbf{I} + \bar{\boldsymbol{\tau}}, \tag{5.2.79}$$

where $p = -(1/3)\,\mathrm{tr}\,\boldsymbol{\tau}$. Show that

$$\epsilon = \hat{\epsilon} + \bar{\epsilon}, \tag{5.2.80}$$

where $\hat{\epsilon}$ and $\bar{\epsilon}$ are given by

$$\hat{\epsilon} = \frac{3(1-2\nu)}{2E}p^2 = \frac{1-2\nu}{6E}(\sigma_1 + \sigma_2 + \sigma_3)^2, \tag{5.2.81}$$

$$\bar{\epsilon} = \frac{1}{12\mu}\{(\sigma_1 - \sigma_2)^2 + (\sigma_2 - \sigma_3)^2 + (\sigma_3 - \sigma_1)^2\}. \tag{5.2.82}$$

Here, $\hat{\epsilon}$ is the strain energy due to volume change and $\bar{\epsilon}$ is the strain energy due to shear distortion. Complete the solution.

5.3 Problems

5.1 Show that $C_{ijkl} = C_{ijlk} = C_{jikl}$

References

[1] A. L. Cauchy. Recherches sur l'équilibre et le mouvement intérieur des corps solides ou fluides, élastiques ou non élastiques. 1822.

[2] G. Green. On the laws of reflexion and refraction of light at the common surface of two noncrystallized media (1837). Trans Cambr Phil Soc 1839; 7(1839–1842): 1-24. Mathematical Papers of the Late George Green, edited by N. M. Ferris, pp. 245–269. MacMillan and Company, London, 1871.

[3] K. R. Rajagopal. Implicit constitutive theories. Applications in Mathematics, 48:279–319, 2003.

[4] K. R. Rajagopal. Conspectus of concepts of elasticity. Mathematics and Mechanics of Solids, 16(5):536–562, 2011.

[5] K. R. Rajagopal and A. Srinivasa. On the response of non-dissipative solids. Proceedings of the Royal Society A: Mathematical, Physical and Engineering Sciences, 463(2078):357–367, 2007.

6 Material symmetry for elastic bodies

Here, we present a rudimentary discussion on material symmetry. The subject matter requires much more careful and detailed discussion but we cannot do that here. We refer the reader to Truesdell [1] for details of the same. Given a material point **X** belonging to the body in a particular reference configuration, material symmetry concerns the set of transformations of the reference configuration with regard to which the response at the point **X** of the body remains the same. Thus, in order to talk about material symmetry, we need to know the class of constitutive relations that we are considering. Here, we are interested in constitutive relations for elastic bodies. There are two schools of thought with regard to the class of transformations which are allowed, the first that subscribes to the transformations being rotations, and the second that allows any unimodular transformation. Truesdell [1] uses the terminology "peer group" to refer to the material symmetry group. We shall follow his approach to material symmetry.

Let $\kappa_1(\mathcal{B})$ and $\kappa_2(\mathcal{B})$ denote two reference configurations of a body. Next, consider the motion $\kappa_1(\mathcal{B}) \to \kappa_t(\mathcal{B})$. In an elastic material, the stress **T** is given by

$$\mathbf{T} = \boldsymbol{f}_{\kappa_1}(\mathbf{F}_{\kappa_1}, \mathbf{X}^{\kappa_1}), \quad \mathbf{F}_{\kappa_1} = \frac{\partial \mathbf{x}}{\partial \mathbf{X}^{\kappa_1}}. \tag{6.0.1}$$

Next, consider the motion $\kappa_2(\mathcal{B}) \longrightarrow \kappa_t(\mathcal{B})$. Now the stress is given by

$$\mathbf{T} = \boldsymbol{f}_{\kappa_2}(\mathbf{F}_{\kappa_2}, \mathbf{X}^{\kappa_2}), \quad \mathbf{F}_{\kappa_2} = \frac{\partial \mathbf{x}}{\partial \mathbf{X}^{\kappa_2}}. \tag{6.0.2}$$

Let us suppose that $\lambda : \kappa_1(\mathcal{B}) \to \kappa_2(\mathcal{B})$ is such that the transformation from $\kappa_1(\mathcal{B})$ to $\kappa_2(\mathcal{B})$ is an orthogonal transformation. Since we are considering the same elastic body, we need

$$\boldsymbol{f}_{\kappa_1}(\mathbf{F}_{\kappa_1}, \mathbf{X}^{\kappa_1}) = \boldsymbol{f}_{\kappa_2}(\mathbf{F}_{\kappa_2}, \mathbf{X}^{\kappa_2}). \tag{6.0.3}$$

Since by chain rule

$$\frac{\partial \mathbf{x}}{\partial \mathbf{X}^{\kappa_1}} = \frac{\partial \mathbf{x}}{\partial \mathbf{X}^{\kappa_2}} \frac{\partial \mathbf{X}^{\kappa_2}}{\partial \mathbf{X}^{\kappa_1}} \tag{6.0.4}$$

and

$$\mathbf{F}_{\kappa_1} = \mathbf{F}_{\kappa_2}\mathbf{P}, \quad \mathbf{P} \equiv \frac{\partial \mathbf{X}^{\kappa_2}}{\partial \mathbf{X}^{\kappa_1}} \tag{6.0.5}$$

$$\mathbf{F}_{\kappa_2} = \mathbf{F}_{\kappa_1}\mathbf{P}^{\mathrm{T}}, \tag{6.0.6}$$

we obtain

$$\boldsymbol{f}_{\kappa_1}(\mathbf{F}_{\kappa_1}, \mathbf{X}^{\kappa_1}) = \boldsymbol{f}_{\kappa_2}(\mathbf{F}_{\kappa_1}\mathbf{P}^{\mathrm{T}}, \mathbf{X}^{\kappa_2}). \tag{6.0.7}$$

https://doi.org/10.1515/9783110789515-006

6.1 Definition of material symmetry for an elastic body

Let $\kappa_1(\mathcal{B})$ and $\kappa_2(\mathcal{B})$ be two reference configurations such that the mapping $\kappa_1(\mathcal{B}) \longrightarrow \kappa_2(\mathcal{B})$ is an invertible C^1 mapping λ. Let $\kappa_1(\mathcal{B}) \longrightarrow \kappa_t(\mathcal{B})$ and $\kappa_2(\mathcal{B}) \longrightarrow \kappa_t(\mathcal{B})$ be such that

$$\mathbf{F}_{\kappa_1} = \nabla_{\kappa_1}\mathbf{x} = \mathbf{F}_{\kappa_2} = \nabla_{\kappa_2}\mathbf{x} = \hat{\mathbf{F}}, \tag{6.1.1}$$

where $\nabla_{\kappa_1}\mathbf{x} = \frac{\partial \mathbf{x}}{\partial \mathbf{X}_{\kappa_1}}$ and $\nabla_{\kappa_2}\mathbf{x} = \frac{\partial \mathbf{x}}{\partial \mathbf{X}_{\kappa_2}}$ at the abstract material point $\mathcal{P}$.

Definition 6.1.1. If with

$$\rho_{\kappa_1}(\mathbf{X}^{\kappa_1}) = \rho_{\kappa_2}(\mathbf{X}^{\kappa_2}) \tag{6.1.2}$$

we also have

$$\mathbf{T} = \boldsymbol{f}_{\kappa_1}(\hat{\mathbf{F}}, \mathbf{X}^{\kappa_1}) = \boldsymbol{f}_{\kappa_2}(\hat{\mathbf{F}}, \mathbf{X}^{\kappa_2}), \tag{6.1.3}$$

then the body is said to possess material symmetry at $\mathcal{P}$ with respect to $\lambda(\cdot)$.

Thus, if the body is materially symmetric at $\mathcal{P}$ (or say $\mathbf{X}_{\kappa_1}$) with respect to λ, then

$$\boldsymbol{f}_{\kappa_1}(\hat{\mathbf{F}}, \mathbf{X}^{\kappa_1}) = \boldsymbol{f}_{\kappa_2}(\hat{\mathbf{F}}, \mathbf{X}^{\kappa_2}). \tag{6.1.4}$$

Thus by Eq. (6.1.3), (6.0.7) and (6.1.4)

$$\boldsymbol{f}_{\kappa_1}(\mathbf{F}_{\kappa_1}, \mathbf{X}^{\kappa_1}) = \boldsymbol{f}_{\kappa_1}(\mathbf{F}_{\kappa_1}\mathbf{P}^{\mathrm{T}}, \mathbf{X}^{\kappa_1}) \tag{6.1.5}$$

Hence, if a particle $\mathcal{P}$ (or $\mathbf{X}^{\kappa_1}$) is materially symmetric with respect to λ, then Eq. (6.1.5) is true for all nonsingular $\mathbf{F}_{\kappa_1}$. If, Eq. (6.1.5) is true for a set $\mathcal{G}_\lambda$ of orthogonal transformations then $\mathcal{P}$ (or $\mathbf{X}^{\kappa_1}$) is said to be materially symmetric with respect to $\mathcal{G}_\lambda$. If the above is true for all particles, the body is said to be materially symmetric with respect to $\mathcal{G}_\lambda$. It can be shown that $\mathcal{G}_\lambda$ forms a group (see Truesdell [1]).

6.2 Isotropic solid

The elastic body is said to be isotropic at the particle $\mathcal{P}$ (or $\mathbf{X}^{\kappa_1} \in \kappa_1(\mathcal{B})$) if the symmetry group is $\mathcal{O}$, where $\mathcal{O}$ is the orthogonal group. Then,

$$\boldsymbol{f}(\mathbf{F}, \mathbf{X}) = \boldsymbol{f}(\mathbf{F}\mathbf{Q}^{\mathrm{T}}, \mathbf{X}) \ \forall\ \mathbf{Q} \in \mathcal{O} \text{ and } \forall \text{ nonsingular } \mathbf{F}. \tag{6.2.1}$$

The body is said to be isotropic if the above is true for all points $\mathcal{P}$ belonging to the body. For isotropic elastic bodies, one can show that (see Truesdell [1])

$$\boldsymbol{\beta}(\mathbf{Q}\mathbf{E}\mathbf{Q}^{\mathrm{T}}, \mathbf{X}) = \mathbf{Q}\boldsymbol{\beta}(\mathbf{E}, \mathbf{X})\mathbf{Q}^{\mathrm{T}} \ \forall\ \mathbf{Q} \in \mathcal{O}, \tag{6.2.2}$$

where $\mathbf{E}$ is the Green-St. Venant strain tensor.

6.3 Linearization of the isotropy condition

Linearization of the above isotropy condition implies that $\mathcal{C}$ is an isotropic fourth-order tensor. This then implies that the components of the elasticity tensor has to satisfy

$$C_{ijkl} = C_{mnrs} Q_{mi} Q_{nj} Q_{rk} Q_{sl} \ \forall \ \mathbf{Q} \in \mathcal{O}. \tag{6.3.1}$$

We saw in Chapter 5 that the components of $\mathcal{C}$ satisfy

$$C_{ijkl} = C_{jikl} = C_{ijlk} \tag{6.3.2}$$

and it follows from a representation theorem for isotropic tensors that

$$C_{ijkl} = \lambda \delta_{ij} \delta_{kl} + \mu(\delta_{ik}\delta_{jl} + \delta_{jk}\delta_{il}). \tag{6.3.3}$$

Thus,

$$T_{ij} = C_{ijkl}\varepsilon_{kl} \tag{6.3.4}$$

$$= [\lambda \delta_{ij}\delta_{kl} + \mu(\delta_{ik}\delta_{jl} + \delta_{jk}\delta_{il})]\varepsilon_{kl} \tag{6.3.5}$$

$$= \lambda \varepsilon_{kk}\delta_{ij} + \mu(\varepsilon_{ij} + \varepsilon_{ji}) \tag{6.3.6}$$

$$= \lambda \varepsilon_{kk}\delta_{ij} + 2\mu\varepsilon_{ij}, \tag{6.3.7}$$

or

$$\mathbf{T} = \lambda(\mathrm{tr}\,(\boldsymbol{\varepsilon}))\mathbf{I} + 2\mu\boldsymbol{\varepsilon}, \tag{6.3.8}$$

where $\lambda = \lambda(\mathbf{x})$ is the Lame modulus and $\mu = \mu(\mathbf{x})$ is the shear modulus. (In the case of a linearized elastic solid, we do not make any distinction between the Lagrangian and Eulerian representations as the deformations are small.)

If $\det \mathcal{C} \neq 0$, then

$$\boldsymbol{\varepsilon} = \mathbf{K}[\mathbf{T}]. \tag{6.3.9}$$

Now, if we apply the trace operator to Eq. (6.3.8) we obtain:

$$\mathrm{tr}\,\mathbf{T} = 3\lambda(\mathrm{tr}\,\boldsymbol{\varepsilon}) + 2\mu(\mathrm{tr}\,\boldsymbol{\varepsilon}) \tag{6.3.10}$$

$$= (3\lambda + 2\mu)(\mathrm{tr}\,\boldsymbol{\varepsilon}), \tag{6.3.11}$$

where

$$\mathrm{tr}\,\boldsymbol{\varepsilon} = \frac{\mathrm{tr}\,\mathbf{T}}{3\lambda + 2\mu}. \tag{6.3.12}$$

It follows from Eq. (6.3.8) and Eq. (6.3.12) that

$$\varepsilon = \frac{\mathbf{T}}{2\mu} - \frac{\lambda}{2\mu(3\lambda + 2\mu)}(\operatorname{tr}\mathbf{T})\mathbf{I}. \tag{6.3.13}$$

1. Notice that in simple shear

$$T_{xy} = 2\mu\varepsilon_{xy}. \tag{6.3.14}$$

   We expect ε_{xy} and T_{xy} to have the same sign. This implies that $\mu > 0$.
2. Consider a uniaxial state of stress, i. e.,

$$\mathbf{T} = \begin{bmatrix} T_{xx} & 0 & 0 \\ 0 & 0 & 0 \\ 0 & 0 & 0 \end{bmatrix}. \tag{6.3.15}$$

   Then,

$$\varepsilon_{xx} = \frac{1}{E}[(1+\nu)T_{xx} - \nu T_{xx}] = \frac{T_{xx}}{E}. \tag{6.3.16}$$

   Since we expect ε_{xx} and T_{xx} to have the same sign, we conclude $E > 0$.
3. In the case of uniaxial state of stress, it follows that:

$$\frac{\varepsilon_{yy}}{\varepsilon_{xx}} = \frac{-\nu T_{xx}/E}{T_{xx}/E} = -\nu. \tag{6.3.17}$$

   Since we expect ε_{xx} and ε_{yy} to have opposite signs we conclude that $\nu > 0$. It is now common to come across statements that Poisson's ratio can be negative for what is referred to as meta-materials. It is far from clear if meta-materials can be thought of as continua or whether they are structured materials that have "gaps" that would not qualify them to be continua. These meta-materials also are said to possess rather strange characteristics such as negative density. We shall not discuss these issues here.
4. Consider a hydrostatic state of stress at a point:

$$\mathbf{T} = -P\mathbf{I}, \quad P > 0 \tag{6.3.18}$$

$$\operatorname{tr}\mathbf{T} = -3P = (3\lambda + 2\mu)\operatorname{tr}\varepsilon \tag{6.3.19}$$

$$P = -\left(\lambda + \frac{2}{3}\mu\right)\operatorname{tr}\varepsilon. \tag{6.3.20}$$

   Recall, $\det\mathbf{F} \approx 1 + \operatorname{tr}\varepsilon$. Thus, we expect $\operatorname{tr}\varepsilon$ to be negative (< 0). Hence, $(\lambda + \frac{2}{3}\mu) = k > 0$.

6.4 Response of linearized anisotropic elastic bodies

We shall find it convenient to study the anisotropic response of a linearized elastic solid using a notation due to Voigt [2]. Since the stress and strain are symmetric and have only 6 independent components, and since the elasticity tensor has only 36 independent components, we can identify the stress and strain components in the following manner

$$T_{11} = T_1, \quad T_{22} = T_2, \quad T_{33} = T_3, \tag{6.4.1}$$

$$T_{23} = T_4, \quad T_{31} = T_5, \quad T_{12} = T_6, \tag{6.4.2}$$

$$\varepsilon_{11} = \varepsilon_1, \quad \varepsilon_{22} = \varepsilon_2, \quad \varepsilon_{33} = \varepsilon_3, \tag{6.4.3}$$

$$2\varepsilon_{23} = \varepsilon_4, \quad 2\varepsilon_{31} = \varepsilon_5, \quad 2\varepsilon_{12} = \varepsilon_6, \tag{6.4.4}$$

and express Eq. (6.3.4) as

$$\mathbf{T} = \hat{\boldsymbol{C}}\boldsymbol{\varepsilon}, \tag{6.4.5}$$

where $\hat{\boldsymbol{C}}$ is a linear transformation which assigns to six dimensional vectors other six dimensional vectors, that is the matrix $\hat{\boldsymbol{C}}$ is 6×6, it has 36 elements. In indicial notation, Eq. (6.4.5) can be expressed as

$$T_\alpha = \hat{C}_{\alpha\beta}\varepsilon_\beta, \quad \alpha, \beta = 1, 2, \ldots, 6. \tag{6.4.6}$$

The elasticity tensor in its contracted form has the matrix

$$(\hat{\boldsymbol{C}}) = \begin{pmatrix} \hat{C}_{11} & \hat{C}_{12} & \hat{C}_{13} & \hat{C}_{14} & \hat{C}_{15} & \hat{C}_{16} \\ \hat{C}_{21} & \hat{C}_{22} & \hat{C}_{23} & \hat{C}_{24} & \hat{C}_{25} & \hat{C}_{26} \\ \cdot & \cdot & \cdot & \cdot & \cdot & \cdot \\ \cdot & \cdot & \cdot & \cdot & \cdot & \cdot \\ \cdot & \cdot & \cdot & \cdot & \cdot & \cdot \\ \hat{C}_{61} & \hat{C}_{62} & \hat{C}_{63} & \hat{C}_{64} & \hat{C}_{65} & \hat{C}_{66} \end{pmatrix}. \tag{6.4.7}$$

It can be shown if the material is a Green elastic material, that is if one can associate the notion of a stored energy with the deformations of the body, then it follows that $\hat{\boldsymbol{C}}$ is symmetric and hence it will only have 21 independent components, that is the elasticity tensor $\boldsymbol{C}$ will also satisfy $C_{ijkl} = C_{klij}$, and

$$\hat{C}_{\alpha\beta} = \hat{C}_{\beta\alpha}. \tag{6.4.8}$$

Thus,

$$(\hat{\mathcal{C}}_G) = \begin{pmatrix} \hat{C}_{11} & \hat{C}_{12} & \hat{C}_{13} & \hat{C}_{14} & \hat{C}_{15} & \hat{C}_{16} \\ \cdot & \hat{C}_{22} & \hat{C}_{23} & \hat{C}_{24} & \hat{C}_{25} & \hat{C}_{26} \\ \cdot & \cdot & \hat{C}_{33} & \hat{C}_{34} & \hat{C}_{35} & \hat{C}_{36} \\ \cdot & \cdot & \cdot & \hat{C}_{44} & \hat{C}_{45} & \hat{C}_{46} \\ \cdot & \cdot & \cdot & \cdot & \hat{C}_{55} & \hat{C}_{56} \\ \cdot & \cdot & \cdot & \cdot & \cdot & \hat{C}_{66} \end{pmatrix}, \tag{6.4.9}$$

the other terms are given by the symmetry condition.

The elements of the contracted matrix of $\hat{\mathcal{C}}$ can be identified with the elements of the elasticity tensor $\mathcal{C}$, but we shall not do so here (see Ting [3] for the relationship between the components of $\mathcal{C}$ and the components of $\hat{\mathcal{C}}_G$), where $\hat{\mathcal{C}}_G$ denotes the elasticity tensor for a Green elastic solid.

6.4.1 One plane of material symmetry

We will first consider the case when the body has one plane of symmetry, say the $X_1\, X_2$ plane. In this case the contracted tensor $\hat{\mathcal{C}}$ has a matrix whose entries are (see Sokolnikoff [4] for details)

$$(\hat{\mathcal{C}}) = \begin{pmatrix} \hat{C}_{11} & \hat{C}_{12} & \hat{C}_{13} & 0 & 0 & \hat{C}_{16} \\ \hat{C}_{21} & \hat{C}_{22} & \hat{C}_{23} & 0 & 0 & \hat{C}_{26} \\ \hat{C}_{31} & \hat{C}_{32} & \hat{C}_{33} & 0 & 0 & \hat{C}_{36} \\ 0 & 0 & 0 & \hat{C}_{44} & \hat{C}_{45} & 0 \\ 0 & 0 & 0 & \hat{C}_{54} & \hat{C}_{55} & 0 \\ \hat{C}_{61} & \hat{C}_{62} & \hat{C}_{63} & 0 & 0 & \hat{C}_{66} \end{pmatrix}. \tag{6.4.10}$$

We notice that there are 20 independent elasticities. However, if the elastic body is Green elastic, then we will have 13 independent constants, since $\hat{\mathcal{C}}$ is symmetric:

$$(\hat{\mathcal{C}}_G) = \begin{pmatrix} \hat{C}_{11} & \hat{C}_{12} & \hat{C}_{13} & 0 & 0 & \hat{C}_{16} \\ \cdot & \hat{C}_{22} & \hat{C}_{23} & 0 & 0 & \hat{C}_{26} \\ \cdot & \cdot & \hat{C}_{33} & 0 & 0 & \hat{C}_{36} \\ \cdot & \cdot & \cdot & \hat{C}_{44} & \hat{C}_{45} & 0 \\ \cdot & \cdot & \cdot & \cdot & \hat{C}_{55} & 0 \\ \cdot & \cdot & \cdot & \cdot & \cdot & \hat{C}_{66} \end{pmatrix}. \tag{6.4.11}$$

6.4.2 Orthotropic elastic body

Let us consider an elastic body when the body has reflectional symmetries about three mutually perpendicular directions (three orthogonal planes of symmetry). Such bodies are said to be orthotropic. In this case, the contracted elasticity tensor $\hat{\mathcal{C}}$ has a matrix given by

$$(\hat{\mathcal{C}}) = \begin{pmatrix} \hat{C}_{11} & \hat{C}_{12} & \hat{C}_{13} & 0 & 0 & 0 \\ \hat{C}_{21} & \hat{C}_{22} & \hat{C}_{23} & 0 & 0 & 0 \\ \hat{C}_{31} & \hat{C}_{32} & \hat{C}_{33} & 0 & 0 & 0 \\ 0 & 0 & 0 & \hat{C}_{44} & 0 & 0 \\ 0 & 0 & 0 & 0 & \hat{C}_{55} & 0 \\ 0 & 0 & 0 & 0 & 0 & \hat{C}_{66} \end{pmatrix}. \tag{6.4.12}$$

Thus, there are 12 independent constants that characterize a Cauchy linearized elastic body. However, if the material is a Green elastic body, we see below since the elasticity tensor has additional symmetry, it needs fewer constants to characterize the body.

In the case of a Green linearized elastic body the contracted elasticity tensor $\hat{\mathcal{C}}$ takes the form

$$(\hat{\mathcal{C}}_G) = \begin{pmatrix} \hat{C}_{11} & \hat{C}_{12} & \hat{C}_{13} & 0 & 0 & 0 \\ \hat{C}_{12} & \hat{C}_{22} & \hat{C}_{23} & 0 & 0 & 0 \\ \hat{C}_{13} & \hat{C}_{23} & \hat{C}_{33} & 0 & 0 & 0 \\ 0 & 0 & 0 & \hat{C}_{44} & 0 & 0 \\ 0 & 0 & 0 & 0 & \hat{C}_{55} & 0 \\ 0 & 0 & 0 & 0 & 0 & \hat{C}_{66} \end{pmatrix}. \tag{6.4.13}$$

We note that a Green linearized elastic orthotropic body is characterized by nine independent constants.

6.4.3 Transverse isotropy

In this short subsection, we shall merely introduce the notion of transverse isotropy and discuss briefly the representation for the elasticity tensor for transversely isotropic elastic solids.

Let $\mathbf{R}_{\mathbf{e}}^{\varphi}$ denote the rotation by an angle φ about an axis which is along the direction of the unit vector $\mathbf{e}$ at the point $\mathbf{X}$ belonging to the body. Let us consider the group $\mathcal{G}$ that consists of the identity tensor and the set of all rotations $\mathbf{R}_{\mathbf{e}}^{\varphi}$, $0 < \varphi < 2\pi$ at the point $\mathbf{X} \in \kappa_R(\mathcal{B})$. If the response of the body at the point $\mathbf{X} \in \kappa_R(\mathcal{B})$ is the same for all the transformations belonging to $\mathcal{G}$, then the body is said to be transversely isotropic at $\mathbf{X} \in \kappa_R(\mathcal{B})$. If the above is true for all particles $\mathbf{X} \in \kappa_R(\mathcal{B})$, the body $\mathcal{B}$ is said to be transversely isotropic.

As in the case of isotropic elastic bodies, the number of material constants that characterize a transversely isotropic body is far fewer than 21. It can be shown (see Ting [3]) that in the case of a Green elastic body there are five independent material constants that determine the elasticity tensor $\mathcal{C}$ in the case of transverse isotropy.

In the case of a transversely isotropic homogeneous Cauchy elastic solid, the elasticity tensor has the form:

$$(\hat{\boldsymbol{C}}) = \begin{pmatrix} C_{11} & C_{12} & C_{13} & 0 & 0 & 0 \\ C_{21} & C_{11} & C_{23} & 0 & 0 & 0 \\ C_{31} & C_{32} & C_{33} & 0 & 0 & 0 \\ 0 & 0 & 0 & C_{44} & 0 & 0 \\ 0 & 0 & 0 & 0 & C_{55} & 0 \\ 0 & 0 & 0 & 0 & 0 & (C_{11}-C_{12})/2 \end{pmatrix}. \tag{6.4.14}$$

In the case of a transversely Green elastic solid, the contracted tensor $\hat{\boldsymbol{C}}$ has the matrix representation

$$(\hat{\boldsymbol{C}}_G) = \begin{pmatrix} \hat{C}_{11} & \hat{C}_{12} & \hat{C}_{13} & 0 & 0 & 0 \\ \hat{C}_{21} & \hat{C}_{11} & \hat{C}_{13} & 0 & 0 & 0 \\ \hat{C}_{13} & \hat{C}_{13} & \hat{C}_{33} & 0 & 0 & 0 \\ 0 & 0 & 0 & \hat{C}_{44} & 0 & 0 \\ 0 & 0 & 0 & 0 & \hat{C}_{44} & 0 \\ 0 & 0 & 0 & 0 & 0 & (\hat{C}_{11}-\hat{C}_{12})/2 \end{pmatrix}. \tag{6.4.15}$$

Thus, there are five independent constants C_{11}, C_{12}, C_{13}, C_{33}, and C_{44}, and the constitutive relation takes the form

$$\mathbf{T} = \boldsymbol{C}\varepsilon, \tag{6.4.16}$$

with $\boldsymbol{C}$ being given by Eq. (6.4.15).

The reader is referred to the book by Lekhnitskii [5] for a detailed treatment of anisotropic elastic bodies.

6.5 Problems

6.1 Given that

$$\mathbf{T} = \lambda(\operatorname{tr}\varepsilon)\mathbf{I} + 2\mu\varepsilon, \tag{6.5.1}$$

show that

$$\varepsilon = \frac{\mathbf{T}}{2\mu} - \frac{\lambda}{2\mu(3\lambda+2\mu)}(\operatorname{tr}\mathbf{T})\mathbf{I}. \tag{6.5.2}$$

References

[1] C. A. Truesdell. A First Course in Rational Continuum Mechanics V1, 1992.

[2] W. Voigt. Lehrbuch der kristallphysik: (mit ausschluss der kristalloptik), Vol. 34. BG Teubner, 1910.
[3] T. C. T. Ting. Anisotropic Elasticity. Theory and Applications. Oxford University Press, 1996.
[4] I. S. Sokolnikoff. Mathematical Theory of Elasticity. McGraw Hill, New York, 2nd edition, 1956.
[5] S. G. Lekhnitskii. Theory of Elasticity of Anisotropic Elastic Body, translated by P. Fern, edited by J. J. Brandstatter, Golden-Day Series in Mathematical Physics, San Francisco, 1963.

7 Elasto-statics

In this chapter we will discuss static deformations of linearized elastic bodies. In this case the equations of motion reduce to

$$\operatorname{div}\mathbf{T} + \rho\mathbf{b} = \mathbf{0}. \tag{7.0.1}$$

Since the body is at rest, the above equations are called the equations of equilibrium. We shall consider four classes of problems, anti-plane strain, plane-strain and plane-stress, as well as the problem of generalized plane stress, within the context of Cartesian coordinates. In this chapter we will merely derive the governing equations for the different classes, we shall not solve any specific problem, that will be carried out in later chapters.

7.1 Anti-plane strain

We start by defining the class of problems that falls within the purview of anti-plane strain. Let us consider a cylindrical body whose cross-section is such that the z-axis is normal to it. Let us assume furthermore that the body is

(i) uniform in the z-direction.

(ii) is a cylinder of infinite length
and

(iii) loading is along the z-direction and is uniform with only shearing traction being applied along the z-direction.

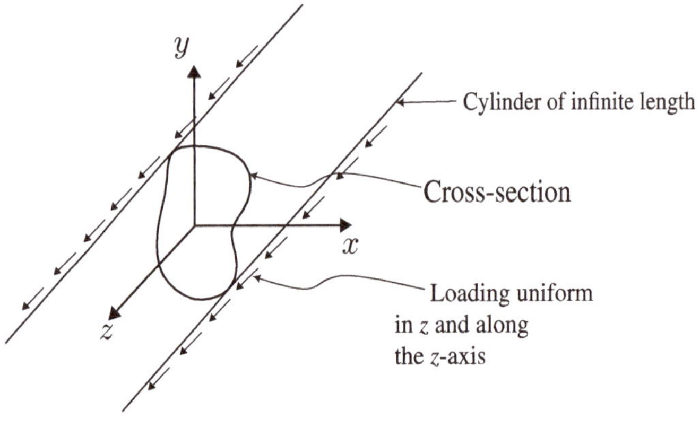

Figure 7.1: An elastic body in a state of anti-plane strain.

The above assumptions imply that it would be reasonable to assume that the displacement field $\mathbf{u}(x,y,z) = u(x,y,z)\mathbf{i} + v(x,y,z)\mathbf{j} + w(x,y,z)\mathbf{k}$ is such that it will have the following components (see Figure 7.1):

https://doi.org/10.1515/9783110789515-007

$$u = 0, \quad v = 0, \quad w = w(x, y),^{1} \tag{7.1.1}$$

and thus the linearized strain has the matrix

$$\boldsymbol{\varepsilon} = \begin{bmatrix} 0 & 0 & \frac{1}{2}\frac{\partial w}{\partial x} \\ 0 & 0 & \frac{1}{2}\frac{\partial w}{\partial y} \\ \frac{1}{2}\frac{\partial w}{\partial x} & \frac{1}{2}\frac{\partial w}{\partial y} & 0 \end{bmatrix}. \tag{7.1.2}$$

Notice that the tr $\boldsymbol{\varepsilon} = 0$, which implies that the motion is isochoric. The stress tensor for a linearized elastic solid becomes

$$\mathbf{T} = 2\mu\boldsymbol{\varepsilon} = \mu \begin{bmatrix} 0 & 0 & \frac{\partial w}{\partial x} \\ 0 & 0 & \frac{\partial w}{\partial y} \\ \frac{\partial w}{\partial x} & \frac{\partial w}{\partial y} & 0 \end{bmatrix}. \tag{7.1.3}$$

The equation of equilibrium

$$\operatorname{div} \mathbf{T} + \rho \mathbf{b} = \mathbf{0}, \tag{7.1.4}$$

which on assuming that $\mathbf{b} = \mathbf{0}$ implies that

$$\operatorname{div} \mathbf{T} = \mathbf{0}. \tag{7.1.5}$$

Thus, we find that

$$\frac{\partial^2 w}{\partial x^2} + \frac{\partial^2 w}{\partial y^2} = 0. \tag{7.1.6}$$

Thus, w is a harmonic function. In order to have a well-defined problem, we need to prescribe the boundary conditions and the appropriate *boundary conditions* are as follows. Recall that the traction vector is expressed as (since the stress is symmetric)

$$\mathbf{t} = \mathbf{T}\mathbf{n}, \tag{7.1.7}$$

where $\mathbf{n} = n_x\mathbf{e}_1 + n_y\mathbf{e}_2$ is the normal to a surface. It then follows from Eq. (7.1.3) and (7.1.7) that

$$\mathbf{t} = \mathbf{T}\mathbf{n} = \mu \begin{bmatrix} 0 & 0 & \frac{\partial w}{\partial x} \\ 0 & 0 & \frac{\partial w}{\partial y} \\ \frac{\partial w}{\partial x} & \frac{\partial w}{\partial y} & 0 \end{bmatrix} \begin{pmatrix} n_x \\ n_y \\ 0 \end{pmatrix} = \mu \begin{pmatrix} 0 \\ 0 \\ n_x\frac{\partial w}{\partial x} + n_y\frac{\partial w}{\partial y} \end{pmatrix}. \tag{7.1.8}$$

1 From a mathematical standpoint, we can use (7.1.1) as the definition of anti-plane strain.

Thus,

$$t_z = \mu\left(n_x \frac{\partial w}{\partial x} + n_y \frac{\partial w}{\partial y}\right) = \mu \frac{\partial w}{\partial n}, \tag{7.1.9}$$

indicating that we need to solve

$$\frac{\partial^2 w}{\partial x^2} + \frac{\partial^2 w}{\partial y^2} = 0 \tag{7.1.10}$$

subject to the boundary condition boundary condition

$$\frac{\partial w}{\partial n} = \frac{t_z}{\mu} \tag{7.1.11}$$

7.2 Plane strain

A body is said to be in a state of plane strain (with regard to the $x - y$ plane) if there is no component of the displacement field perpendicular to the $(x - y)$ plane and the in plane displacements depend only on (x, y). Such a state would be appropriate for the following geometry and the following loading. We shall assume (see Figure 7.2)

(i) The elastic body under consideration is assumed to be an infinitely long cylinder whose axis coincides with the z-axis.

(ii) the loading is uniform in z.

(iii) the loading has no component in the z direction (except possibly at infinity).

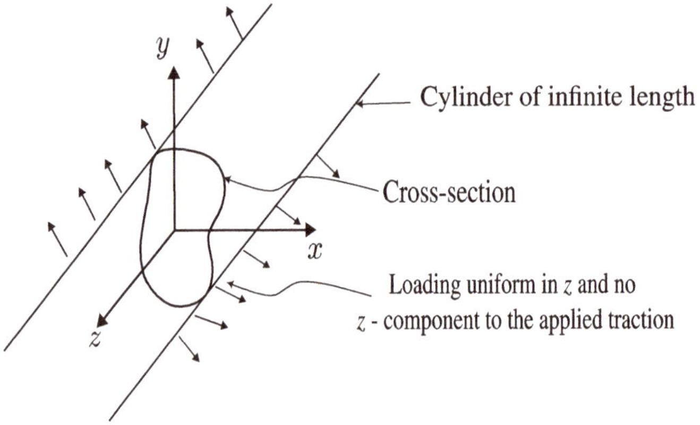

Figure 7.2: A cylinder of infinite length subjected to a Plane Strain condition.

It follows that the components of the displacement u and v in the x and y directions, respectively, would only depend on x and y (a consequence of the second assumption). Thus, $u = u(x,y)$, $v = v(x,y)$, and $w = 0$ leads to[2]

$$\varepsilon = \frac{1}{2}\begin{bmatrix} 2\frac{\partial u}{\partial x} & \frac{\partial u}{\partial y} + \frac{\partial v}{\partial x} & 0 \\ \frac{\partial u}{\partial y} + \frac{\partial v}{\partial x} & 2\frac{\partial v}{\partial y} & 0 \\ 0 & 0 & 0 \end{bmatrix} \tag{7.2.1}$$

Now,

$$\mathbf{T} = \lambda(\mathrm{tr}\,\varepsilon)\mathbf{I} + 2\mu\varepsilon. \tag{7.2.2}$$

On substituting the expression for the stress in the equations of equilibrium

$$\mathrm{div}\,\mathbf{T} + \rho\mathbf{b} = \mathbf{0} \tag{7.2.3}$$

and expressing $\hat{\mathbf{b}} = \rho\mathbf{b}$, we obtain from Eq. (7.2.3), that

$$T_{xx,x} + T_{xy,y} + T_{xz,z} + \hat{b}_x = 0, \tag{7.2.4}$$

where

$$T_{xx,x} = \frac{\partial}{\partial x}\left[\lambda\left(\frac{\partial u}{\partial x} + \frac{\partial v}{\partial y}\right)\right] + \frac{\partial}{\partial x}\left(2\mu\frac{\partial u}{\partial x}\right), \tag{7.2.5}$$

where $\frac{\partial u}{\partial x} = \varepsilon_{xx}$. Similarly we can get expression for the other derivatives of the stress in (7.2.4). Let us define, for convenience, $\theta = \frac{\partial u}{\partial x} + \frac{\partial v}{\partial y}$, allowing us to write

$$T_{xx} = \lambda\theta + 2\mu\frac{\partial u}{\partial x}, \tag{7.2.6}$$

$$T_{yy} = \lambda\theta + 2\mu\frac{\partial v}{\partial y}, \tag{7.2.7}$$

$$T_{zz} = \lambda\theta \tag{7.2.8}$$

$$T_{xy} = \mu\left(\frac{\partial u}{\partial y} + \frac{\partial v}{\partial x}\right) \tag{7.2.9}$$

$$T_{xz} = T_{yz} = 0. \tag{7.2.10}$$

Thus, in the x-direction, we obtain

$$T_{xx,x} + T_{xy,y} + T_{xz,z} + \hat{b}_x = 0 \tag{7.2.11}$$

2 From a mathematical standpoint, we could view a body undergoing the displacement field given is in a state of plane strain.

$$\frac{\partial}{\partial x}\left[\lambda\theta + 2\mu\frac{\partial u}{\partial x}\right] + \frac{\partial}{\partial y}\left[\mu\left(\frac{\partial u}{\partial y} + \frac{\partial v}{\partial x}\right)\right] + \hat{b}_x = 0 \tag{7.2.12}$$

$$\frac{\partial}{\partial x}[\lambda\theta] + 2\mu\left[\frac{\partial^2 u}{\partial x^2}\right] + \mu\left[\frac{\partial^2 u}{\partial y^2}\right] + \mu\frac{\partial^2 v}{\partial x \partial y} + \hat{b}_x = 0 \tag{7.2.13}$$

$$\frac{\partial}{\partial x}[(\lambda + \mu)\theta] + \mu\left[\frac{\partial^2 u}{\partial x^2} + \frac{\partial^2 u}{\partial y^2}\right] + \hat{b}_x = 0. \tag{7.2.14}$$

Similarly in the y-direction we obtain from the equilibrium equation that

$$\frac{\partial}{\partial y}[(\lambda + \mu)\theta] + \mu\left[\frac{\partial^2 v}{\partial x^2} + \frac{\partial^2 v}{\partial y^2}\right] + \hat{b}_y = 0. \tag{7.2.15}$$

In the z-direction, we have

$$\hat{b}_z = 0. \tag{7.2.16}$$

Now, we can pose the question: can $\hat{\mathbf{b}}$ be a function of x, y, z? The answer is a clear no as the above equations imply that $\hat{\mathbf{b}} = \hat{\mathbf{b}}(x, y)$. We have, for two-dimensions,

$$\nabla_2[(\lambda + \mu)\theta] + \mu\nabla_2^2\mathbf{u} + \hat{\mathbf{b}} = \mathbf{0} \tag{7.2.17}$$

with

$$\mathbf{u} = u(x, y)\mathbf{e}_1 + v(x, y)\mathbf{e}_2 \tag{7.2.18}$$

and ∇_2 and ∇_2^2 are the two-dimensional operators for the gradient and the Laplacian. Note that

$$T_{xx} + T_{yy} = 2(\lambda + \mu)\theta = \frac{2(\lambda + \mu)}{\lambda}T_{zz}. \tag{7.2.19}$$

Recall that

$$E = \frac{\mu(3\lambda + 2\mu)}{\lambda + \mu}, \tag{7.2.20}$$

$$\nu = \frac{\lambda}{2(\lambda + \mu)}, \tag{7.2.21}$$

$$k = \lambda + \frac{2}{3}\mu, \tag{7.2.22}$$

$$\mu = G \quad \text{(shear modulus)}. \tag{7.2.23}$$

Thus,

$$T_{zz} = \nu(T_{xx} + T_{yy}). \tag{7.2.24}$$

Since the normal component along the z direction is not zero, this implies that plane strain does not imply plane stress (a state of stress defined in the next section). Recall from the equations of equilibrium that

$$T_{xx,x} + T_{xy,y} = -\hat{b}_x, \tag{7.2.25}$$

$$T_{xy,x} + T_{yy,y} = -\hat{b}_y. \tag{7.2.26}$$

On taking the partial derivative of Eq. (7.2.25) with respect to x and partial derivative of Eq. (7.2.26) with respect to y and adding we find that

$$2T_{xy,yx} = -(\hat{b}_{x,x} + \hat{b}_{y,y}) - T_{xx,xx} - T_{yy,yy}. \tag{7.2.27}$$

Let us now recall that the compatibility condition, that

$$\epsilon_{irs}\epsilon_{jkl}\varepsilon_{rk,sl} = 0, \tag{7.2.28}$$

which reduces to just the one equation

$$2\frac{\partial^2 \varepsilon_{xy}}{\partial x \partial y} = \frac{\partial^2 \varepsilon_{xx}}{\partial y^2} + \frac{\partial^2 \varepsilon_{yy}}{\partial x^2}. \tag{7.2.29}$$

Since,

$$\varepsilon = \frac{1}{E}\{(1+\nu)\mathbf{T} - \nu(\operatorname{tr}\mathbf{T})\mathbf{I}\}, \tag{7.2.30}$$

on substituting Eq. (7.2.30) into Eq. (7.2.29), we find that

$$\frac{2}{E}[(1+\nu)T_{xy,xy}] = \frac{\partial^2}{\partial y^2}\left[\frac{(1+\nu)}{E}T_{xx} - \frac{\nu}{E}(T_{xx} + T_{yy})\right] + \frac{\partial^2}{\partial x^2}\left[\frac{(1+\nu)}{E}T_{yy} - \frac{\nu}{E}(T_{xx} + T_{yy})\right]. \tag{7.2.31}$$

It follows from Eq. (7.2.24), (7.2.27), and (7.2.31) that

$$\nabla_2^2(T_{xx} + T_{yy}) = \frac{-2(\lambda + \mu)}{\lambda + 2\mu}\operatorname{div}_2 \hat{\mathbf{b}} = \frac{-1}{(1-\nu)}\operatorname{div}_2 \hat{\mathbf{b}}. \tag{7.2.32}$$

We need to prescribe the boundary conditions that go with the above equation. However, we shall defer this to later chapters where we reduce the above equations in terms of a single function, referred to as the Airy's stress function.

7.3 Plane stress

A body is in a state of plane stress relative to the $x-y$ plane if there are no shear stresses (T_{xz}, T_{yz}) and no normal stress (T_{zz}), with the stresses depending only on x and y. We

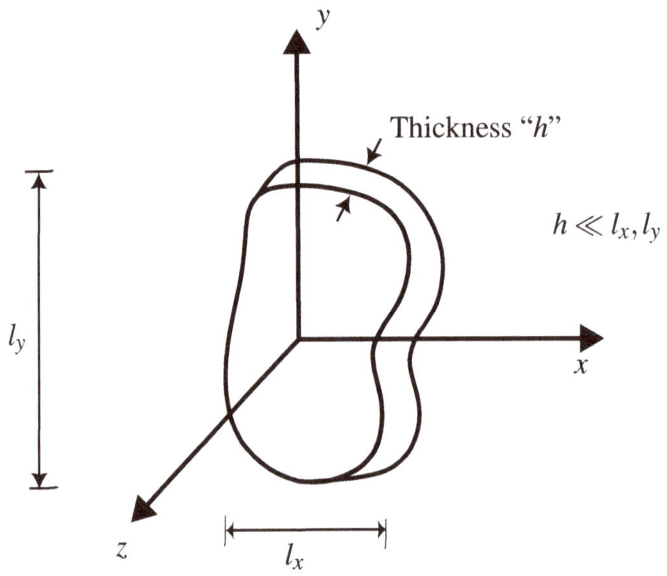

Figure 7.3: A body subjected to a Plane Stress condition.

will see later if T_{zz} is very small, then we can generalize the treatment followed below and such a problem is called the generalized plane stress problem. Let us consider a body (see Fig. 7.3) where the thickness dimension; h is very small in comparison to a representative diameter of the body. Further, suppose that the

(i) loading is uniform in z.
(ii) loading has no z-component.

Since

$$\mathbf{T} = \lambda(\text{tr}\,\varepsilon)\mathbf{I} + 2\mu\varepsilon, \tag{7.3.1}$$

it follows that

$$T_{zz} = 0 = \lambda(\varepsilon_{xx} + \varepsilon_{yy} + \varepsilon_{zz}) + 2\mu\varepsilon_{zz} \tag{7.3.2}$$

$$\varepsilon_{zz} = \frac{-\lambda}{(\lambda + 2\mu)}(\varepsilon_{xx} + \varepsilon_{yy}). \tag{7.3.3}$$

Thus,

$$\varepsilon_{zz} \neq 0. \tag{7.3.4}$$

That is, plane stress does not imply plane strain and plane strain does not imply plane stress. We now present the equations of equilibrium. It follows from Eq. (7.2.2) that

$$T_{xx} = \lambda(\operatorname{tr} \varepsilon) + 2\mu\varepsilon_{xx}, \tag{7.3.5}$$

$$T_{yy} = \lambda(\operatorname{tr} \varepsilon) + 2\mu\varepsilon_{yy}, \tag{7.3.6}$$

and thus

$$\frac{\partial T_{xx}}{\partial x} + \frac{\partial T_{xy}}{\partial y} + \hat{b}_x = 0, \tag{7.3.7}$$

$$\frac{\partial T_{xy}}{\partial x} + \frac{\partial T_{yy}}{\partial y} + \hat{b}_y = 0, \tag{7.3.8}$$

$$\hat{b}_z = 0. \tag{7.3.9}$$

The two-dimensional compatibility equation is given by

$$2\frac{\partial^2 \varepsilon_{xy}}{\partial x \partial y} = \frac{\partial^2 \varepsilon_{xx}}{\partial x^2} + \frac{\partial^2 \varepsilon_{yy}}{\partial y^2}. \tag{7.3.10}$$

It follows from Eq. (7.3.5)-(7.3.10) that

$$\left(\frac{2\lambda\mu}{\lambda + 2\mu} + \mu\right)\nabla_2(\operatorname{tr} \varepsilon) + \mu\nabla_2^2\mathbf{u} + \hat{\mathbf{b}} = \mathbf{0}. \tag{7.3.11}$$

The compatibility equation reduces to

$$\nabla_2^2(T_{xx} + T_{yy}) = \frac{(3\lambda + 2\mu)}{2(\lambda + \mu)}\operatorname{div}_2 \hat{\mathbf{b}} = -(1 + \nu)\operatorname{div}_2 \hat{\mathbf{b}}. \tag{7.3.12}$$

We notice that the only difference between the equation governing plane strain and plane stress is the coefficient multiplying the body force $\hat{\mathbf{b}}$ is different.

As before, we shall not discuss the issue of boundary conditions here but defer it to a later chapter.

7.4 Generalized plane stress

Let us consider an example of a cylinder of thickness h, where h is much smaller than the diameter of the cross-section such that the two faces of the cylinder are at $z = \pm h/2$. Suppose we apply loads only on the lateral surfaces and also assume that there is no z component. Let us neglect body forces. We shall now define the average values for the displacement field in the following manner:

$$\tilde{u}_i(x, y) = \frac{1}{h}\int_{-h/2}^{h/2} u_i(x, y, z)\, dz, \tag{7.4.1}$$

where $i = 1, 2, 3$. In virtue of the assumption that there is no loading in the z-direction, we shall assume that, $\hat{u}_z = 0$ (if the loading is uniform in z). Also,

$$T_{xz}(x, y, \pm h/2) = T_{yz}(x, y, \pm h/2) = T_{zz}(x, y, \pm h/2) = 0, \tag{7.4.2}$$

since there are no loads on the cross-section at $z = \pm h/2$. Consider the equilibrium equation in the z-direction:

$$T_{xz,x} + T_{zy,y} + T_{zz,z} = 0. \tag{7.4.3}$$

Thus,

$$T_{xz,x}(x, y, \pm h/2) + T_{zy,y}(x, y, \pm h/2) + T_{zz,z}(x, y, \pm h/2) = 0. \tag{7.4.4}$$

Thus,

$$T_{zz,z}(x, y, \pm h/2) = 0. \tag{7.4.5}$$

This gives some credence to the assumption that T_{zz} would be very small and hence can be neglected. The other equilibrium equations are

$$T_{i1,1} + T_{i2,2} + T_{i3,3} = 0, \tag{7.4.6}$$

where $i = 1, 2$. Thus

$$\frac{1}{h} = \int_{-h/2}^{h/2} [T_{i1,1} + T_{i2,2} + T_{i3,3}] \, \mathrm{d}z = 0. \tag{7.4.7}$$

Consider

$$\int_{-h/2}^{h/2} T_{i3,3} \, \mathrm{d}z = T_{i3,3}(x, y, h/2) - T_{i3}(x, y, -h/2) = 0. \tag{7.4.8}$$

Thus, we find

$$\frac{1}{h} \int_{-h/2}^{h/2} (T_{i1,1} + T_{i2,2}) \, \mathrm{d}z = 0. \tag{7.4.9}$$

Hence,

$$\tilde{T}_{i1,1} + \tilde{T}_{i2,2} = 0. \tag{7.4.10}$$

If we write down the constitutive equations for linearized elasticity in an average sense,

$$\tilde{\mathbf{T}} = \lambda(\mathrm{tr}\,\tilde{\varepsilon})\mathbf{I} + \mu(\tilde{\varepsilon}), \tag{7.4.11}$$

it then follows (as in the plane stress problem) that the equations of equilibrium (in the average) reduce to

$$\mu\nabla_2^2\tilde{\mathbf{u}} + \nabla_2\left[\left(\frac{2\lambda\mu}{\lambda+2\mu} + \mu\right)(\mathrm{tr}\,\tilde{\varepsilon})\right] = \mathbf{0}. \tag{7.4.12}$$

If you have body forces, the average equation of equilibrium is

$$\mu\nabla_2^2\tilde{\mathbf{u}} + \nabla_2\left[\left(\frac{2\lambda\mu}{\lambda+2\mu} + \mu\right)(\mathrm{tr}\,\tilde{\varepsilon})\right] + \tilde{\mathbf{b}} = \mathbf{0}, \tag{7.4.13}$$

where $\tilde{\mathbf{b}}$ is the average body force.

8 Stress functions

In Chapter 8, we developed the governing equations for plane stress and plane strain, both of which were in terms of the stress components T_{xx} and T_{yy}. For the moment, let us ignore the body forces. Then, plane stress and plane strain problems are governed by a single partial differential equation, the two-dimensional Laplacian operating on the sum of two unknown normal stresses in the x and y directions. This equation can be transformed to an equation for a single variable by introducing a function (potential) in terms of which the stress components T_{xx}, T_{yy}, T_{xy} can be expressed. The potential is referred to as the Airy stress function, since Airy introduced it in 1863 [1]. The Airy stress function was generalized to three-dimensional problems by Maxwell [2] and later by Morera [3]. Morera introduced three stress functions from which the six components of the stress are defined. The stress functions of Maxwell and Morera are special cases of the even more general stress function studied by Beltrami [4] and Mitchell [5].

The reader will find a detailed discussion of stress functions in Love. The reader will find a discussion of use of stress function in elastodynamics in the paper by Teodorescu [6].

8.1 Airy's stress function

We will now discuss how the equation (7.2.32) and (7.3.12) for the two normal stresses T_{xx} and T_{yy} can be reduced to a single equation for a potential that is referred to as the Airy stress function. Let us suppose that the body force is conservative, that is $\mathbf{b} = -\nabla\gamma$. Let us introduce a function ϕ through

$$T_{xx} = \frac{\partial^2\phi}{\partial y^2} + \gamma, \quad T_{xy} = -\frac{\partial^2\phi}{\partial x\partial y}, \quad T_{yy} = \frac{\partial^2\phi}{\partial x^2} + \gamma. \tag{8.1.1}$$

We notice that Eq. (8.1.1) automatically satisfies the equilibrium equations in two dimensions. Next, we determine the consequences of the compatibility equation for plane strain. In this case, it follows from Eq. (7.2.32) and Eq. (8.1.1) that we obtain

$$\nabla_2^4\phi + \frac{1-2\nu}{1-\nu}\nabla^2\gamma = 0, \tag{8.1.2}$$

where ∇_2^4 is the biharmonic operator given by $\partial^4/(\partial x^4) + 2\partial^4/(\partial x^2\partial y^2) + \partial^4/(\partial y^4)$. In the case of plane stress, it follows from Eq. (7.3.12) and Eq. (8.1.1) that

$$\nabla_2^4\phi + (1-\nu)\nabla^2\gamma = 0. \tag{8.1.3}$$

https://doi.org/10.1515/9783110789515-008

We notice that in the absence of body forces, the equations governing the Airy stress function are the same for both plane strain and plane stress. Also, the solution for the Airy's stress function does not depend on the material parameters. Moreover, the strains calculated for the plane stress and plane strain problems determined from knowing the stresses will not be the same, as the form of the stress-strain relation for the two problems are different.

In order to solve equations Eq. (8.1.2) and Eq. (8.1.3), we need to specify boundary conditions. We can consider the domains which are multiply connected. However, we shall require that the boundaries of the domain be smooth. We shall denote by **n** the unit outward normal to the boundary. The tangent vector $\boldsymbol{t}$ is considered positive when it is such that it points in the direction which would be transversed on the boundary so that the domain is always to the left. We are given the traction **t** on $\partial\Omega$

$$\mathbf{t} = \mathbf{T}^{\mathrm{T}}\mathbf{n} = \mathbf{T}\mathbf{n}, \tag{8.1.4}$$

or, in the x-direction

$$t_x = T_{xx}n_x + T_{xy}n_y \tag{8.1.5}$$

$$= \frac{\partial^2 \phi}{\partial y^2}n_x - \frac{\partial^2 \phi}{\partial x \partial y}n_y. \tag{8.1.6}$$

Similarly for the y-direction,

$$t_y = -\frac{\partial^2 \phi}{\partial x \partial y}n_x + \frac{\partial^2 \phi}{\partial x^2}n_y. \tag{8.1.7}$$

The solutions for plane problems involving a biharmonic operator can be obtained by knowing the solution ψ_1, ψ_2, ψ_3 to three harmonic problems, namely

$$\phi = x\psi_1(x,y) + y\psi_2(x,y) + \psi_3(x,y), \tag{8.1.8}$$

and

$$\frac{\partial \psi_1}{\partial x} = \frac{\partial \psi_2}{\partial y}; \quad \frac{\partial \psi_1}{\partial y} = \frac{-\partial \psi_2}{\partial x} \tag{8.1.9}$$

We shall solve specific boundary value problems using the Airy stress function later.

8.2 Westergaard potential

It follows from Eq. (8.1.8) and the definition of the stresses in terms of the Airy potential, that

$$T_{xy} = -x\frac{\partial^2 \psi_1}{\partial x \partial y} - y\frac{\partial^2 \psi_2}{\partial x \partial y} - \frac{\partial^2 \psi_3}{\partial x \partial y}. \tag{8.2.1}$$

In many problems, we encounter a plane of symmetry which we can take to be $y = 0$. Also, there are problems wherein the shear stress $T_{xy} = 0$ on a plane (which we can assume to be the $y = 0$ plane). For such problems, one can introduce a potential, called the Westergaard Potential that greatly simplifies the solution of the problem.

Let $y = 0$, then consider

$$T_{xy} = -x\frac{\partial^2\psi_1}{\partial x\partial y} - \frac{\partial^2\psi_3}{\partial x\partial y} = 0. \tag{8.2.2}$$

Integrating over x yields

$$\frac{\partial\psi_3}{\partial y} = -x\frac{\partial\psi_1}{\partial y} + \int_x \frac{\partial\psi_1}{\partial y}\,dx + \tilde{g}(y), \tag{8.2.3}$$

but $\partial\psi_1/\partial y = -\partial\psi_2/\partial x$. Thus,

$$\left.\frac{\partial\psi_3}{\partial y}\right|_{(x,y)=(x,0)} = -x\frac{\partial\psi_1(x,0)}{\partial y} - \psi_2(x,0) + \tilde{g}(y). \tag{8.2.4}$$

This implies that

$$\psi_2(x,0) = -x\left.\frac{\partial\psi_1}{\partial y}\right|_{(x,y)=(x,0)} - \left.\frac{\partial\psi_3}{\partial y}\right|_{(x,y)=(x,0)} + \tilde{g}(y). \tag{8.2.5}$$

Next, recall that $T_{xx} = \phi_{yy}$ and $T_{yy} = \phi_{xx}$. Equation (8.2.3) implies that

$$\psi_3 = -x\psi_1 - \int_y \psi_2\,dy + g(y). \tag{8.2.6}$$

Now, using Eq. (8.2.6), we see that

$$T_{xx} = \phi_{,yy} = x\psi_{1,yy} + y\psi_{2,yy} + 2\psi_{2,y} + \psi_{3,yy} \tag{8.2.7}$$
$$= y\psi_{2,yy} + \psi_{2,y} + G(y). \tag{8.2.8}$$

Similarly, for T_{yy}, we have that

$$T_{yy} = \phi_{,xx} = x\psi_{1,xx} + 2\psi_{1,x} + y\psi_{2,xx} + \psi_{3,xx} \tag{8.2.9}$$
$$= \int_y \psi_{2,xx}\,dy = y\psi_{2,xx} + \psi_{2,y} + F(x). \tag{8.2.10}$$

Finally, for T_{xy}, we have

$$T_{xy} = -\phi_{xy} = -x\psi_{1,xy} - y\psi_{2,xy} - \psi_{3,xy} \tag{8.2.11}$$
$$= -y\phi_{2,xy}. \tag{8.2.12}$$

Thus, in summary we have determined

$$T_{xx} = y\psi_{2,yy} + \psi_{2,y} + G(y) \tag{8.2.13}$$

$$T_{yy} = -y\psi_{2,yy} + \psi_{2,y} + F(x) \tag{8.2.14}$$

$$T_{xy} = -y\psi_{2,xy}. \tag{8.2.15}$$

Now, we use compatibility to check if our derivation is correct (i. e., to determine $G(y)$ and $F(x)$). Consider

$$\nabla^2(T_{xx} + T_{yy}) = \cancelto{0}{\nabla^2 \frac{\partial \psi_2}{\partial y}} + \nabla^2 G(y) + \nabla^2 F(x) = \nabla^2 G(y) + \nabla^2 F(x) = 0. \tag{8.2.16}$$

Equation (8.2.16) implies that

$$G''(y) + F''(x) = 0, \tag{8.2.17}$$

or

$$F''(x) = -a^*, \quad G''(y) = a^*, \tag{8.2.18}$$

where prime denotes derivative with respect to the argurment and a^* is a constant. Thus, it follows that

$$F = ax + b - \frac{1}{2}a^* x^2 \tag{8.2.19}$$

$$G = cy + d + \frac{1}{2}a^* y^2, \tag{8.2.20}$$

where a, b, c, and d are constants. Let us define $\psi \equiv \partial\psi_2/\partial y$. Then, the final form of Eq. (8.2.13) becomes

$$T_{xx} = y\psi_{,y} + \psi + a^* \frac{y^2}{2} + cy + d, \tag{8.2.21}$$

$$T_{yy} = -y\psi_{,y} + \psi - a^* \frac{x^2}{2} + ax + b, \tag{8.2.22}$$

$$T_{xy} = -y\psi_{,x}, \tag{8.2.23}$$

where

$$\nabla^2 \psi = 0, \tag{8.2.24}$$

and ψ is the Westergaard stress function. It is important to note that the Westergaard stress function ψ is not complete in that there exist stress fields that do not admit representation as a Westergaard solution.

8.2.1 Problems using the Westergaard potential

Plate with uniform bi-axial loading

Consider a plate with uniform bi-axial loading (i. e., uniform tension or compression in the x and y directions). We can take $\psi \equiv 0$, which implies that

$$a^* = a = c = 0. \tag{8.2.25}$$

Then,

$$T_{xx} = d, \quad T_{yy} = b, \quad T_{xy} = 0. \tag{8.2.26}$$

Note, we can always superimpose a constant shear stress $T_{xy} = \tau_0$ on a Westergaard field. Equilibrium and compatibility equations are satisfied.

First we must solve the problem with the Westergaard function as if $T_{xy} = 0$ on $y = 0$. Then, we add to that solution $T_{xy} = \tau_0$. Now, with $\psi = 0$, we immediately obtain the solution for a rectangular plate with (a) a constant loading distribution, (b) a linear loading distribution, and (c) a parabolic loading distribution for T_{xx} and T_{yy}. Parabolic distributions would not be independent if $\psi = 0$.

Bending of a cantilever beam

For the bending of a cantilever beam, we have the following boundary conditions (see Figure 8.1):

$$T_{yy}(x,0) = 0, \quad T_{yy}(x,h) = 0, \quad T_{xy}(x,0) = 0, \quad T_{xy}(x,h) = 0, \quad T_{xx}(0,y) = 0, \tag{8.2.27}$$

and

$$\int_0^h T_{xy}(0,y)\,dy = -P. \tag{8.2.28}$$

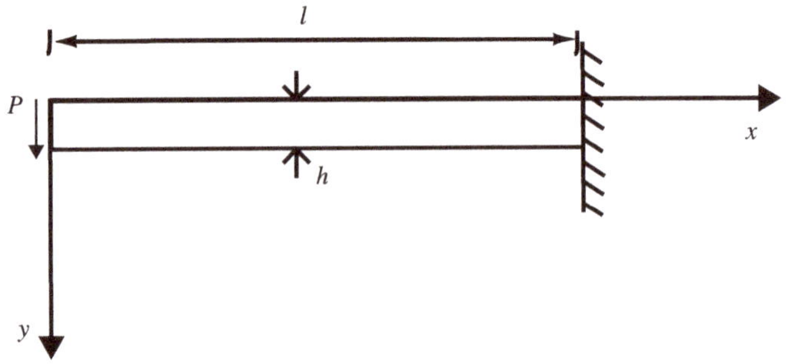

Figure 8.1: Cantilever Beam loaded at the end by a concentrated shear force.

The key boundary condition for the Westergaard potential problem is $T_{xy}(x,0) = 0$. Take $a^* = c = d = 0$ and assume that $T_{yy} = 0$, which satisfies boundary conditions $T_{yy}(x,0) = 0$ and $T_{yy}(x,h) = 0$. This implies that

$$T_{yy} = -y\psi_{,y} + \psi + ax + b = 0, \tag{8.2.29}$$

or

$$-y^2 \frac{\partial}{\partial y}\left(\frac{\psi}{y}\right) = -(ax+b) \implies \frac{\partial}{\partial y}\left(\frac{\psi}{y}\right) = \frac{(ax+b)}{y^2}. \tag{8.2.30}$$

Therefore,

$$\frac{\psi}{y} = -\frac{ax+b}{y} + f_1(x) \implies \psi = -(ax+b) + f_1(x)y. \tag{8.2.31}$$

Also, we have that (using boundary condition $T_{xy}(x,h) = 0$)

$$T_{xy} = -y\psi_x = -y[-a + yf_1'(x)] \implies -h[-a + hf_1'(x)] = 0. \tag{8.2.32}$$

Thus, $f_1'(x) = a/h$ and $f_1 = (a/h)x + a$. Now,

$$\psi = -ax - b + \frac{a}{h}xy + a_1 y. \tag{8.2.33}$$

Using the boundary condition $T_{xx}(0,y) = 0$ yields

$$T_{xx} = y\psi_{,y} + \psi \implies -b + 2a_1 y = 0, \tag{8.2.34}$$

or

$$a_1 = b = 0. \tag{8.2.35}$$

Therefore,

$$\psi = -ax + \frac{a}{h}xy, \tag{8.2.36}$$

giving us

$$T_{xx} = ax\left(\frac{2y}{h} - 1\right) \tag{8.2.37}$$

$$T_{yy} = 0 \tag{8.2.38}$$

$$T_{xy} = ay\left(1 - \frac{y}{h}\right). \tag{8.2.39}$$

Note, if T_{xy} is distributed over the end of the beam in the form of the parabola (Eq. (8.2.39)), then we would have the value of "a" and an exact solution for the stresses is obtained. However, usually the net load P is given and we do not know the actual shear stress distribution. So, for an approximation, we use the average condition, i. e., $\int_0^h T_{xy}(0,y)\,dy = -P$:

$$a\int_0^h\left(y-\frac{y^2}{h}\right)dy = a\left(\frac{h^2}{2}-\frac{h^2}{3}\right) = \frac{ah^2}{6} = -P \implies a = -\frac{6P}{h^2} = -\frac{P}{h^2/6} = -\frac{Ph}{2I}, \quad (8.2.40)$$

where I is the moment of inertia of the cross section. This average condition yields

$$T_{xx} = \frac{Pxh}{2I}\left(1-2\frac{y}{h}\right) \quad (8.2.41)$$

$$T_{yy} = 0 \quad (8.2.42)$$

$$T_{xy} = \frac{Pyh}{2I}\left(\frac{y}{h}-1\right). \quad (8.2.43)$$

Let us shift to the standard coordinate system (i. e., change y to $y + h/2$). This shift yields the expressions as:

$$T_{xx} = \frac{-Pxy}{I} \quad (8.2.44)$$

$$T_{yy} = 0 \quad (8.2.45)$$

$$T_{xy} = \frac{-P}{2I}\left(\frac{h^2}{4}-y^2\right). \quad (8.2.46)$$

Next, we examine the displacements and strains. For the strain in the x direction, we have

$$\varepsilon_{xx} = \frac{\partial u}{\partial x} = \frac{1}{E}(T_{xx} - \nu \overset{0}{\cancel{T_{yy}}}) = -\frac{Pxy}{EI}. \quad (8.2.47)$$

Integrating to find the displacement yields

$$u = -\frac{Px^2y}{2EI} + \varphi_1(y). \quad (8.2.48)$$

Similarly, the strain in the y direction is

$$\varepsilon_{yy} = \frac{\partial v}{\partial y} = \frac{1}{E}(\overset{0}{\cancel{T_{yy}}} - \nu T_{xx}) = \frac{\nu Pxy}{EI}. \quad (8.2.49)$$

We integrate this to find

$$v = \frac{\nu Pxy^2}{2EI} + \varphi_2(x). \quad (8.2.50)$$

Notice that

$$T_{xy} = G\gamma_{xy} = G\left(\frac{\partial u}{\partial y}+\frac{\partial v}{\partial x}\right) = \frac{E}{2(1+\nu)}\left(\frac{\partial u}{\partial y}+\frac{\partial v}{\partial x}\right). \quad (8.2.51)$$

Therefore,

$$-\frac{P(1+\nu)}{EI}\left(\frac{h^2}{4}-y^2\right)=-\frac{Px^2}{2EI}+\varphi_1'(y)+\frac{\nu Py^2}{2EI}\varphi_2'(x), \tag{8.2.52}$$

or

$$\varphi_2'(x)+\varphi_1'(y)=\frac{Px^2}{2EI}+\frac{Py^2}{2EI}(2+\nu)-\frac{Ph^2}{4EI}(1+\nu). \tag{8.2.53}$$

Equation (8.2.53) implies that

$$\varphi_2'(x)=\frac{Px^2}{2EI}-\frac{Ph^2}{4EI}(1+\nu)+A \tag{8.2.54}$$

$$\varphi_1'(y)=\frac{Py^2}{2EI}(2+\nu)-A. \tag{8.2.55}$$

Integrating Eq. (8.2.54) with respect to x and Eq. (8.2.55) with respect to y yields

$$\varphi_2(x)=\frac{Px^3}{6EI}-(1+\nu)\frac{Ph^2x}{6EI}+Ax+B \tag{8.2.56}$$

$$\varphi_1(y)=\frac{Py^3}{4EI}(2+\nu)-Ay+C. \tag{8.2.57}$$

Therefore, the displacements become

$$u=-\frac{Px^2y}{2EI}+\frac{Py^3}{6EI}(2+\nu)-Ay+C \tag{8.2.58}$$

$$v=\frac{\nu Pxy^2}{2EI}+\frac{Px^3}{6EI}-(1+\nu)\frac{Ph^2x}{4EI}+Ax+B. \tag{8.2.59}$$

However, if we evaluate the expressions at $x=l$, we see that

$$u(l,y)=-\frac{Pl^2y}{2EI}+\frac{Py^3}{6EI}(2+\nu)-Ay+C \tag{8.2.60}$$

$$v(l,y)=\frac{\nu Ply^2}{2EI}+\frac{Pl^3}{6EI}-(1+\nu)\frac{Ph^2l}{4EI}+Al+B, \tag{8.2.61}$$

which cannot satisfy $u=v=0$ at $x=l$. Let us require the center line to be held fixed at the wall:

$$u(l,0)=C=0,\quad v(l,0)=\frac{Pl^3}{6EI}-(1+\nu)\frac{Ph^2l}{4EI}+Al+B=0. \tag{8.2.62}$$

Now, our expressions for displacements u and v become

$$u(x,y)=-\frac{Px^2y}{2EI}+\frac{Py^3}{6EI}(2+\nu)-Ay \tag{8.2.63}$$

$$v(x,y) = \frac{\nu Pxy^2}{2EI} + \frac{P}{6EI}(x^3 - l^3) - (1+\nu)\frac{Ph^2}{4EI}(x-l) + A(x-l), \tag{8.2.64}$$

but

$$\frac{\partial v}{\partial x}(l,0) = 0 \implies \frac{Pl^2}{2EI} - (1+\nu)\frac{Ph^2}{4EI} + A = 0. \tag{8.2.65}$$

Finally, we have

$$u(x,y) = \frac{Px^2y}{2EI} + (2+\nu)\frac{Py^3}{6EI} + \frac{Pl^2y}{2EI} - (1+\nu)\frac{Ph^2y}{4EI} \tag{8.2.66}$$

$$v(x,y) = \frac{\nu Pxy^2}{2EI} + \frac{P}{6EI}(x^3 + 2l^3 - 3l^2x), \tag{8.2.67}$$

for the solution to the displacements and

$$v(x,0) = \frac{P}{6EI}(x^3 - 3l^2x + 2l^3), \tag{8.2.68}$$

for the beam tip deflection, which is the same as the Euler-Bernoulli result.

Example 8.2.1. Use the condition

$$\left.\frac{\partial u}{\partial y}\right|_{y=0} = 0 \tag{8.2.69}$$

to determine the constant A in Eq. (8.2.63). Then, find the usual tip deflection $v(x,0)$.

Beam with uniformly distributed load

Consider a beam with a uniformly distributed constant load, w. The boundary conditions for this problem are (see Figure 8.2)

$$T_{xy}(x,0) = 0, \quad T_{xy}(x,h) = 0, \quad T_{xy}(0,y) = 0, \quad T_{yy}(x,0) = -w, \quad T_{yy}(x,h) = 0, \tag{8.2.70}$$

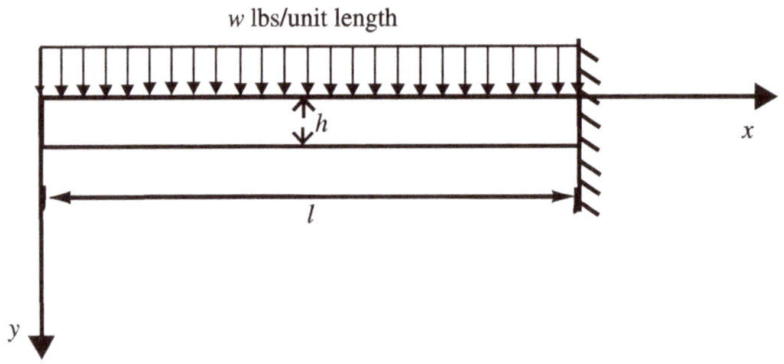

Figure 8.2: Cantilever Beam with a load that is uniformly distributed.

and

$$T_{xx}(0,y) = 0, \tag{8.2.71}$$

which we will see cannot be enforced pointwise, but we can meet it in an average sense. Next, the stresses in terms of the Westergaard function are

$$T_{xx} = y\psi_{,y} + \psi + \frac{a^*}{2}y^2 + cy + d \tag{8.2.72}$$

$$T_{yy} = -y\psi_{,y} + \psi - \frac{a^*}{2}x^2 + ax + b \tag{8.2.73}$$

$$T_{xy} = -y\psi_{,x}. \tag{8.2.74}$$

Suppose that

$$T_{xy} = Axy(y-h), \tag{8.2.75}$$

which satisfies boundary conditions $T_{xy}(x,0) = 0$, $T_{xy}(x,h) = 0$, and $T_{xy}(0,y) = 0$. Therefore,

$$\psi_{,x} = -Ax(y-h). \tag{8.2.76}$$

Integrating Eq. (8.2.76) yields

$$\psi = -\frac{A}{2}x^2(y-h) + g_1(y). \tag{8.2.77}$$

Now,

$$\nabla^2\psi = -Ay + Ah + g_1''(y) = 0, \tag{8.2.78}$$

implies that

$$g_1 = \frac{A}{6}y^3 - \frac{Ah}{2}y^2 + \alpha_1 y + \alpha. \tag{8.2.79}$$

Thus,

$$\psi = -\frac{A}{2}x^2(y-h) + \frac{A}{6}y^3 - \frac{Ah}{2}y^2 + \alpha_1 y + \alpha_2. \tag{8.2.80}$$

Using Eq. (8.2.80), we see that

$$T_{yy} = -\frac{A}{3}y^3 + \frac{Ah}{2}y^2 + \frac{A}{2}x^2 h + \alpha_2 - \frac{a'}{2}x^2 + ax + b. \tag{8.2.81}$$

By applying boundary condition $T_{yy}(x,0) = -w$, we see that

$$\frac{A}{2}x^2h + \alpha_2 - \frac{a^*}{2}x^2 + ax + b = -w, \tag{8.2.82}$$

which implies that

$$Ah = a^*, \quad a = 0, \quad \alpha_2 + b = -w. \tag{8.2.83}$$

Thus, we now have

$$T_{yy} = -\frac{A}{3}y^3 + \frac{Ah}{2}y^2 - w, \tag{8.2.84}$$

whereby using boundary condition $T_{yy}(x, h) = 0$ gives

$$A = \frac{6w}{h^3}. \tag{8.2.85}$$

So, for the stress in the y direction, we have

$$T_{yy} = w\left(\frac{3y^2}{h^2} - \frac{2y^3}{h^3} - 1\right), \tag{8.2.86}$$

and for the shear stress

$$T_{xy} = \frac{6wx}{h^3}(y^2 - yh). \tag{8.2.87}$$

Using the Westergaard function to determine the stress in the x direction yields

$$T_{xx} = \frac{6w}{h^3}\left(-x^2y. + \frac{2}{3}y^3 - hy^2 + \frac{1}{2}x^2h\right) + (2\alpha_1 + c)y + (\alpha_2 + d), \tag{8.2.88}$$

which cannot satisfy $T_{xx} = 0$ at $x = 0$. Instead, we require

$$\int_0^h T_{xx}(0, y)\, dy = 0, \tag{8.2.89}$$

which implies zero axial force, and

$$\int_0^h yT_{xx}(0, y)\, dy = 0, \tag{8.2.90}$$

which implies zero end couple. Now, let

$$\beta_1 = 2\alpha_1 + c, \quad \beta_2 = \alpha_2 + d, \tag{8.2.91}$$

such that

$$\frac{h}{2}\beta_1 + \beta_2 = w, \tag{8.2.92}$$

$$\frac{h}{3}\beta_1 + \frac{1}{2}\beta_2 = \frac{7}{10}w. \tag{8.2.93}$$

Solving Eq. (8.2.92) for β_1 and β_2 yields

$$\beta_1 = \frac{12}{5h}w, \quad \beta_2 = -\frac{1}{5}w. \tag{8.2.94}$$

Finally, we obtain the following expressions for the stresses

$$T_{xx} = \frac{w}{I}\left(-\frac{1}{2}x^2y + \frac{1}{3}y^3 - \frac{1}{2}hy^2 + \frac{1}{4}hx^2 + \frac{1}{5}h^2y - \frac{1}{60}h^3\right), \tag{8.2.95}$$

$$T_{yy} = w\left(\frac{3y^2}{h^2} - \frac{2y^3}{h^3} - 1\right), \tag{8.2.96}$$

$$T_{xy} = \frac{wx}{2I}(y^2 - yh), \tag{8.2.97}$$

or, by shifting to standard coordinates,

$$T_{xx} = -\frac{w}{2I}x^2y + \frac{w}{2I}\left(\frac{2}{3}y^3 - \frac{1}{10}h^2y\right), \tag{8.2.98}$$

$$T_{yy} = -\frac{w}{2I}\left(\frac{1}{3}y^3 - \frac{1}{4}yh^2 + \frac{1}{12}h^3\right), \tag{8.2.99}$$

$$T_{xy} = \frac{wx}{2I}\left(y^2 - \frac{h^2}{4}\right). \tag{8.2.100}$$

Now, let

$$F_1(y) = \left(\frac{2}{3}y^3 - \frac{1}{10}h^2y\right)\frac{w}{2I}, \tag{8.2.101}$$

such that

$$F_1'(y) = \frac{w}{2I}\left(2y^2 - \frac{1}{10}h^2\right) = 0, \tag{8.2.102}$$

at $y = Ih/\sqrt{w}$. This implies that

$$F_{1,\max} = \frac{w}{5\sqrt{5}}. \tag{8.2.103}$$

The maximum stress in the x direction from the Euler-Bernoulli[1] Theory is given by

1 The student has not yet been introduced to the Euler-Bernoulli beam theory. However, the comparison is provided with the Euler-Bernoulli beam theory will become meaningful when the student will be made conversant with the theory later in the course.

$$T_{xx,\max} = \frac{3wl^2}{h^2}, \tag{8.2.104}$$

resulting in an error between the two approaches of

$$\%\ \text{error} = \frac{h^2}{15\sqrt{5}l^2}, \tag{8.2.105}$$

which is very small for $h/l \ll 1$.

We now examine the displacements. Consider the following expressions for displacements in the x and y directions (u and v):

$$E\frac{\partial u}{\partial x} = T_{xx} - \nu T_{yy}, \tag{8.2.106}$$

$$E\frac{\partial v}{\partial y} = T_{yy} - \nu T_{xx}, \tag{8.2.107}$$

which implies

$$Eu = -\frac{w}{6I}x^3y + \frac{wx}{2I}\left(\frac{2}{3}y^3 - \frac{1}{10}h^2y\right) + \frac{\nu wx}{2I}\left(\frac{1}{3}y^3 - \frac{1}{4}yh^2 + \frac{1}{12}h^3\right) + g(y), \tag{8.2.108}$$

$$Ev = -\frac{w}{2I}\left(\frac{1}{12}y^4 - \frac{1}{8}y^2h^2 + \frac{1}{12}h^3y\right) + \frac{\nu wx^2y^2}{4I} - \frac{\nu w}{2I}\left(\frac{1}{6}y^4 - \frac{1}{20}h^2y^2\right) + f(x). \tag{8.2.109}$$

Recall the relationship between the displacements and the shear stress is given by

$$T_{xy} = \frac{E}{2(1+\nu)}\left(\frac{\partial u}{\partial y} + \frac{\partial v}{\partial x}\right), \tag{8.2.110}$$

which yields

$$f(x) = \frac{wx^4}{24I} - \frac{wx^2h^2}{4I}\left(\frac{2}{5} + \frac{\nu}{4}\right) + Ax + B, \tag{8.2.111}$$

$$g(y) = -Ay + C. \tag{8.2.112}$$

As before, we require that

$$u(l,0) = 0, \quad v(l,0) = 0, \quad \frac{dv}{dx}(l,0) = 0. \tag{8.2.113}$$

Using Eq. (8.2.113), the constants A, B, and C can be determined as

$$A = \frac{wlh^2}{2I}\left(\frac{2}{5} + \frac{\nu}{4}\right) - \frac{wl^3}{6I} \tag{8.2.114}$$

$$B = \frac{wl^4}{8I} - \frac{wl^2h^2}{4I}\left(\frac{2}{5} + \frac{\nu}{4}\right) \tag{8.2.115}$$

$$C = -\frac{wlh^3 \nu}{24I}. \tag{8.2.116}$$

Thus, for the end deflection, we have

$$v(0,0) = B = \frac{wl^4}{8EI}\left[1 - \frac{2h^2}{l^2}\left(\frac{2}{5} + \frac{\nu}{4}\right)\right], \tag{8.2.117}$$

where the term $wl^4/8EI$ corresponds to the Euler-Bernoulli result and the second term is the correction due to shear. We can express the term

$$\frac{wl^2h^2}{4EI}\left(\frac{2}{5} + \frac{\nu}{4}\right) = \frac{3}{2}\frac{wl^2}{AG}\left(\frac{2}{5} + \frac{\nu}{4}\right)\left(\frac{1}{1+\nu}\right), \tag{8.2.118}$$

which implies the shear factor is

$$k = \frac{2}{3}\frac{1+\nu}{(\frac{2}{5} + \frac{\nu}{4})}, \tag{8.2.119}$$

with

$$\text{shear correction} = \frac{wl^2}{kAG}. \tag{8.2.120}$$

8.3 Problems

8.1 Show that the coefficient $\frac{-2(\lambda+\mu)}{\lambda+2\mu}$ follows from Eq. (7.2.24), (7.2.27), and (7.2.31).

8.2 Using the compatibility equation, show that

$$\nabla_2^4 \phi = 0, \tag{8.3.1}$$

where ∇_2^4 is the biharmonic operator such that

$$\nabla_2^4 \phi = \nabla_2^2(\nabla_2^2 \phi). \tag{8.3.2}$$

References

[1] G. B. Airy. On the strains in the interior of beams. Phil. Trans. Roy. Soc. London, 153:49–80, 1863.

[2] J. C. Maxwell. On reciprocal diagrams in space and their relation to Airy's function of stress. Proc. Roy. London Math. Soc., 2:1865–1969, 1868.

[3] G. Morera. Soluzione genrale della equazioni indefinite dell'equilibrio de un corpo continuo. Atti Accad. Lincei Rend., 1:137–141, 1892. Appendice, 1, 233–234, 1892.

[4] E. Beltrami. Osservazioni sulla nota precedente. Atti Accad. Lincei Rend., Series (5), 1:141–142, 1892.

[5] J. H. Mitchell. On the direct determination of stress in an elastic solid with applications to the theory of plates. Proc. London. Math. Soc., 100–124, 1899.

[6] P. P. Teodorescu. Stress functions in three dimensional elastodynamics. Acta Mechanica, 14:103–118, 1972.

9 Axi-symmetric problems

In this chapter we consider problems wherein the body has an axis of symmetry. We will derive the appropriate equations governing such problems with regard to a cylindrical polar co-ordinate system. We shall be interested in studying plane stress problems, the plane of interest being the $x - y$ plane, and hence we will ignore the displacement in the z-direction. We shall derive the equations of equilibrium in polar coordinates. In order to do that, we need to express the gradient and the divergence operator in cylindrical polar coordinate system. In the cylindrical polar coordinate system, the gradient operator is expressed as

$$\nabla = \mathbf{e}_r \frac{\partial}{\partial r} + \mathbf{e}_\theta \frac{1}{r}\frac{\partial}{\partial \theta} + \mathbf{e}_z \frac{\partial}{\partial z}, \tag{9.0.1}$$

where $\mathbf{e}_r$, $\mathbf{e}_\theta$, $\mathbf{e}_z$ are the unit vectors in the r, θ, and z directions, respectively. In general, displacement $\mathbf{u}$ is given by

$$\mathbf{u} = u_r \mathbf{e}_r + u_\theta \mathbf{e}_\theta + u_z \mathbf{e}_z, \tag{9.0.2}$$

and since we will be interested in plane problems, we assume $u_z = 0$ and we make the further assumption that u_r and u_θ are functions of r and θ. In order to be able to record the equations of equilibrium, we need to determine the gradient of the displacement field, substitute the same into the constitutive relation for the stress and then find the divergence of the stress. The divergence of a tensor $\mathbf{T}$ in polar co-ordinate system is expressed as (we are assuming that there is no dependence in the z-direction in view of the assumptions):

$$\operatorname{div} \mathbf{T} = \left\{ \frac{1}{r}\frac{\partial}{\partial r}(rT_{rr}) + \frac{1}{r}\frac{\partial}{\partial \theta}(T_{\theta r}) - \frac{1}{r}T_{\theta\theta} \right\} \mathbf{e}_r + \left\{ \frac{1}{r}\frac{\partial}{\partial \theta}(T_{\theta\theta}) + \frac{1}{r}\frac{\partial}{\partial r}(rT_{\theta r}) + \frac{1}{r}T_{\theta r} \right\} \mathbf{e}_\theta. \tag{9.0.3}$$

Thus, using (9.0.3) we can write down the equations of equilibrium in polar coordinates.

It follows from Eq. (7.0.1) and Eq. (9.0.3) that in the absence of body forces the equations of equilibrium reduces to

$$T_{rr} + r\frac{\partial}{\partial r}(T_{rr}) + \frac{\partial}{\partial \theta}(T_{r\theta}) - T_{\theta\theta} = 0, \tag{9.0.4}$$

$$r\frac{\partial (T_{r\theta})}{\partial r} + 2T_{r\theta} + \frac{\partial}{\partial \theta}(T_{\theta\theta}) = 0. \tag{9.0.5}$$

We need to substitute the constitutive expression for the stress into the above to get the governing equation in terms of the displacement. We will defer this till we discuss the compatibility condition. Recall for plane problems without body forces that

$$\nabla^2(T_{xx} + T_{yy}) = 0. \tag{9.0.6}$$

https://doi.org/10.1515/9783110789515-009

The expressions for T_{rr} and $T_{\theta\theta}$ in terms of T_{xx} and T_{yy} implies that (see homework Problem **9.1**) $T_{rr} + T_{\theta\theta} = T_{xx} + T_{yy}$, and thus the compatibility equation in polar coordinates is

$$\nabla^2(T_{rr} + T_{\theta\theta}) = 0, \tag{9.0.7}$$

where the Laplacian in polar coordinate is

$$\nabla^2(\cdot) = \frac{1}{r}\frac{\partial}{\partial r}\left(r\frac{\partial(\cdot)}{\partial r}\right) + \frac{1}{r^2}\frac{\partial^2(\cdot)}{\partial\theta^2}. \tag{9.0.8}$$

For plane problems in polar coordinates, we have to ensure Eq. (9.0.4), (9.0.5), and (9.0.7) are met.

We will assume that geometry and loading do not vary with θ. In this case, Eq. (9.0.4) and (9.0.5) simplify to

$$T_{rr} + r\frac{\partial}{\partial r}(T_{rr}) - T_{\theta\theta} = 0, \tag{9.0.9}$$

$$r\frac{\partial T_{r\theta}}{\partial r} + 2T_{r\theta} = 0. \tag{9.0.10}$$

While Eq. (9.0.4) and Eq. (9.0.5) are coupled, Eq. (9.0.9) and Eq. (9.0.10) are uncoupled. Thus, we can solve Eq. (9.0.10) for the shear stress, without having to solve Eq. (9.0.9) for the normal stresses. Equation (9.0.10) can be rewritten as

$$\frac{1}{r}\frac{\mathrm{d}}{\mathrm{d}r}(r^2 T_{r\theta}) = 0 \implies \frac{\mathrm{d}}{\mathrm{d}r}(r^2 T_{r\theta}) = 0 \tag{9.0.11}$$

or

$$r^2 T_{r\theta} = c \implies T_{r\theta} = \frac{c}{r^2}. \tag{9.0.12}$$

Example. Consider the problem shown in Fig. 9.1. From the previous discussion, we can conclude that

$$T_0 = \frac{c}{b^2}, \quad T_1 = \frac{c}{a^2}. \tag{9.0.13}$$

Thus,

$$\frac{T_0}{T_1} = \left(\frac{a}{b}\right)^2. \tag{9.0.14}$$

It follows that T_0 and T_1 are not arbitrary. They have to satisfy a particular ratio for the body to be in equilibrium. Also, for finite bodies, the shear stress is either *always zero or never zero*. So if no shear traction is applied on the boundary, there are no shear stresses.

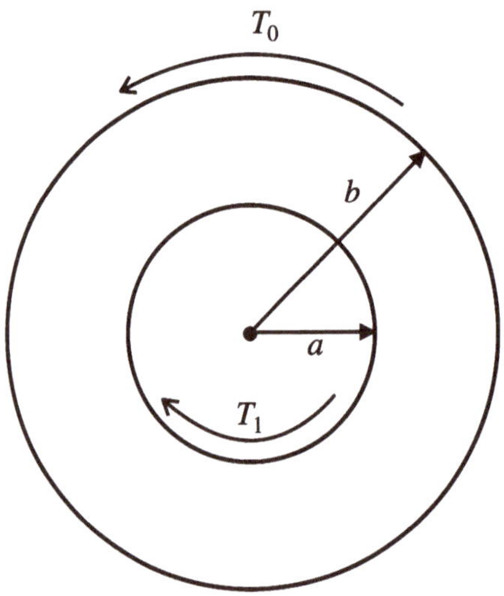

Figure 9.1: Domain used for discussion on polar coordinates.

To find the normal stresses, we need to solve Eq. (9.0.7) and Eq. (9.0.9) simultaneously. We first note that, for axis-symmetric problems,

$$\nabla^2(\cdot) = \frac{1}{r}\frac{\mathrm{d}}{\mathrm{d}r}\left(r\frac{\mathrm{d}(\cdot)}{\mathrm{d}r}\right). \tag{9.0.15}$$

Thus, we have

$$\frac{1}{r}\frac{\mathrm{d}}{\mathrm{d}r}\left(r\frac{\mathrm{d}}{\mathrm{d}r}(T_{rr} + T_{\theta\theta})\right) = 0 \implies \frac{\mathrm{d}}{\mathrm{d}r}\left(r\frac{\mathrm{d}}{\mathrm{d}r}(T_{rr} + T_{\theta\theta})\right) = 0, \tag{9.0.16}$$

and

$$r\frac{\mathrm{d}}{\mathrm{d}r}(T_{rr} + T_{\theta\theta}) = c_1 \implies \frac{\mathrm{d}}{\mathrm{d}r}(T_{rr} + T_{\theta\theta}) = \frac{c_1}{r}, \tag{9.0.17}$$

which on integrating yields

$$T_{rr} + T_{\theta\theta} = c_1 \ln r + c_2 \tag{9.0.18}$$

or

$$T_{\theta\theta} = c_1 \ln r + c_2 - T_{rr}. \tag{9.0.19}$$

However, by Eq. (9.0.9),

$$T_{\theta\theta} = r\frac{\mathrm{d}T_{rr}}{\mathrm{d}r} + T_{rr}. \tag{9.0.20}$$

Thus,

$$r\frac{\mathrm{d}T_{rr}}{\mathrm{d}r} + 2T_{rr} - c_1 \ln r - c_2 = 0 \tag{9.0.21}$$

$$\frac{\mathrm{d}}{\mathrm{d}r}(r^2 T_{rr}) = c_1 r \ln r + c_2 r. \tag{9.0.22}$$

Now, we see that

$$r^2 T_{rr} = \frac{c_1}{2} r^2 \ln r - \frac{c_1}{4} r^2 + \frac{c_2}{2} r^2 + c_3, \tag{9.0.23}$$

or

$$T_{rr} = \frac{c_1}{2} \ln r - \frac{c_1}{4} + \frac{c_2}{2} + \frac{c_3}{r^2} \tag{9.0.24}$$

or

$$T_{rr} = (B - A) \ln r + A + \frac{D}{r^2}, \tag{9.0.25}$$

where $A \equiv \frac{c_2}{2} - \frac{c_1}{4}$, $B \equiv \frac{c_2}{2} + \frac{c_1}{4}$, and $D \equiv c_3$.

Note. The student is asked to show in problem **9.3** that when $u_\theta \equiv 0$, $A - B = 0$. Thus, for instance, in the following example $A = B$. However, if $u_\theta \neq 0$, then A is not generally equal to B. Thus, in problems wherein we expect $u_\theta \equiv 0$ and since $A = B$,

$$T_{rr} = A + \frac{D}{r^2} \tag{9.0.26}$$

$$T_{\theta\theta} = A - \frac{D}{r^2} \tag{9.0.27}$$

Next, consider an infinitely long hollow cylinder with internal and external pressures as shown in Fig. 9.2. The appropriate boundary conditions are

$$T_{rr}(a) = -P_1 \tag{9.0.28}$$

$$T_{rr}(b) = -P_0. \tag{9.0.29}$$

Therefore, on enforcing the boundary conditions Eq. (9.0.28) and Eq. (9.0.29) we obtain the equations

$$A + \frac{D}{a^2} = -P_1 \tag{9.0.30}$$

$$A + \frac{D}{b^2} = -P_0, \tag{9.0.31}$$

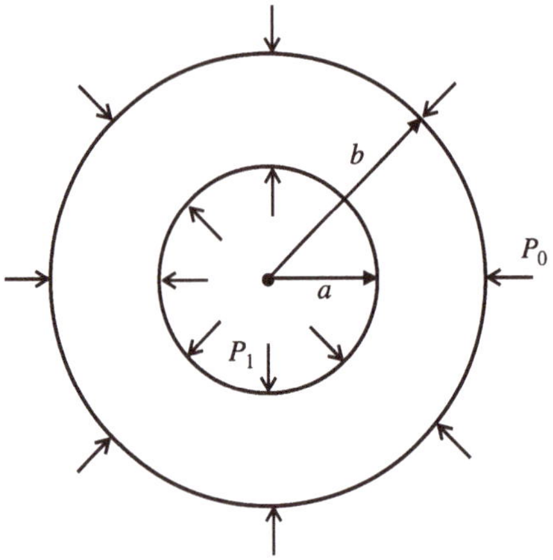

Figure 9.2: Hollow cylinder with internal and external pressures P_1 and P_0, respectively.

for A and D, yielding

$$D = \frac{a^2b^2}{(b^2 - a^2)}(P_0 - P_1) \tag{9.0.32}$$

$$A = \frac{a^2P_1 - b^2P_0}{(b^2 - a^2)}. \tag{9.0.33}$$

Thus, we have

$$T_{rr} = \frac{a^2P_1 - b^2P_0}{(b^2 - a^2)} + \frac{a^2b^2}{(b^2 - a^2)}\frac{(P_0 - P_1)}{r^2} \tag{9.0.34}$$

$$T_{\theta\theta} = \frac{a^2P_1 - b^2P_0}{(b^2 - a^2)} - \frac{a^2b^2}{(b^2 - a^2)}\frac{(P_0 - P_1)}{r^2}. \tag{9.0.35}$$

In this problem, $T_{r\theta} = 0$ (i. e., there is no applied shear). Therefore, T_{rr} and $T_{\theta\theta}$ are principal stresses. We will now consider states of the normal stresses corresponding to various values for a, b, P_0 and P_1

(i) $P_0 > P_1$: this implies that $b^2P_0 > a^2P_1$, since $b > a$ (see Fig. 9.2).

(ii) $P_0 = P_1 = P$: Then we have the following:

$$T_{rr} = -P, \quad T_{\theta\theta} = -P. \tag{9.0.36}$$

(iii) $P_1 > P_0$: For this case, we consider two subcases, that is, Subcase A where $a^2P_1 > b^2P_0$ and Subcase B where $a^2P_1 < b^2P_0$ (see Fig. 9.2). (1) This happens if $2a^2P_1 > (a^2 + b^2)P_0$. (2) occurs if $1 < \sqrt{\frac{P_1 - P_0}{P_0 - \frac{a^2}{b^2}P_1}} < \frac{b}{a}$. (3) This happens if $(a^2 + b^2)P_1 < 2b^2P_0$.

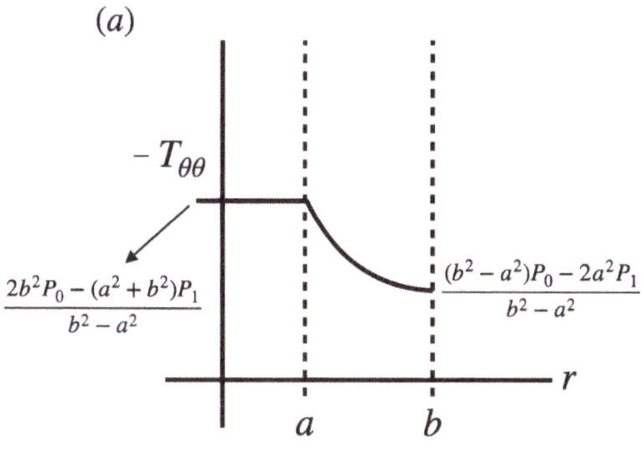

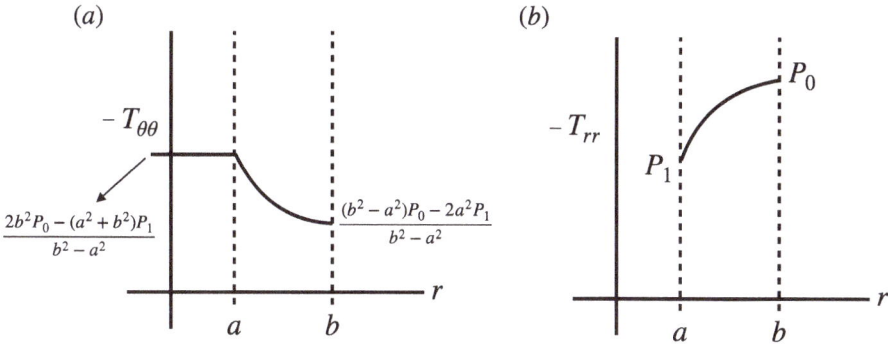

Figure 9.3: Stress plots as a function of r for Case (i).

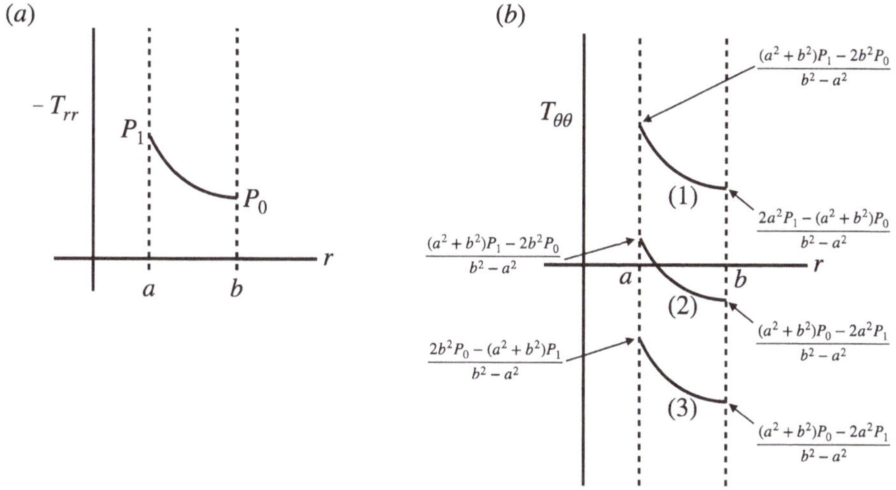

Figure 9.4: Stress plots as a function of r for Case (iii).

Special cases.

1. Suppose $a = 0$ (solid cylinder). Then

$$T_{rr} = T_{\theta\theta} = -P_0. \tag{9.0.37}$$

2. No outer pressure ($P_0 = 0$). Then

$$T_{rr} = -\frac{a^2}{r^2}P_1\frac{(b^2-r^2)}{(b^2-a^2)}, \quad T_{\theta\theta} = \frac{a^2}{r^2}P_1\frac{(b^2+r^2)}{(b^2-a^2)}. \tag{9.0.38}$$

3. Suppose $P_0 = 0$ and $b \longrightarrow \infty$ (hole in an infinite plane with internal pressure). Then,

$$T_{rr} = -\frac{a^2}{r^2}P_1, \quad T_{\theta\theta} = \frac{a^2}{r^2}P_1. \tag{9.0.39}$$

4. $P_1 = 0$ and $b \longrightarrow \infty$ (hole in an infinite plane with uniform tensile loading at ∞). Then,

$$T_{rr} = -P_0\left(1 - \frac{a^2}{r^2}\right), \quad T_{\theta\theta} = -P_0\left(1 + \frac{a^2}{r^2}\right). \tag{9.0.40}$$

Notice $T_{\theta\theta}$ at $a = -2P_0$:

$$k_f = \frac{T_{\theta\theta}\text{ (with hole)}}{T_{\theta\theta}\text{ (without hole)}} = \frac{-2P_0}{-P_0} = 2, \tag{9.0.41}$$

where k_f is the stress concentration factor.

Stress concentration refers to a change in the stress due to an inhomogeneity (inclusion, cavity, etc.). The *stress concentration factor* is the ratio of the stress at a point where there is an inhomogeneity, due to the inhomogeneity, and the stress at the same point when there is no such inhomogeneity.

9.1 Three-dimensional stress functions

In the beginning of the chapter, mention was made of the generalizations of the Airy stress function due to Maxwell, Morera, and Beltrami. We provide a brief discussion of the generalizations of the Airy stress function to study three-dimensional problems. While Airy introduced one stress function (one potential) in terms of which stresses in plane problems can be expressed, Maxwell and Morera introduced three stress functions (three potentials) from which the different components of the stress can be obtained. Beltrami recognized that the three dimensional stress functions due to Maxwell and Morera are special cases of a more general triplet of stress functions that he introduced.

We note that if the stress $\mathbf{T}$ can be expressed as

$$\mathbf{T} = \operatorname{curl}(\operatorname{curl}\mathbf{G}), \quad T_{ij} = \epsilon_{ilm}\epsilon_{jnk}G_{ln,mk}, \tag{9.1.1}$$

where $\mathbf{G}$ is a symmetric, sufficiently smooth tensor, then

$$\mathbf{T} = \mathbf{T}^{\mathrm{T}}, \tag{9.1.2}$$

and $\mathbf{T}$ automatically satisfies

$$\operatorname{div}\mathbf{T} = \mathbf{0}. \tag{9.1.3}$$

The Airy stress function corresponds to the special form of $\mathbf{G}$, whose matrix is

$$(\mathbf{G}) = \begin{pmatrix} 0 & 0 & 0 \\ 0 & 0 & 0 \\ 0 & 0 & \varphi \end{pmatrix}, \tag{9.1.4}$$

where $\varphi = \varphi(x, y)$. It follows from Eq. (9.1.1)

$$T_{11} = \epsilon_{132}\epsilon_{132}\varphi_{,22}, \quad T_{12} = \epsilon_{132}\epsilon_{231}\varphi_{,21}, \quad T_{22} = \epsilon_{231}\epsilon_{231}\varphi_{,11}, \tag{9.1.5}$$

i. e.,

$$T_{xx} = \frac{\partial^2\varphi}{\partial y^2}, \quad T_{yy} = \frac{\partial^2\varphi}{\partial x^2}, \quad T_{xy} - \frac{\partial^2\varphi}{\partial x \partial y}. \tag{9.1.6}$$

For the triplet of stress functions proposed by Maxwell, the matrix of **G** takes the form

$$(\mathbf{G}) = \begin{pmatrix} \tilde{\varphi}_1 & 0 & 0 \\ 0 & \tilde{\varphi}_2 & 0 \\ 0 & 0 & \tilde{\varphi}_3 \end{pmatrix}, \tag{9.1.7}$$

where $\tilde{\varphi}_i$, $i = 1, 2, 3$ are functions of x, y, and z.

It immediately follows that

$$T_{xx} = \frac{\partial^2\tilde{\varphi}_3}{\partial y^2} + \frac{\partial^2\tilde{\varphi}_2}{\partial z^2}, \quad T_{xy} = T_{yx} = -\frac{\partial^2\tilde{\varphi}_3}{\partial x \partial y}, \tag{9.1.8}$$

$$T_{yy} = \frac{\partial^2\tilde{\varphi}_1}{\partial z^2} + \frac{\partial^2\tilde{\varphi}_3}{\partial x^2}, \quad T_{yz} = T_{zy} = -\frac{\partial^2\tilde{\varphi}_1}{\partial y \partial z}, \tag{9.1.9}$$

$$T_{zz} = \frac{\partial^2\tilde{\varphi}_2}{\partial x^2} + \frac{\partial^2\tilde{\varphi}_1}{\partial y^2}, \quad T_{xz} = T_{zx} = -\frac{\partial^2\tilde{\varphi}_2}{\partial x \partial z}. \tag{9.1.10}$$

It is trivial to verify the above components for the stress satisfy the equations of equilibrium, in the absence of body forces.

The triplet of stress functions ψ_i, $i = 1, 2, 3$ considered by Morera, take the following form in the Beltrami representation

$$(\mathbf{G}) = \begin{pmatrix} 0 & \psi_3 & \psi_2 \\ \psi_3 & 0 & \psi_1 \\ \psi_2 & \psi_1 & 0 \end{pmatrix}, \tag{9.1.11}$$

where ψ_i, $i = 1, 2, 3$ are functions of x, y, and z. It immediately follows that

$$T_{xx} = -2\frac{\partial^2\psi_1}{\partial y \partial z}, \quad T_{xy} = T_{yx} = \frac{\partial}{\partial z}\left(\frac{\partial\psi_1}{\partial x} + \frac{\partial\psi_2}{\partial y} - \frac{\partial\psi_3}{\partial z}\right), \tag{9.1.12}$$

$$T_{yy} = -2\frac{\partial^2\psi_2}{\partial x \partial z}, \quad T_{yz} = T_{zy} = \frac{\partial}{\partial x}\left(\frac{\partial\psi_2}{\partial y} + \frac{\partial\psi_3}{\partial z} - \frac{\partial\psi_1}{\partial x}\right), \tag{9.1.13}$$

$$T_{zz} = -2\frac{\partial^2 \psi_3}{\partial x \partial y}, \quad T_{zx} = T_{xz} = \frac{\partial}{\partial y}\left(\frac{\partial \psi_3}{\partial z} + \frac{\partial \psi_1}{\partial x} - \frac{\partial \psi_2}{\partial y}\right). \tag{9.1.14}$$

The above stress components automatically satisfy the equilibrium equations, in the absence of body forces. In the cases considered above, while the stress functions satisfy the equilibrium equations in the absence of body forces, we have to ensure that the stress functions so chosen satisfy the Beltrami-Mitchell compatibility equations. We saw that in the case of the Airy stress function, we obtain the bi-harmonic equation. We have to substitute the stress function in the Beltrami-Mitchell compatibility equation and then solve them using appropriate boundary conditions.

Thus far, we have discussed the components of the stress given in terms of the second derivatives of a potential. It is possible to introduce the stresses that are given in terms of higher derivatives, which of course require the potential to have greater smoothness (C^4). This is due to V. I. Blokh and Ju. A. Krutkov.

Suppose a, b, and c are constants. Furthermore, suppose

$$T_{xx} = (b + c)\frac{\partial^4 F}{\partial y^2 \partial z^2}, \quad T_{xy} = T_{yx} = -c\frac{\partial^4 F}{\partial x \partial y \partial z^2}, \tag{9.1.15}$$

$$T_{yy} = (c + a)\frac{\partial^4 F}{\partial z^2 \partial x^2}, \quad T_{yz} = T_{zy} = -a\frac{\partial^4 F}{\partial y \partial z \partial x^2}, \tag{9.1.16}$$

$$T_{zz} = (a + b)\frac{\partial^4 F}{\partial x^2 \partial y^2}, \quad T_{xz} = T_{zx} = -b\frac{\partial^4 F}{\partial z \partial x \partial y^2}. \tag{9.1.17}$$

The above stress components automatically satisfy the equation of equilibrium, in the absence of body forces. We can also generalize Eq. (9.1.15)-(9.1.17) in the following manner. Suppose a_m, b_m, and c_m, $i = 1, \ldots,$ are constants. Furthermore, suppose

$$T_{xx} = \sum_{m=1}^{\infty}(b_m + c_m)\frac{\partial^4 F_m}{\partial y^2 \partial z^2}, \quad T_{xy} = T_{yx} = -\sum_{m=1}^{\infty} c_m \frac{\partial^4 F_m}{\partial x \partial y \partial z^2}, \tag{9.1.18}$$

$$T_{yy} = \sum_{m=1}^{\infty}(c_m + a_m)\frac{\partial^4 F_m}{\partial z^2 \partial x^2}, \quad T_{yz} = T_{zy} = -\sum_{m=1}^{\infty} a_m \frac{\partial^4 F_m}{\partial y \partial z \partial x^2}, \tag{9.1.19}$$

$$T_{zz} = \sum_{m=1}^{\infty}(a_m + b_m)\frac{\partial^4 F_m}{\partial x^2 \partial y^2}, \quad T_{xz} = T_{zx} = -\sum_{m=1}^{\infty} b_m \frac{\partial^4 F_m}{\partial z \partial x \partial y^2}, \tag{9.1.20}$$

where a_m, b_m, and c_m are arbitrary constants. Once again, the above stresses satisfy the equations of equilibrium, in the absence of body forces.

We notice that if

$$\tilde{\varphi}_1 = \sum_{m=1}^{\infty} a_m \frac{\partial^2 F_m}{\partial x^2}, \quad \tilde{\varphi}_2 = \sum_{m=1}^{\infty} b_m \frac{\partial^2 F_m}{\partial y^2}, \quad \tilde{\varphi}_3 = \sum_{m=1}^{\infty} c_m \frac{\partial^2 F_m}{\partial z^2}, \tag{9.1.21}$$

then $\tilde{\varphi}$, $i = 1, 2, 3$ are precisely the stress functions introduced by Maxwell. Since a_m, b_m, and c_m are arbitrary constants, and since $F_m(x, y, z)$ are arbitrary functions, $\tilde{\varphi}_1$, $\tilde{\varphi}_2$, and

$\tilde{\varphi}_3$ are arbitrary. In the above definition, a_m, b_m, and c_m have to be such that the series converges.

Next, suppose the functions a_m, b_m, and c_m are such that

$$b_m + c_m = \alpha_m, \quad c_m + a_m = \beta_m, \quad a_m + b_m = \gamma_m, \tag{9.1.22}$$

and

$$a_m = \frac{1}{2}(\alpha_m + \beta_m + \gamma_m) = b_m = c_m. \tag{9.1.23}$$

Suppose

$$\sum_{m=1}^{\infty} \alpha_m \frac{\partial^2 F_m}{\partial y \partial z} = \psi_1(x,y,z), \tag{9.1.24}$$

$$\sum_{m=1}^{\infty} \beta_m \frac{\partial^2 F_m}{\partial x \partial z} = \psi_2(x,y,z), \tag{9.1.25}$$

$$\sum_{m=1}^{\infty} \gamma_m \frac{\partial^2 F_m}{\partial x \partial y} = \psi_3(x,y,z), \tag{9.1.26}$$

then ψ_i, $i = 1, 2, 3$ are the stress functions defined by Morera.

In the above discussion a_m, b_m, c_m, α_m, β_m, and γ_m have to be so chosen that the series defined above converges.

9.2 Problems

9.1 Show that the following relations hold

$$T_{rr} = T_{xx}\cos^2\theta + T_{yy}\sin^2\theta + T_{xy}\sin 2\theta, \tag{9.2.1}$$

$$T_{\theta\theta} = T_{xx}\sin^2\theta + T_{yy}\cos^2\theta - T_{xy}\sin 2\theta, \tag{9.2.2}$$

$$T_{r\theta} = \frac{T_{yy} - T_{xx}}{2}\sin 2\theta + T_{xy}\cos 2\theta. \tag{9.2.3}$$

Thus,

$$T_{rr} + T_{\theta\theta} = T_{xx} + T_{yy}. \tag{9.2.4}$$

9.2 Show that Eq. (9.0.22) leads to the expression

$$r^2 T_{rr} = \frac{c_1}{2} r^2 \ln r - \frac{c_1}{4} r^2 + \frac{c_2}{2} r^2 + c_3, \tag{9.2.5}$$

or

$$T_{rr} = \frac{c_1}{2}\ln r - \frac{c_1}{4} + \frac{c_2}{2} + \frac{c_3}{r^2} \tag{9.2.6}$$

or

$$T_{rr} = (B - A) \ln r + A + \frac{D}{r^2}, \tag{9.2.7}$$

where $A \equiv \frac{c_2}{2} - \frac{c_1}{4}$, $B \equiv \frac{c_2}{2} + \frac{c_1}{4}$, and $D \equiv c_3$.

9.3 Show that if $u_\theta = 0$, as it is in axi-symmetrix problems, then $A = B$.

10 Torsion of prismatic cylinders

In the historical introduction to his treatise on elasticity, Love [1] mentions that interest in the problem of torsion can be traced back to the investigations of Galileo. He also discusses the early work of Coulomb which inspired St. Venant's study of torsion of prismatic cylinders (see St. Venant [2]). The study here follows the semi-inverse approach adopted by St. Venant. We will now study the problem of the torsion of a cylinder.[1]

We will assume that the cylinder is of constant cross-section and is twisted due to the application of end moments about the axis of the cylinder, which we shall assume coincides with the z-coordinate direction (see Figure 10.1).

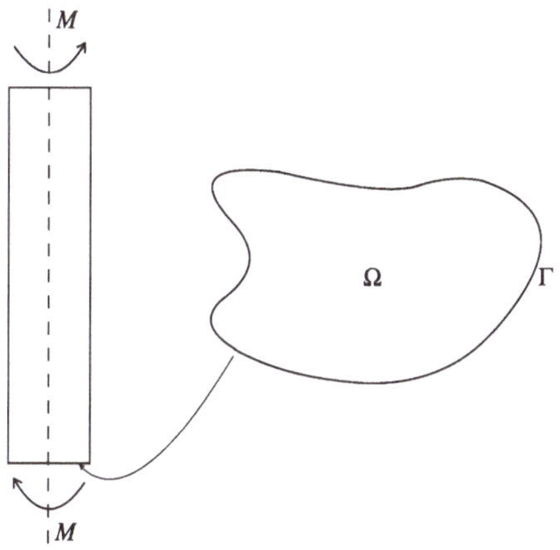

Figure 10.1: Cylinder and the cross-sectional domain.

10.1 Semi-inverse solution

Let us assume a displacement field solution of the following form:

$$u = \alpha yz \tag{10.1.1}$$

$$v = \alpha xz \tag{10.1.2}$$

$$w = \alpha\varphi(x,y), \tag{10.1.3}$$

where u, v, and w are the components of displacement in the x, y, and z directions, respectively. The function φ is known as the warping function and denotes the displacement perpendicular to the cross-section that takes place due to the applied twist. Also, α is the angle of twist per unit length. When one has a circular cross-section, there is no displacement in the axial direction due to the application of a twist.

1 A much more detailed treatment of torsion can be found in Timoshenko and Goodier [3].

https://doi.org/10.1515/9783110789515-010

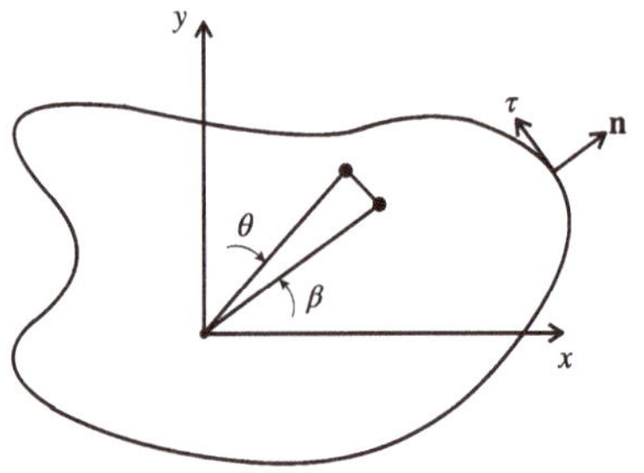

Figure 10.2: Cylinder cross-section with normal and tangent vectors.

We now motivate the assumption that leads to us seeking a solution of the form Eq. (10.1.1)-(10.1.3). Let the total angle of twist be θ (see Figure 10.2). Then,

$$u = r\cos(\theta+\beta) - r\cos(\beta) \tag{10.1.4}$$
$$= r[\cos\theta\cos\beta - \sin\theta\sin\beta] - r\cos\beta \tag{10.1.5}$$
$$\approx -r\theta\sin\beta \quad (\text{since } \theta \text{ is small, } \cos\theta\approx 1, \sin\theta\approx\theta) \tag{10.1.6}$$
$$= -y\theta = -y\alpha z, \tag{10.1.7}$$

where α is the twist per unit length. Thus,

$$u = -\alpha yz. \tag{10.1.8}$$

Similarly for the displacement in the y-direction, we have

$$v = r[\sin(\theta+\beta) - \sin\beta] = r[\sin\theta\cos\beta + \cos\theta\sin\beta - \sin\beta] \tag{10.1.9}$$
$$\approx \theta x = \alpha zx \tag{10.1.10}$$
$$= \alpha xz. \tag{10.1.11}$$

Hence, the strain tensor ε is given by

$$\varepsilon = \frac{1}{2}\begin{bmatrix} 0 & 0 & -\alpha y + \alpha\frac{\partial\varphi}{\partial x} \\ 0 & 0 & \alpha x + \alpha\frac{\partial\varphi}{\partial y} \\ -\alpha y + \alpha\frac{\partial\varphi}{\partial x} & \alpha x + \alpha\frac{\partial\varphi}{\partial y} & 0 \end{bmatrix}. \tag{10.1.12}$$

Recall the linearized rotation tensor:

$$\omega = \begin{bmatrix} 0 & -\alpha z & -\frac{1}{2}\alpha(y+\frac{\partial\varphi}{\partial x}) \\ \alpha z & 0 & \frac{1}{2}\alpha(x-\frac{\partial\varphi}{\partial y}) \\ \frac{1}{2}\alpha(y+\frac{\partial\varphi}{\partial x}) & -\frac{1}{2}\alpha(x-\frac{\partial\varphi}{\partial y}) & 0 \end{bmatrix}. \tag{10.1.13}$$

Thus, $w_{xy,z} = -\alpha$. This implies that α is the angle of twist per unit length. It follows from Hooke's law (Eq. (6.3.8)) that if $(\operatorname{tr} \varepsilon) = 0$,

$$\mathbf{T} = \begin{bmatrix} 0 & 0 & \alpha\mu(\frac{\partial\varphi}{\partial x} - y) \\ 0 & 0 & \alpha\mu(\frac{\partial\varphi}{\partial y} + x) \\ \alpha\mu(\frac{\partial\varphi}{\partial x} - y) & \alpha\mu(\frac{\partial\varphi}{\partial y} + x) & 0 \end{bmatrix}. \tag{10.1.14}$$

Let us assume $\mathbf{b} = \mathbf{0}$. Thus, the equilibrium equation reduces to $\operatorname{div} \mathbf{T} = \mathbf{0}$ and this leads to

$$\frac{\partial}{\partial x}\left[\alpha\mu\left(\frac{\partial\varphi}{\partial x} - y\right)\right] + \frac{\partial}{\partial y}\left[\alpha\mu\left(\frac{\partial\varphi}{\partial y} + x\right)\right] = 0, \tag{10.1.15}$$

or

$$\nabla^2\varphi = 0 \tag{10.1.16}$$

in the domain Ω.

10.1.1 Boundary conditions

The lateral surface of the cylinder is free of traction. Recall the traction $\mathbf{t}$ is given by Eq. (8.1.4). Thus,

$$t_x = 0 \tag{10.1.17}$$

$$t_y = 0 \tag{10.1.18}$$

$$t_z = 0. \tag{10.1.19}$$

Now,

$$t_x = \cancelto{0}{T_{xx}} n_x + \cancelto{0}{T_{xy}} n_y + T_{xz} \cancelto{0}{n_z} \tag{10.1.20}$$

$$t_y = \cancelto{0}{T_{yx}} n_x + \cancelto{0}{T_{yy}} n_y + T_{yz} \cancelto{0}{n_z} \tag{10.1.21}$$

$$t_z = T_{zx} n_x + T_{zy} n_y + T_{zz} \cancelto{0}{n_z}, \tag{10.1.22}$$

where we have used the fact that the normal to the lateral surface has no z component so $n_z = 0$. Thus,

$$T_{zx} n_x + T_{zy} n_y = 0 \tag{10.1.23}$$

$$\implies \alpha\mu\left(\frac{\partial\varphi}{\partial x} - y\right) n_x + \alpha\mu\left(\frac{\partial\varphi}{\partial y} + x\right) n_y = 0, \tag{10.1.24}$$

which can also be written as

$$\frac{\partial \varphi}{\partial x}n_x + \frac{\partial \varphi}{\partial y}n_y = \frac{\partial \varphi}{\partial n} = yn_x - xn_y \quad \text{on } \Gamma. \tag{10.1.25}$$

We have to solve Eq. (10.1.16) and (10.1.25).

$$\nabla\varphi = \frac{\partial \varphi}{\partial n}\mathbf{n} + \frac{\partial \varphi}{\partial \tau}\boldsymbol{\tau} \implies \nabla\varphi \cdot \mathbf{n} = \frac{\partial \varphi}{\partial n}, \tag{10.1.26}$$

since the unit normal and unit tangent are orthogonal to one-another, and where

$$\nabla\varphi \cdot \mathbf{n} = \frac{\partial \varphi}{\partial n} = \left(\frac{\partial \varphi}{\partial x}\mathbf{e}_x + \frac{\partial \varphi}{\partial y}\mathbf{e}_y\right) \cdot (n_x\mathbf{e}_x + n_y\mathbf{e}_y) = \frac{\partial \varphi}{\partial x}n_x + \frac{\partial \varphi}{\partial y}n_y. \tag{10.1.27}$$

The forces acting in the x and y directions on any cross section of the cylinder should be zero. We will check this now. The force acting along the x-direction, denoted by f_x, is given by

$$f_x = \int_\Omega t_x \, da = \int_\Omega T_{xz} n_z \, da, \tag{10.1.28}$$

and since $n_z = 1$, we have

$$f_x = \int_\Omega T_{xz} \, da. \tag{10.1.29}$$

Notice that

$$\frac{\partial}{\partial x}\left[x\left(\frac{\partial \varphi}{\partial x} - y\right)\right] + \frac{\partial}{\partial y}\left[x\left(\frac{\partial \varphi}{\partial y} + x\right)\right] = \left(\frac{\partial \varphi}{\partial x} - y\right) + x\frac{\partial^2 \varphi}{\partial x^2} + x\frac{\partial^2 \varphi}{\partial y^2} \tag{10.1.30}$$

$$= \left[\frac{\partial \varphi}{\partial x} - y\right] + x\cancelto{0}{\nabla^2\varphi}. \tag{10.1.31}$$

The right hand side multiplied by α and μ is the shear stress T_{xz}. Recall the divergence theorem:

$$\int_\Omega \operatorname{div} \mathbf{v} \, da = \int_\Gamma \mathbf{v} \cdot \mathbf{n} \, ds. \tag{10.1.32}$$

Thus,

$$\int_\Omega \left(\frac{\partial \varphi}{\partial x} - y\right) da = \int_\Gamma \left\{x\left(\frac{\partial \varphi}{\partial x} - y\right)n_x + x\left(\frac{\partial \varphi}{\partial y} + x\right)n_y\right\} ds \tag{10.1.33}$$

$$= \int_\Gamma x\left[\frac{\partial \varphi}{\partial x}n_x + \frac{\partial \varphi}{\partial y}n_y - yn_x + xn_y\right] ds \tag{10.1.34}$$

$$= \int_{\Gamma} x\left[\frac{\partial\varphi}{\partial n} - yn_x + xn_y\right] ds, \tag{10.1.35}$$

which in virtue of Eq. (10.1.25) leads to

$$\int_{\Omega}\left(\frac{\partial\varphi}{\partial x} - y\right) da = \int_{\Gamma} x\left[\frac{\partial\varphi}{\partial n} - yn_x + xn_y\right] ds = 0. \tag{10.1.36}$$

Thus, we have shown that the force acting in the x-direction is zero. Similarly, we can show the force acting in the y-direction is zero. Next, the moment M is given by

$$M = \int_{\Omega} (xT_{yz} - yT_{xz})\, da. \tag{10.1.37}$$

If we substitute the expressions for T_{yz} and T_{xz}, we find

$$M = \alpha\mu \int_{\Omega}\left(x^2 + y^2 + x\frac{\partial\varphi}{\partial y} - y\frac{\partial\varphi}{\partial x}\right) da, \tag{10.1.38}$$

where

$$K = \mu \int_{\Omega}\left(x^2 + y^2 + x\frac{\partial\varphi}{\partial y} - y\frac{\partial\varphi}{\partial x}\right) da, \tag{10.1.39}$$

is referred to as the torsional rigidity of the cylinder.

Let ψ denote the conjugate harmonic of φ and let

$$F(z) = \varphi + i\psi. \tag{10.1.40}$$

Then,

$$\frac{\partial\varphi}{\partial x} = \frac{\partial\psi}{\partial y}, \quad \frac{\partial\varphi}{\partial y} = -\frac{\partial\psi}{\partial x}, \tag{10.1.41}$$

which are known as the Cauchy-Riemann Conditions. Also,

$$\nabla^2\varphi = 0, \quad \text{in } \Omega \tag{10.1.42}$$

$$\nabla^2\psi = 0, \quad \text{in } \Omega. \tag{10.1.43}$$

Now,

$$T_{xz} = \alpha\mu\left(\frac{\partial\psi}{\partial y} - y\right) \tag{10.1.44}$$

$$T_{yz} = -\alpha\mu\left(\frac{\partial\psi}{\partial x} - x\right). \tag{10.1.45}$$

Thus, Eq. (10.1.25) becomes

$$\left(\frac{\partial\psi}{\partial y}-y\right)n_x-\left(\frac{\partial\psi}{\partial x}-x\right)n_y=0,\quad \text{on }\Gamma. \tag{10.1.46}$$

However,

$$n_x=\frac{\partial y}{\partial s},\quad n_y=-\frac{\partial x}{\partial s}. \tag{10.1.47}$$

Thus,

$$\frac{\partial}{\partial s}\left[\psi-\frac{1}{2}(x^2+y^2)\right]=0,\quad \text{on }\Gamma. \tag{10.1.48}$$

Hence,

$$\psi-\frac{1}{2}(x^2+y^2)=\text{const.}=c,\quad \text{on }\Gamma. \tag{10.1.49}$$

Let us introduce Φ (Prandtl introduced this stress function in his paper in 1903[2]) such that

$$\Phi=\psi-\frac{1}{2}(x^2+y^2). \tag{10.1.50}$$

Then,

$$\nabla^2\Phi=-2\quad \text{in }\Omega \tag{10.1.51}$$

$$\Phi=c,\quad \text{on }\Gamma. \tag{10.1.52}$$

The original torsion problem has been reduced to solving Eq. (10.1.51) subject to Eq. (10.1.52).

For simply connected domains Ω, we can set c to be a single constant and solve the problem. However, for multiply connected domains Ω with several holes, we cannot use a single constant c to solve the problem. Let us first consider the case when the domain is simply connected.

10.2 Torsion of an elliptic shaft

Consider an elliptic cross section of the form

$$\frac{x^2}{a^2}+\frac{y^2}{b^2}=1. \tag{10.2.1}$$

2 L. Prandtl, Zur Torsion von prismatischen Stäben (Torsion of prismatic rods), Phys. Z. 4, 758–759 (1903).

We first choose $\Phi = c[\frac{x^2}{a^2} + \frac{y^2}{b^2}]$ so that it is constant on the boundary. Next, we need to ensure that Φ satisfies Eq. (10.1.51). Thus,

$$2c\left[\frac{1}{a^2} + \frac{1}{b^2}\right] = -2 \implies c = -\frac{a^2b^2}{a^2+b^2}. \tag{10.2.2}$$

Hence,

$$T_{xz} = \alpha\mu\left(\frac{\partial\phi}{\partial x} - y\right) = \alpha\mu\left(\frac{\partial\psi}{\partial y} - y\right) \tag{10.2.3}$$

$$= \alpha\mu\frac{\partial\Phi}{\partial y} \tag{10.2.4}$$

$$= -\frac{2\alpha\mu a^2}{a^2+b^2}y. \tag{10.2.5}$$

Similarly,

$$T_{yz} = -\alpha\mu\frac{\partial\Phi}{\partial x} = \frac{2\alpha\mu b^2}{a^2+b^2}x. \tag{10.2.6}$$

The magnitude of the tangential stress τ at any point on the boundary is given by

$$\tau = [T_{xz}^2 + T_{yz}^2]^{1/2} = \frac{2\mu ab}{a^2+b^2}\left[\frac{b^2x^2}{a^2} + \frac{a^2y^2}{b^2}\right]^{1/2}. \tag{10.2.7}$$

10.3 Torsion of a cylinder with rectangular cross-section

Often times, using the equations for the boundary to determine the stress function, as we did in the case of a shaft with elliptic cross-section, will not work. A simple example is a cylinder with a rectangular cross-section such as that shown in Fig. 10.3. If we use the same method to pick the function as in the case of the elliptic cross-section, then

$$\Phi = (y-b)(y+b)(x-a)(x+a) = (y^2-b^2)(x^2-a^2), \tag{10.3.1}$$

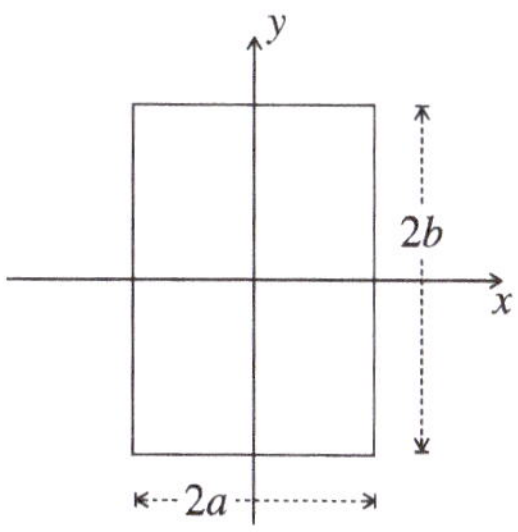

Figure 10.3: Cross section of a rectangular cylinder.

and

$$\nabla^2\Phi = 2[(y^2 - b^2) + (x^2 - a^2)] \quad \text{in } \Omega. \tag{10.3.2}$$

Thus, we see that

$$\nabla^2\Phi \neq -2 \quad \text{in } \Omega. \tag{10.3.3}$$

Since we are dealing with a linear operator, we can pick Φ such that $\Phi = \Phi_1 + \Phi_2$ where Φ_1 satisfies the Laplace equation (Harmonic equation) instead of the Poisson equation. Here, Φ_2 is chosen so that its Laplacian is -2, but does not satisfy the zero boundary condition. On the other hand, we require that Φ_1 satisfy the negative of what Φ_2 satisfies on the boundary so that $\Phi_1 + \Phi_2$ is zero on the boundary. Let us pick

$$\Phi_2 = (a^2 - x^2). \tag{10.3.4}$$

We note that $\nabla^2\Phi_2 = -2$. Thus,

$$\nabla^2\Phi = \nabla^2(\Phi_1 + \Phi_2) = -2 \implies \nabla^2\Phi_1 = 0. \tag{10.3.5}$$

Since $\Phi_1|_\Gamma = -\Phi_2|_\Gamma$, Φ_1 satisfies $\Phi_1(\pm a, y) = 0$ and $\Phi_1(x, \pm b) = 0$. So, we have to solve

$$\nabla^2\Phi_1 = 0 \tag{10.3.6}$$

$$\Phi_1(\pm a, y) = 0, \; \Phi_1(x, \pm b) = 0. \tag{10.3.7}$$

We seek a solution using the method of separation of variables. Let

$$\Phi_1(x, y) = F(x)G(y). \tag{10.3.8}$$

Substituting Eq. (10.3.8) into Eq. (10.3.6), we obtain

$$F''(x) + \lambda^2 F(x) = 0 \tag{10.3.9}$$

$$G''(y) - \lambda^2 G(y) = 0. \tag{10.3.10}$$

Thus, the solution to Eq. (10.3.9) and Eq. (10.3.10) are

$$F(x) = A \sin \lambda x + B \cos \lambda x \tag{10.3.11}$$

$$G(y) = C \sinh \lambda y + D \cosh \lambda y. \tag{10.3.12}$$

Due to the symmetry of the expected solution, $A = C = 0$. In general, any $\lambda = \frac{n\pi}{2a}$, $n = 1, 2, 3, \ldots$ will meet Eq. (10.3.9) and Eq. (10.3.10). However, in view of Eq. (10.3.7), λ is restricted to $\lambda = \frac{n\pi}{2a}$, $n = 1, 3, 5, \ldots$. The solution for $\Phi_1(x, y)$ is given by

$$\Phi_1(x,y) = \sum_{n=1}^{\infty} E_n \cos\frac{n\pi x}{2a} \cosh\frac{n\pi y}{2a}. \tag{10.3.13}$$

Since Φ_1 has to satisfy the boundary condition given by Eq. (10.3.7), we need

$$(a^2 - x^2) = \sum_{n=1}^{\infty} H_n \cos\frac{n\pi x}{2a}, \tag{10.3.14}$$

where

$$H_n = E_n \cosh\frac{n\pi h}{2a}. \tag{10.3.15}$$

Under appropriate conditions, a function $f(x)$ can be expressed as

$$f(x) = \frac{\alpha_0}{2} + \sum_{n=1}^{\infty} (\alpha_n \sin nx + \beta_n \cos nx), \tag{10.3.16}$$

where the Fourier coefficients α_0, α_n, and β_n are given by

$$\alpha_0 = \frac{1}{\pi}\int_{-\pi}^{\pi} f(x)\,\mathrm{d}x, \tag{10.3.17}$$

$$\alpha_n = \frac{1}{\pi}\int_{-\pi}^{\pi} f(x)\sin nx\,\mathrm{d}x, \tag{10.3.18}$$

$$\beta_n = \frac{1}{\pi}\int_{-\pi}^{\pi} f(x)\cos nx\,\mathrm{d}x. \tag{10.3.19}$$

For our problem, $\alpha_0 = 0$, $\alpha_n = 0 \;\forall\, n = 1, 2, 3, \ldots$, and

$$\beta_n = \frac{1}{a}\int_{-a}^{a} (a^2 - x^2)\cos\frac{n\pi x}{2a}\,\mathrm{d}t. \tag{10.3.20}$$

Thus,

$$E_n = \frac{32(-1)^{(n-1)/2}}{n^3\pi^3\cosh\frac{n\pi b}{a}}. \tag{10.3.21}$$

It then follows from $\Phi = \Phi_1 + \Phi_2$, Eq. (10.3.4), (10.3.13), and (10.3.21) that

$$\Phi = \underbrace{\frac{-32a^2}{\pi^3}\sum_{n=1}^{\infty}\left[\frac{(-1)^{(n-1)/2}}{n^3\cosh\frac{n\pi b}{a}}\cos\frac{n\pi}{2a}x\cosh\frac{n\pi}{2a}y\right]}_{\Phi_1} + \underbrace{(a^2 - x^2)}_{\Phi_2}. \tag{10.3.22}$$

Once the stress function has been determined, one can find the stresses T_{xz} and T_{yz} and the torsional rigidity.

10.4 Torsion of multiply connected domains

In our analysis of torsion, we have thus far only considered simply-connected domains (i. e., where any closed curve in the domain can be continuously reduced to a point). In this section, we will investigate the situation when the cross sections are multiply connected as that shown in Fig. 10.4. Additional information and examples of torsion of cylinders with multiply connected cross-sections may be found in [3].

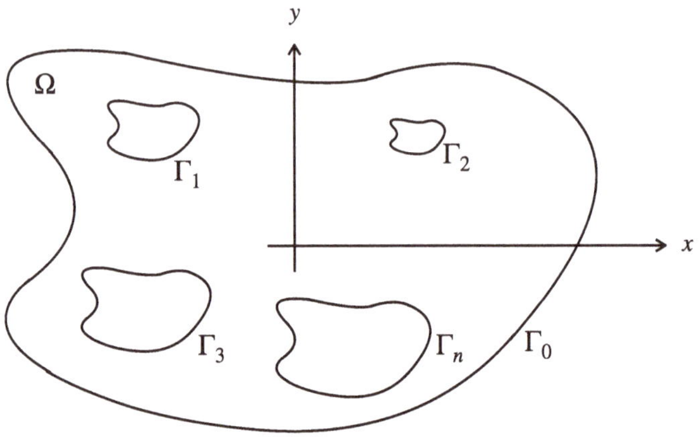

Figure 10.4: Generalized multiply-connected domain containing arbitrarily-shaped holes.

Let Γ_i denote the ith boundary, Γ_0 refer to the outermost boundary, and let Ω denote the domain of the cross-section of the cylinder being considered. We will assume that all lateral surfaces of the cylinder corresponding to each boundary are to be traction free such that

$$\Phi = \hat{\Phi}_i, \tag{10.4.1a}$$

and

$$\frac{d\phi}{dn} = (yn_x - xn_y), \tag{10.4.1b}$$

on the ith boundary Γ_i, and where $\hat{\Phi}_i$ are the values taken by the stress function on the boundaries that are specified to be constant. Although the stress function on the boundary Γ_i can be chosen to be any arbitrary constant, we shall choose $\hat{\Phi}_0 = 0$ whenever possible. In doing so, the constants $\hat{\Phi}_i$ where $i = 1, 2 - -n$, cannot be equal to zero. Instead, the constants $\hat{\Phi}_i$ on the interior boundaries, signified here as Γ_i, $i = 1, 2 - -n$, are determined by requiring that the displacement is single-valued. This requirement implies that

$$\oint_{\Gamma_i} d\phi(x, y) = 0. \tag{10.4.2}$$

Since

$$d\phi = \frac{\partial \phi}{\partial x} dx + \frac{\partial \phi}{\partial y} dy, \tag{10.4.3}$$

the left hand side of Eq. (10.4.2) can be expressed as

$$\oint_{\Gamma_i} d\phi(x, y) = \oint_{\Gamma_i} \left(\frac{\partial \phi}{\partial x} dx + \frac{\partial \phi}{\partial y} dy \right). \tag{10.4.4}$$

Recall that the shear stresses can be expressed in terms of the displacement, $w(x, y)$, via

$$T_{xz} = \alpha\mu \frac{\partial \phi}{\partial x} - y\alpha\mu, \quad T_{yz} = \alpha\mu \frac{\partial \phi}{\partial y} + x\alpha\mu. \tag{10.4.5}$$

Using Eqs. (10.4.4) and (10.4.5), one can show that

$$\oint_{\Gamma_i} d\phi(x, y) = \frac{1}{\alpha\mu} \oint_{\Gamma_i} (T_{xz} dx + T_{yz} dy) - \frac{\alpha}{\mu} \oint_{\Gamma_i} (x dy - y dx). \tag{10.4.6}$$

It follows from Green's theorem that

$$\oint_{\Gamma_i} (x dy - y dx) = 2 \iint_{A_i} dx dy = 2A_i, \tag{10.4.7}$$

where A_i is the area enclosed by the ith boundary Γ_i. Recalling that

$$T_{xz} dx + T_{yz} dy = \bar{\tau} ds, \tag{10.4.8}$$

where $\bar{\tau}$ is the resultant shear stress, Eqs. (10.4.2), (10.4.6), and (10.4.7) imply that

$$\oint_{\Gamma_i} \bar{\tau}\, ds = 2A_i, \tag{10.4.9}$$

where, as previously mentioned, i is an integer greater than zero. Finally, the resultant moment on the shaft is defined as

$$M = \iint_{\Omega} (xT_{yz} - yT_{xz})\, dxdy = -\alpha\mu \iint_{\Omega} \left(x\frac{\partial \Phi}{\partial x} + y\frac{\partial \Phi}{\partial y} \right) dxdy, \tag{10.4.10}$$

which on invoking Green's theorem can equivalently be written as

$$M = 2\alpha\mu \iint_{\Omega} \Phi\, dxdy - \alpha\mu \oint_{\Gamma_i} (xn_x + yn_y)\Phi\, ds. \tag{10.4.11a}$$

Thus, one arrives at the expression

$$M = 2\alpha\mu \iint_{\Omega} \Phi\, dxdy + 2\alpha\mu \sum_{i=1}^{n} \hat{\Phi}_i A_i, \tag{10.4.12}$$

where n corresponds to the number of holes in the cross section.

10.5 The torsion membrane analogy

A powerful tool in developing solutions to problems is making use of analogies that arise in very distinct physical problems giving rise to the same class of governing equations. For instance, two disparate problems such as a certain class of flows of fluids are governed by the same partial differential equations as a class of deformations of a linearized elastic solid, allowing one to use similar mathematical techniques in solving the two equations. This is also true even with regard to different classes of deformations of a linearized elastic solid. In this section, we show how one could gainfully exploit the analogy that exists between the torsion problem and that of the deformation of a membrane due to the application of pressure. Ludwig Prandtl[3] recognized that the equations governing the torsion problem resemble those for the static deflection of a thin elastic membrane under uniform pressure, and tension at the edges, the cross-section of the bar being twisted and the membrane being the same, such as that shown in Fig. 10.5a. Recall that the governing equation for the torsion problem defined in terms of the Prandtl stress function, $\Phi = \Phi(x, y)$, is

$$\nabla^2 \Phi = \frac{\partial^2 \Phi}{\partial x^2} + \frac{\partial^2 \Phi}{\partial y^2} = -2 \quad \text{in } \Omega, \tag{10.5.1}$$

with boundary condition

$$\frac{\partial \Phi}{\partial s} = 0 \quad \text{on } \Gamma, \tag{10.5.2}$$

3 L. Prandtl, Zur Torsion von prismatischen Stäben (Torsion of prismatic rods), Phys. Z. 4, 758-759 (1903).

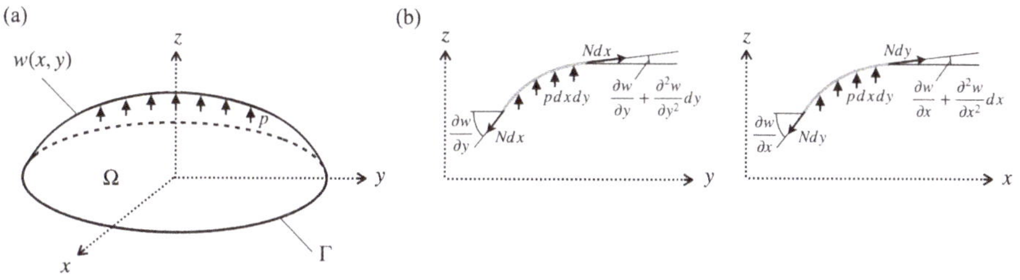

Figure 10.5: The static deflection of a membrane with uniform tension: (a) the representation of the deflected membrane with domain Ω in the xy-plane and boundary Γ and (b) the equilibrium representation of an arbitrary infinitesimal membrane element.

where μ is the shear modulus, α is the angle of twist per unit length, and

$$T_{xz} = \alpha\mu \frac{\partial \Phi}{\partial y}, \quad T_{yz} = -\alpha\mu \frac{\partial \Phi}{\partial x}. \tag{10.5.3}$$

Next, consider the thin elastic membrane shown in Fig. 10.5. The membrane is initially stretched over a region defined as Ω with boundary Γ. We shall assume that this stretching induces a uniform tension, N, throughout the membrane. The membrane is then subjected to a uniform transverse pressure, p, that induces a transverse deflection, $w = w(x, y)$. We assume the applied pressure leads to the tension at the edges to be uniform. Suppose the membrane is in equilibrium (Fig. 10.5b). Now consider an increase in the deflection, w of the membrane. The equation of equilibrium in the z-direction acting on the infinitesimal element is

$$\frac{\partial^2 w}{\partial x^2} + \frac{\partial^2 w}{\partial y^2} = -\frac{p}{N} \tag{10.5.4}$$

We shall assume that the domain defining the membrane is fixed at the boundary such that transverse deflection is prohibited. That is,

$$w(x, y) = 0 \quad \text{on } \Gamma. \tag{10.5.5}$$

By comparing the membrane problem of Eq. (10.5.4) with the torsion problem of Eq. (10.5.1), a few things are immediately obvious. On comparing the equation governing torsion and that which has been derived for the membrane under pressure, we conclude the solution for the torsion problem can be obtained from the solution of the membrane problem, by replacing p/N by $2\mu\alpha$. Furthermore, since Φ is analogous to w and the volume enclosed by the membrane can be expressed as

$$V = \iint_{\Omega} w(x, y)\, dxdy, \tag{10.5.6}$$

the resultant torque on a given cylinder is analogous to double the volume, or $M = 2V$. To see the analogy for the shear stresses, consider the only nonzero stresses defined in terms of the stress function

$$T_{xz} = \alpha\mu\frac{\partial\Phi}{\partial y} = \frac{\partial w}{\partial y}, \quad T_{yz} = -\alpha\mu\frac{\partial\Phi}{\partial x} = \frac{\partial w}{\partial x}. \tag{10.5.7}$$

Thus, the stress is given by the slope of the deflection of the membrane. The condition that defines a contour line (w = constant) on the membrane surface equivalently defines a line of constant Φ for the torsion problem. Recalling that

$$n_x = \frac{\partial y}{\partial s}, \quad n_y = \frac{\partial x}{\partial s}, \tag{10.5.8}$$

along the contour line

$$\frac{\partial w}{\partial s} = \alpha\mu\frac{\partial\Phi}{\partial s} = \alpha\mu\left(\frac{\partial\Phi}{\partial x}\frac{\partial x}{\partial s} + \frac{\partial\Phi}{\partial y}\frac{\partial y}{\partial s}\right) = T_{zn} = 0, \tag{10.5.9}$$

where s denotes the tangential coordinate, n is the normal coordinate, and T_{zn} is the stress normal to the contour line. It then follows that

$$\bar{\tau} = T_{zt} = -\alpha\mu\left(\frac{\partial\Phi}{\partial y}\frac{\partial y}{\partial n} + \frac{\partial\Phi}{\partial x}\frac{\partial x}{\partial n}\right) = -\alpha\mu\frac{\partial\Phi}{\partial n} = -\frac{\partial w}{\partial n}. \tag{10.5.10}$$

Thus, the resultant shear stress in the domain of the torsion problem is given by the negative of the slope of the membrane in the direction normal to the contour line. This implies that the maximum shear stress will occur at the location corresponding to the largest membrane slope.

10.6 Problems

10.1 Repeat the process shown in Eq. (10.1.28)-(10.1.36) to show that $f_y = 0$.

10.2 Show that the maximum of tangential stress, τ, occurs at the point where the semi-minor axes intersect the boundary, Γ, as shown in Fig. 10.6.

10.3 Find the torsional rigidity of the elliptic cylinder.

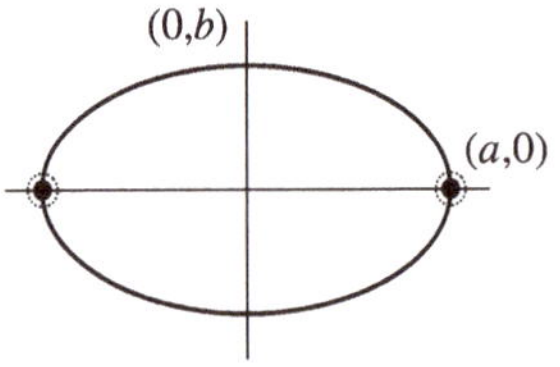

Figure 10.6: Elliptic cylinder with maximum tangential stress indicated where semi-minor axes intersect the boundary.

References

[1] A. E. H. Love. A Treatise on the Mathematical Theory of Elasticity. Dover Publications, New York, 1944.
[2] A. B. St. Venant. Me'm des Savants e'strangers, 14, 1855.
[3] S. Timoshenko and J. N. Goodier. Theory of Elasticity. McGraw-Hill, New York, etc., 1987.

11 Problems in cylindrical polar coordinates

In Chapter 9, we studied problems using a Polar coordinate system. The Cylindrical polar coordinate system is particularly well suited to analyze a large class of problems that are technologically important. Such problems include problems of revolution, where a section/surface is revolved around an axis defining the natural axis of the problem. These problems can further be simplified under the assumption of axial symmetry (e. g., cylinder, sphere, disk, etc.). Under the appropriate conditions, some problems can be reduced to the study of planar problems. That is, they can be cast as two-dimensional problems and analysed using polar coordinates. The section will begin by considering a problem in cylindrical coordinates and as the problem satisfies certain assumptions it can be treated as a two dimensional problem and can be studied within the context of polar coordinates.

One can express all the equations of linearized elasticity: the constitutive relation, the strain-displacement equations, the balance of linear momentum, and the compatibility equations with regard to any basis. That is, we can document the equations in component form with respect to the Cartesian base vectors $(\mathbf{i}, \mathbf{j}, \mathbf{k})$ in terms of the Cartesian coordinate system or the base vectors $(\mathbf{e}_r, \mathbf{e}_\theta, \mathbf{e}_z)$ in terms of the cylindrical polar coordinate system, or for that matter any other coordinate system. But this requires knowledge of curvilinear tensor analysis. On page 181 of Sokolnikoff ([1]) one can find a succinct documentation of all the appropriate equations valid for coordinate systems using covariant and contravariant components and derivatives of the relevant quantities (see Sokolnikoff ([1]) for a detailed discussion of tensor analysis and the development of equations for a continuum. See also Malvern ([2]) and Jaunzemis ([3])). In the development here, though it is not the optimal mathematical way to arrive at the field equations in cylindrical polar co-ordinates, we shall use a rather cumbersome procedure starting with the results in a Cartesian basis and then using the relation between the Cartesian basis and the basis for the cylindrical polar coordinate system to obtain the relevant equations as we do not assume the reader be conversant with curvilinear tensor analysis.

Let $(\mathbf{e}_r, \mathbf{e}_\theta, \mathbf{e}_z)$ denote the basis vectors in a cylindrical coordinate system. Then, these basis vector can be are expressed in terms of the Cartesian basis vectors $(\mathbf{e}_x, \mathbf{e}_y, \mathbf{e}_z)$ as

$$\mathbf{e}_r = \cos\theta\mathbf{e}_x + \sin\theta\mathbf{e}_y, \quad \mathbf{e}_\theta = -\sin\theta\mathbf{e}_x + \cos\theta\mathbf{e}_y, \quad \mathbf{e}_z = \mathbf{e}_z. \tag{11.0.1}$$

We recall that

$$x = r\cos\theta, \quad y = r\sin\theta, \quad r = \sqrt{x^2 + y^2}, \quad \theta = \arctan(y/x). \tag{11.0.2}$$

For many applications, it is also important to know the relation between derivatives and differential operators in Cartesian coordinates and cylindrical coordinates. Here, we will present the operators of interest, such as the gradient, divergence, curl, and

https://doi.org/10.1515/9783110789515-011

Laplacian, expressed in cylindrical coordinates. Using Eq. (11.0.2), the Cartesian partial derivatives of some arbitrary function, ξ, can be expressed as

$$\frac{\partial \xi}{\partial x} = \frac{\partial r}{\partial x}\frac{\partial \hat{\xi}}{\partial r} + \frac{\partial \theta}{\partial x}\frac{\partial \hat{\xi}}{\partial \theta} = \cos\theta \frac{\partial \hat{\xi}}{\partial r} - \frac{\sin\theta}{r}\frac{\partial \hat{\xi}}{\partial \theta}, \tag{11.0.3a}$$

$$\frac{\partial \xi}{\partial y} = \frac{\partial r}{\partial y}\frac{\partial \hat{\xi}}{\partial r} + \frac{\partial \theta}{\partial y}\frac{\partial \hat{\xi}}{\partial \theta} = \sin\theta \frac{\partial \hat{\xi}}{\partial r} + \frac{\cos\theta}{r}\frac{\partial \hat{\xi}}{\partial \theta}, \tag{11.0.3b}$$

$$\frac{\partial \xi}{\partial z} = \frac{\partial \hat{\xi}}{\partial z}, \tag{11.0.3c}$$

where $\hat{\xi}$ is a scalar function expressed in cylindrical coordinates. Furthermore, the second partial derivatives can be found from the above expressions but are excluded here for brevity. It can also be shown using Eqs. (11.0.1) and (11.0.3) that the gradient operator can be expressed as

$$\nabla \xi = \frac{\partial \hat{\xi}}{\partial r}\mathbf{e}_r + \frac{1}{r}\frac{\partial \hat{\xi}}{\partial \theta}\mathbf{e}_\theta + \frac{\partial \hat{\xi}}{\partial z}\mathbf{e}_z. \tag{11.0.4}$$

The gradient of a vector, ζ, yields a second order tensor. The matrix associated with the gradient of a vector in cylindrical coordinates is

$$(\nabla \zeta) = \frac{\partial \zeta_i}{\partial x_j} = \begin{bmatrix} \frac{\partial \zeta_r}{\partial r} & \frac{1}{r}\frac{\partial \zeta_r}{\partial \theta} - \frac{\zeta_\theta}{r} & \frac{\partial \zeta_r}{\partial z} \\ \frac{\partial \zeta_\theta}{\partial r} & \frac{1}{r}\frac{\partial \zeta_\theta}{\partial \theta} + \frac{\zeta_r}{r} & \frac{\partial \zeta_\theta}{\partial z} \\ \frac{\partial \zeta_z}{\partial r} & \frac{1}{r}\frac{\partial \zeta_z}{\partial \theta} & \frac{\partial \zeta_z}{\partial z} \end{bmatrix}. \tag{11.0.5}$$

The divergence of a vector in cylindrical coordinates is

$$\operatorname{div} \zeta = \frac{\partial \zeta_i}{\partial x_i} = \operatorname{tr}(\operatorname{grad} \zeta) = \frac{1}{r}\frac{\partial (r\zeta_r)}{\partial r} + \frac{1}{r}\frac{\partial \zeta_\theta}{\partial \theta} + \frac{\partial \zeta_z}{\partial z}. \tag{11.0.6}$$

Similarly, the divergence of a tensor, $\mathbf{A}$, can be expressed as

$$\begin{aligned} \operatorname{div} \mathbf{A} = \frac{\partial A_{ij}}{\partial x_j}\mathbf{e}_i = &\left[\frac{\partial A_{rr}}{\partial r} + \frac{1}{r}\frac{\partial A_{\theta r}}{\partial \theta} + \frac{\partial A_{zr}}{\partial z} + \frac{1}{r}(A_{rr} - A_{\theta\theta})\right]\mathbf{e}_r \\ &+ \left[\frac{\partial A_{r\theta}}{\partial r} + \frac{1}{r}\frac{A_{\theta\theta}}{\partial \theta} + \frac{\partial A_{z\theta}}{\partial z} + \frac{1}{r}(A_{r\theta} + A_{\theta r})\right]\mathbf{e}_\theta + \left[\frac{\partial A_{rz}}{\partial r} + \frac{1}{r}\frac{\partial A_{\theta z}}{\partial \theta} + \frac{\partial A_{zz}}{\partial z} + \frac{A_{rz}}{r}\right]\mathbf{e}_z, \end{aligned} \tag{11.0.7}$$

which will be crucial in defining the stress equilibrium equations. For the curl of a vector, we have

$$\operatorname{curl} \zeta = \left(\frac{1}{r}\frac{\partial \zeta_z}{\partial \theta} - \frac{\partial \zeta_\theta}{\partial z}\right)\mathbf{e}_r + \left(\frac{\partial \zeta_r}{\partial z} - \frac{\partial \zeta_z}{\partial r}\right)\mathbf{e}_\theta + \frac{1}{r}\left(\frac{\partial (r\zeta_\theta)}{\partial r} - \frac{\partial \zeta_r}{\partial \theta}\right)\mathbf{e}_z. \tag{11.0.8}$$

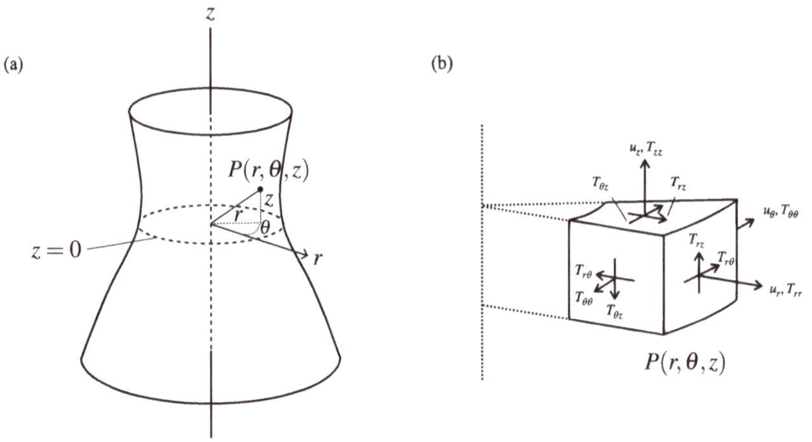

Figure 11.1: Displacements, stresses, and strains in depicted in cylindrical coordinates.

Lastly, the Laplacian of a scalar function expressed in cylindrical coordinates is

$$\nabla^2 \xi = \Delta \xi = \frac{\partial^2 \xi}{\partial x_i^2} = \frac{1}{r}\frac{\partial}{\partial r}\left(r\frac{\partial \hat{\xi}}{\partial r}\right) + \frac{1}{r^2}\frac{\partial^2 \hat{\xi}}{\partial \theta^2} + \frac{\partial^2 \hat{\xi}}{\partial z^2}. \tag{11.0.9}$$

Now that we have the necessary machinery in place, the equilibrium and strain-displacement equations can be presented. Consider an arbitrary body of revolution, such as that shown in Fig. 11.1. The body of interest need not have axially symmetric cross section, but should have a natural axis of revolution for cylindrical coordinates to be advantageous. In the following discussion, we will not assume the cross section is axisymmetric for completeness; however, if one has axisymmetry, $\partial/\partial\theta = 0$. In Fig. 11.1, the displacements, stresses, and strains are shown in cylindrical coordinates. For such a problem, the displacement, **u**, is often denoted by

$$u_1 = u_r, \quad u_2 = u_\theta, \quad \text{and} \quad u_3 = u_z, \tag{11.0.10}$$

where u_r, u_θ, and u_z are the displacements in the radial, tangential, and axial directions, respectively. The stress tensor, **T**, has the following matrix representation:

$$(\mathbf{T}) = \begin{bmatrix} T_{rr} & T_{r\theta} & T_{rz} \\ T_{\theta r} & T_{\theta\theta} & T_{\theta z} \\ T_{zr} & T_{z\theta} & T_{zz} \end{bmatrix}, \tag{11.0.11}$$

and the linearized strain tensor, ε, has the following associated matrix:

$$(\varepsilon) = \begin{bmatrix} \varepsilon_{rr} & \varepsilon_{r\theta} & \varepsilon_{rz} \\ \varepsilon_{\theta r} & \varepsilon_{\theta\theta} & \varepsilon_{\theta z} \\ \varepsilon_{zr} & \varepsilon_{z\theta} & \varepsilon_{zz} \end{bmatrix}. \tag{11.0.12}$$

The displacements, stresses, and strains are shown acting on an infinitesimal element at point $P(r,\theta,z)$ in Fig. 11.1b. Here, it is assumed that that θ is defined positive in the counterclockwise direction when viewing the body from above.

Recall that the general form of the equilibrium equation is

$$\operatorname{div}\mathbf{T} + \rho\mathbf{b} = \mathbf{0}. \tag{11.0.13}$$

Using the definition of the divergence of a tensor in cylindrical coordinates given in Eq. (11.0.7) and the fact that the stress tensor is symmetric, the *equilibrium equations* are

$$\frac{\partial T_{rr}}{\partial r} + \frac{1}{r}\frac{\partial T_{\theta r}}{\partial \theta} + \frac{T_{zr}}{\partial z} + \frac{1}{r}(T_{rr} - T_{\theta\theta}) + b_r = 0 \tag{11.0.14a}$$

$$\frac{\partial T_{r\theta}}{\partial r} + \frac{1}{r}\frac{T_{\theta\theta}}{\partial \theta} + \frac{\partial T_{z\theta}}{\partial z} + \frac{2T_{r\theta}}{r} + b_\theta = 0 \tag{11.0.14b}$$

$$\frac{\partial T_{rz}}{\partial r} + \frac{1}{r}\frac{\partial T_{\theta z}}{\partial \theta} + \frac{\partial T_{zz}}{\partial z} + \frac{T_{rz}}{r} + b_z = 0. \tag{11.0.14c}$$

Now we seek to define the strain-displacement relations in cylindrical coordinates. Recall that the general form of the linearized strain tensor is

$$\varepsilon = \frac{1}{2}(\nabla\mathbf{u} + (\nabla\mathbf{u})^{\mathrm{T}}). \tag{11.0.15}$$

Using the definition of a vector gradient in cylindrical coordinates (Eq. (11.0.5)), the strain tensor components (in terms of the displacement, excluding symmetric components) become

$$\varepsilon_{rr} = \frac{\partial u_r}{\partial r} \qquad \varepsilon_{\theta\theta} = \frac{u_r}{r} + \frac{1}{r}\frac{\partial u_\theta}{\partial \theta} \qquad \varepsilon_{zz} = \frac{\partial u_z}{\partial z} \tag{11.0.16a}$$

$$\varepsilon_{r\theta} = \frac{1}{2}\left(\frac{1}{r}\frac{\partial u_r}{\partial \theta} + \frac{\partial u_\theta}{\partial r} - \frac{u_\theta}{r}\right) \quad \varepsilon_{rz} = \frac{1}{2}\left(\frac{\partial u_r}{\partial z} + \frac{\partial u_z}{\partial r}\right) \quad \varepsilon_{z\theta} = \frac{1}{2}\left(\frac{\partial u_\theta}{\partial z} + \frac{1}{r}\frac{\partial u_z}{\partial \theta}\right). \tag{11.0.16b}$$

We recall that the strain compatibility conditions are defined as

$$\boldsymbol{\Omega} = \operatorname{curl}(\operatorname{curl}\varepsilon) = \mathbf{0}. \tag{11.0.17}$$

The constitutive relation in cylindrical polar coordinates take the form:

$$T_{rr} = \lambda(\varepsilon_{rr} + \varepsilon_{\theta\theta} + \varepsilon_{zz}) + 2\mu\varepsilon_{rr}, \tag{11.0.18a}$$

$$T_{\theta\theta} = \lambda(\varepsilon_{rr} + \varepsilon_{\theta\theta} + \varepsilon_{zz}) + 2\mu\varepsilon_{\theta\theta}, \tag{11.0.18b}$$

$$T_{zz} = \lambda(\varepsilon_{rr} + \varepsilon_{\theta\theta} + \varepsilon_{zz}) + 2\mu\varepsilon_{zz}, \tag{11.0.18c}$$

$$T_{r\theta} = T_{\theta r} = 2\mu\varepsilon_{r\theta}, \tag{11.0.18d}$$

$$T_{rz} = T_{zr} = 2\mu\varepsilon_{rz}, \tag{11.0.18e}$$

$$T_{\theta z} = T_{z\theta} = 2\mu\varepsilon_{\theta z}. \tag{11.0.18f}$$

11.1 Equilibrium equations in cylindrical polar coordinates

Consider an infinitesimal element of material such as that shown in Fig. 11.2. We will assume that the element has length dr in the radial direction, $rd\theta$ in the tangential direction, and dz in the axial direction. On writing the equilibrium equation in the r, θ and z direction, we obtain

$$\begin{aligned}\sum F_r \equiv 0 \implies & \left(T_{rr} + \frac{\partial T_{rr}}{\partial r} dr\right)(r + dr)d\theta dz - T_{rr} r d\theta dz - \sin\frac{d\theta}{2}\left(T_{\theta\theta} + \frac{\partial T_{\theta\theta}}{\partial \theta} d\theta\right)drdz \\ & - \sin\frac{d\theta}{2} T_{\theta\theta} drdz + \cos\frac{d\theta}{2}\left(T_{r\theta} + \frac{\partial T_{r\theta}}{\partial \theta} d\theta\right)drdz - \cos\frac{d\theta}{2} T_{r\theta} drdz \\ & + \left(T_{rz} + \frac{\partial T_{rz}}{\partial z} dz\right)\left(r + \frac{dr}{2}\right)drd\theta - T_{rz}\left(r + \frac{dr}{2}\right)drd\theta b_r = 0. \end{aligned} \tag{11.1.1}$$

For an infinitesimally small element, a small angle approximation can be made such that $\sin\theta \approx \theta$ and $\cos\theta \approx 1$. Substituting these approximations into Eq. (11.1.1), dividing the entire equation by the volume of the element, $rdrd\theta dz$, and letting $dr, d\theta, dz \longrightarrow 0$ yields

$$\frac{\partial T_{rr}}{\partial r} + \frac{1}{r}\frac{\partial T_{\theta r}}{\partial \theta} + \frac{T_{zr}}{\partial z} + \frac{1}{r}(T_{rr} - T_{\theta\theta}) + b_r = 0. \tag{11.1.2}$$

Repeating this process for the tangential and axial directions yields

$$\frac{\partial T_{r\theta}}{\partial r} + \frac{1}{r}\frac{\partial T_{\theta\theta}}{\partial \theta} + \frac{\partial T_{z\theta}}{\partial z} + \frac{2T_{r\theta}}{r} + b_\theta = 0 \tag{11.1.3}$$

$$\frac{\partial T_{rz}}{\partial r} + \frac{1}{r}\frac{\partial T_{\theta z}}{\partial \theta} + \frac{\partial T_{zz}}{\partial z} + \frac{T_{rz}}{r} + b_z = 0, \tag{11.1.4}$$

(a) (b)

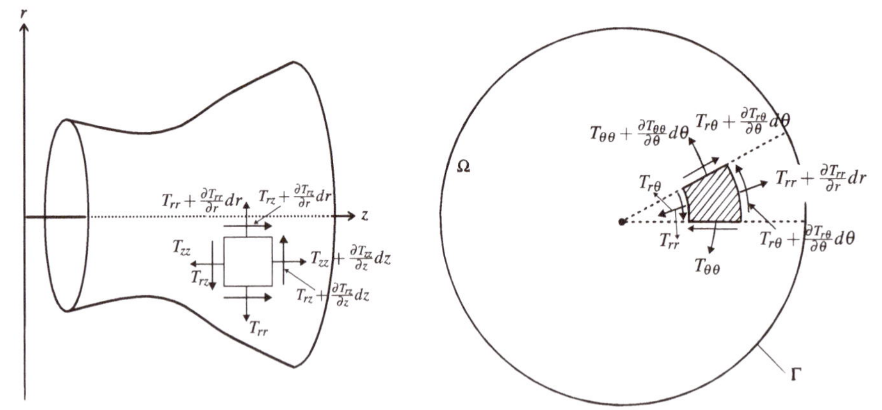

Figure 11.2: Side view and top view of an infinitesimal element of material.

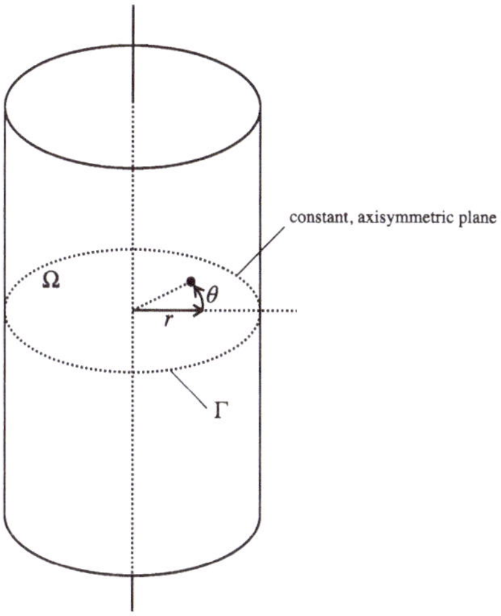

Figure 11.3: Schematic representation of a problem applicable to polar coordinates: example of a right circular cylinder with constant cross section.

respectively. Note that Eq. (11.1.2)-(11.1.4) are identical to Eq. (11.0.14). Next, it will be shown how certain assumptions can lead to a two-dimensional problem in polar coordinates (see Figure 11.3).

The equilibrium equations of Eq. (11.0.14) in polar coordinates become

$$\frac{\partial T_{rr}}{\partial r} + \frac{1}{r}\frac{\partial T_{\theta r}}{\partial \theta} + \frac{1}{r}(T_{rr} - T_{\theta\theta}) + b_r = 0, \tag{11.1.5a}$$

$$\frac{\partial T_{r\theta}}{\partial r} + \frac{1}{r}\frac{\partial T_{\theta\theta}}{\partial \theta} + \frac{2T_{r\theta}}{r} + b_\theta = 0. \tag{11.1.5b}$$

Similarly, if we omit the terms that depend on z in the strain-displacement relations, we see that

$$\varepsilon_{rr} = \frac{\partial u_r}{\partial r}, \tag{11.1.6a}$$

$$\varepsilon_{\theta\theta} = \frac{u_r}{r} + \frac{1}{r}\frac{\partial u_\theta}{\partial \theta}, \tag{11.1.6b}$$

$$\varepsilon_{r\theta} = \frac{1}{2}\left(\frac{1}{r}\frac{\partial u_r}{\partial \theta} + \frac{\partial u_\theta}{\partial r} - \frac{u_\theta}{r}\right). \tag{11.1.6c}$$

Finally, the constitutive relations reduce to

$$T_{rr} = \lambda(\varepsilon_{rr} + \varepsilon_{\theta\theta}) + 2\mu\varepsilon_{rr}, \tag{11.1.7a}$$

$$T_{\theta\theta} = \lambda(\varepsilon_{rr} + \varepsilon_{\theta\theta}) + 2\mu\varepsilon_{\theta\theta}, \tag{11.1.7b}$$

$$T_{r\theta} = T_{\theta r} = 2\mu\varepsilon_{r\theta}. \tag{11.1.7c}$$

Under these two-dimensional assumptions, the strain compatibility equation can be derived directly from the Cartesian expression. One can show that

$$\frac{\partial^2 \xi}{\partial x^2} = \cos^2\theta \frac{\partial^2 \hat{\xi}}{\partial r^2} + \sin^2\theta\left(\frac{1}{r}\frac{\partial \hat{\xi}}{\partial r} + \frac{1}{r^2}\frac{\partial^2 \hat{\xi}}{\partial \theta^2}\right) + \sin 2\theta\left(\frac{1}{r^2}\frac{\partial \hat{\xi}}{\partial \theta} - \frac{1}{r}\frac{\partial^2 \hat{\xi}}{\partial r \partial \theta}\right), \tag{11.1.8a}$$

$$\frac{\partial^2 \xi}{\partial y^2} = \sin^2\theta \frac{\partial^2 \hat{\xi}}{\partial r^2} + \cos^2\theta\left(\frac{1}{r}\frac{\partial \hat{\xi}}{\partial r} + \frac{1}{r^2}\frac{\partial^2 \hat{\xi}}{\partial \theta^2}\right) - \sin 2\theta\left(\frac{1}{r^2}\frac{\partial \hat{\xi}}{\partial \theta} - \frac{1}{r}\frac{\partial^2 \hat{\xi}}{\partial r \partial \theta}\right), \tag{11.1.8b}$$

$$\frac{\partial^2 \xi}{\partial x \partial y} = -\sin\theta\cos\theta\left(-\frac{\partial^2 \hat{\xi}}{\partial r^2} + \frac{1}{r}\frac{\partial \hat{\xi}}{\partial r} + \frac{1}{r^2}\frac{\partial^2 \hat{\xi}}{\partial \theta^2}\right) - \cos 2\theta\left(\frac{1}{r^2}\frac{\partial \hat{\xi}}{\partial \theta} - \frac{1}{r}\frac{\partial^2 \hat{\xi}}{\partial r \partial \theta}\right). \tag{11.1.8c}$$

Recall that

$$u_x = u_r \cos\theta - u_\theta \sin\theta, \quad u_y = u_r \sin\theta + u_\theta \cos\theta. \tag{11.1.9}$$

One can use Eq. (11.0.3) and Eq. (11.1.9) (neglecting terms that depend on z) to arrive at the Cartesian strain components cast in terms of polar coordinates:

$$\varepsilon_{xx} = \cos^2\theta \frac{\partial u_r}{\partial r} + \sin^2\theta\left(\frac{1}{r}\frac{\partial u_\theta}{\partial \theta} + \frac{u_r}{\partial r}\right) - \frac{1}{2}\sin 2\theta\left(\frac{1}{r}\frac{\partial u_r}{\partial \theta} + \frac{\partial u_\theta}{\partial r} - \frac{u_\theta}{r}\right), \tag{11.1.10a}$$

$$\varepsilon_{yy} = \sin^2\theta \frac{\partial u_r}{\partial r} + \cos^2\theta\left(\frac{1}{r}\frac{\partial u_\theta}{\partial \theta} + \frac{u_r}{\partial r}\right) + \frac{1}{2}\sin 2\theta\left(\frac{1}{r}\frac{\partial u_r}{\partial \theta} + \frac{\partial u_\theta}{\partial r} - \frac{u_\theta}{r}\right), \tag{11.1.10b}$$

and

$$\begin{aligned}\varepsilon_{xy} = {} & \frac{1}{2}\cos^2\theta\left(\frac{\partial u_\theta}{\partial r} + \frac{1}{r}\frac{\partial u_r}{\partial \theta} - \frac{u_\theta}{r}\right) - \frac{1}{2}\sin^2\theta\left(\frac{\partial u_\theta}{\partial r} + \frac{1}{r}\frac{\partial u_r}{\partial \theta} - \frac{u_\theta}{r}\right) \\ & + \frac{1}{2}\sin 2\theta\left(\frac{\partial u_r}{\partial r} - \frac{u_r}{r} - \frac{1}{r}\frac{\partial u_\theta}{\partial \theta}\right).\end{aligned} \tag{11.1.11}$$

It is important to note that one can use Eqs. (11.1.10) and (11.1.11) in the strain transformation equations to arrive at the expressions of Eq. (11.1.6). One can show that (after some lengthy algebra) the two-dimensional strain compatibility equation in polar coordinates is

$$\frac{1}{r^2}\frac{\partial^2 \varepsilon_{rr}}{\partial \theta^2} + \frac{\partial^2 \varepsilon_{\theta\theta}}{\partial r^2} - \frac{2}{r}\frac{\partial^2 \varepsilon_{r\theta}}{\partial r \partial \theta} - \frac{1}{r}\frac{\partial \varepsilon_{rr}}{\partial r} + \frac{2}{r}\frac{\partial \varepsilon_{\theta\theta}}{\partial r} - \frac{2}{r^2}\frac{\partial \varepsilon_{r\theta}}{\partial \theta} = 0. \tag{11.1.12}$$

11.2 Stress function in polar coordinates

We now seek to express the stresses in polar coordinates in terms of the stress function Φ. Recall that the stresses in Cartesian coordinates are defined in terms of the stress function as

$$T_{xx} = \frac{\partial^2 \Phi}{\partial y^2}, \quad T_{yy} = \frac{\partial^2 \Phi}{\partial x^2}, \quad T_{xy} = -\frac{\partial^2 \Phi}{\partial x \partial y}. \tag{11.2.1}$$

The stresses in polar coordinates can be cast in terms of those in Cartesian coordinates via the stress transformation equations:

$$T_{rr} = T_{xx}\cos^2\theta + T_{yy}\sin^2\theta + T_{xy}\sin 2\theta \tag{11.2.2a}$$

$$T_{\theta\theta} = T_{xx}\sin^2\theta + T_{yy}\cos^2\theta - T_{xy}\sin 2\theta \tag{11.2.2b}$$

$$T_{r\theta} = \frac{1}{2}\sin 2\theta(T_{yy} - T_{xx}) + T_{xy}\cos 2\theta. \tag{11.2.2c}$$

Substituting Eq. (11.2.1) into Eq. (11.2.2), we see that

$$T_{rr} = \frac{\partial^2\Phi}{\partial y^2}\cos^2\theta + \frac{\partial^2\Phi}{\partial x^2}\sin^2\theta - \frac{\partial^2\Phi}{\partial x\partial y}\sin 2\theta \tag{11.2.3a}$$

$$T_{\theta\theta} = \frac{\partial^2\Phi}{\partial y^2}\sin^2\theta + \frac{\partial^2\Phi}{\partial x^2}\cos^2\theta + \frac{\partial^2\Phi}{\partial x\partial y}\sin 2\theta \tag{11.2.3b}$$

$$T_{r\theta} = \frac{1}{2}\sin 2\theta\left(\frac{\partial^2\Phi}{\partial x^2} - \frac{\partial^2\Phi}{\partial y^2}\right) - \frac{\partial^2\Phi}{\partial x\partial y}\cos 2\theta. \tag{11.2.3c}$$

Here, it is evident that using Eq. (11.1.8) to compute the second derivatives in Eq. (11.2.3) leads to

$$T_{rr} = \frac{1}{r}\frac{\partial\Phi}{\partial r} + \frac{1}{r^2}\frac{\partial^2\Phi}{\partial\theta^2}, \quad T_{\theta\theta} = \frac{\partial^2\Phi}{\partial r^2}, \quad T_{r\theta} = \frac{1}{r^2}\frac{\partial\Phi}{\partial\theta} - \frac{1}{r}\frac{\partial^2\Phi}{\partial r\partial\theta}. \tag{11.2.4}$$

It can further be shown that the expressions of Eq. (11.2.4) satisfy the equilibrium equations (Eq. (11.1.5)). Thus, the governing equation for the two-dimensional problem in polar coordinates is

$$\nabla^4\Phi = \nabla^2(\nabla^2)\Phi = 0, \tag{11.2.5}$$

where

$$\nabla^2 = \Delta = \frac{1}{r}\frac{\partial}{\partial r}\left(r\frac{\partial}{\partial r}\right) + \frac{1}{r^2}\frac{\partial^2}{\partial\theta^2}. \tag{11.2.6}$$

11.3 Torsion of circular shafts of variable diameter

Previously, we have only considered torsional members with constant cross sections. Here, we introduce the problem of a circular shaft with a variable diameter, such as that shown in Fig. 11.4. By considering only a circular-shaped cross section, the problem becomes axisymmetric. For this problem, we will use cylindrical coordinates, introduced in Section 11 to further simplify the analysis. We will again use the semi-inverse method

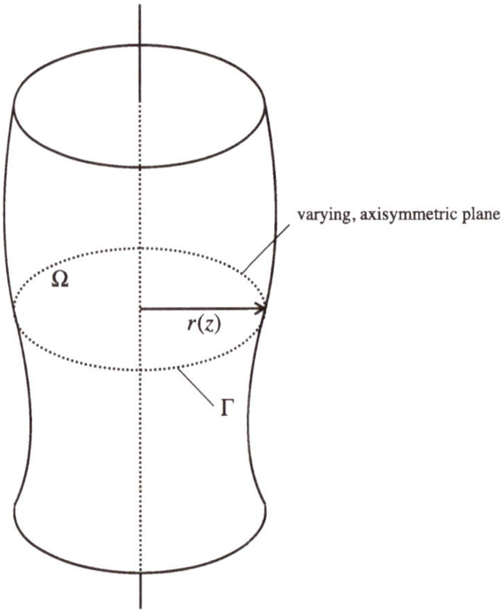

Figure 11.4: Circular cylinder with arbitrarily varying radius.

and begin by establishing the governing differential equation in cylindrical coordinates. In cylindrical coordinates, the strain-displacement relationships are

$$\varepsilon_{rr}=\frac{\partial u_r}{\partial r},\qquad \varepsilon_{\theta\theta}=\frac{u_r}{r}+\frac{1}{r}\frac{\partial u_\theta}{\partial\theta},\qquad \varepsilon_{zz}=\frac{\partial u_z}{\partial z},\tag{11.3.1}$$

$$\varepsilon_{r\theta}=\frac{1}{2}\left(\frac{1}{r}\frac{\partial u_r}{\partial\theta}+\frac{\partial u_\theta}{\partial r}-\frac{u_\theta}{r}\right),\quad \varepsilon_{rz}=\frac{1}{2}\left(\frac{\partial u_r}{\partial z}+\frac{\partial u_z}{\partial r}\right),\quad \varepsilon_{z\theta}=\frac{1}{2}\left(\frac{\partial u_\theta}{\partial z}+\frac{1}{r}\frac{\partial u_z}{\partial\theta}\right).\tag{11.3.2}$$

Under the assumptions of an axisymmetric, circular cross section, $u_r = u_z = 0$ and $\partial u_\theta/\partial\theta = 0$. Thus, the above relations become

$$\varepsilon_{rr}=\varepsilon_{zz}=\varepsilon_{\theta\theta}=\varepsilon_{rz}=0\tag{11.3.3}$$

$$\varepsilon_{r\theta}=\frac{1}{2}\left(\frac{\partial u_\theta}{\partial r}-\frac{u_\theta}{r}\right),\quad \varepsilon_{\theta z}=\frac{1}{2}\frac{\partial u_\theta}{\partial z}.\tag{11.3.4}$$

From these strain-displacement relations, the stress components can be expressed as

$$T_{rr}=T_{zz}=T_{\theta\theta}=T_{rz}=0,\quad T_{r\theta}=\mu\left(\frac{\partial u_\theta}{\partial r}-\frac{u_\theta}{r}\right),\quad T_{\theta z}=\mu\frac{\partial u_\theta}{\partial z}.\tag{11.3.5}$$

The equilibrium equations can be expressed in cylindrical coordinates as

$$\frac{\partial T_{rr}}{\partial r}+\frac{1}{r}\frac{\partial T_{r\theta}}{\partial\theta}+\frac{\partial T_{rz}}{\partial z}+\frac{T_{rr}-T_{\theta\theta}}{r}=0\tag{11.3.6}$$

$$\frac{\partial T_{rz}}{\partial r}+\frac{1}{r}\frac{\partial T_{\theta z}}{\partial\theta}+\frac{\partial T_{zz}}{\partial z}+\frac{T_{rz}}{r}=0\tag{11.3.7}$$

$$\frac{\partial T_{r\theta}}{\partial r} + \frac{1}{r}\frac{\partial T_{\theta\theta}}{\partial \theta} + \frac{\partial T_{\theta z}}{\partial z} + \frac{2T_{r\theta}}{r} = 0 \tag{11.3.8}$$

where, under the assumptions made here, the first two equations are automatically satisfied, and the only remaining equation is

$$\frac{\partial T_{r\theta}}{\partial r} + \frac{\partial T_{\theta z}}{\partial z} + \frac{2T_{r\theta}}{r} = 0. \tag{11.3.9}$$

The above equation can also be written in the form

$$\frac{\partial}{\partial r}(r^2 T_{r\theta}) + \frac{\partial}{\partial z}(r^2 T_{\theta z}) = 0, \tag{11.3.10}$$

where a stress function depending on r and z can be defined such that

$$\frac{\partial \Phi}{\partial z} = -\mu r^3 \frac{\partial}{\partial r}\left(\frac{u_\theta}{r}\right) = -r^2 T_{r\theta}, \tag{11.3.11}$$

$$\frac{\partial \Phi}{\partial r} = \mu r^3 \frac{\partial}{\partial z}\left(\frac{u_\theta}{r}\right) = r^2 T_{\theta z}. \tag{11.3.12}$$

It immediately follows from the above equations that

$$\frac{\partial}{\partial r}\left(\frac{1}{r^3}\frac{\partial \Phi}{\partial r}\right) + \frac{\partial}{\partial z}\left(\frac{1}{r^3}\frac{\partial \Phi}{\partial z}\right) = 0. \tag{11.3.13}$$

The above differential equation can also be expressed as

$$\frac{\partial^2 \Phi}{\partial r^2} - \frac{3}{r}\frac{\partial \Phi}{\partial r} + \frac{\partial^2 \Phi}{\partial z^2} = 0, \tag{11.3.14}$$

which is the governing differential equation for the problem of a shaft with varying cross section. As with the case for a constant circular shaft, the lateral boundary of the cylinder are traction free. Mathematically, this is represented as

$$T_{r\theta} n_r + T_{\theta z} n_z = T_{r\theta}\frac{dz}{ds} - T_{\theta z}\frac{dr}{ds} = 0, \tag{11.3.15}$$

where, as before, ds corresponds to an infinitesimal length along the boundary. Replacing the shear stresses in the above equation with their stress function representations, we see that

$$\frac{\partial \Phi}{\partial z}\frac{dz}{ds} + \frac{\partial \Phi}{\partial r}\frac{dr}{ds} = 0, \tag{11.3.16}$$

which implies

$$\frac{d\Phi}{ds} = 0. \tag{11.3.17}$$

Thus, Φ is constant along the boundary of a given cross section of the shaft. For a given torsional member with circular-shaped cross-section but of varying radius, one must solve Eq. (11.3.14) subject to the boundary condition in Eq. (11.3.17). The resultant torque, M, for a cross section is determined by calculating the moment generated by $T_{\theta z}$. That is,

$$M(z) = \int_{\Omega} r^2 T_{\theta z}\, dA = 2\pi \int_0^{R(z)} r^2 T_{\theta z}\, dr = 2\pi\mu\Phi|_0^{R(z)} \tag{11.3.18}$$

where $R(z)$ denotes the radius of a cross section at a specified z location. For more discussion concerning the torsion of cylinders of non-constant cross section, see Timoshenko and Goodier [4].

Example 11.3.1 (from Timoshenko and Goodier [4]). Consider the problem of a conical shaft (frustum of a cone) with a circular cross-section subjected to end torques. For convenience, cylindrical coordinates will be employed. In doing so, the governing equation is given by Eq. (11.3.14)

$$\frac{\partial^2\Phi}{\partial r^2} - \frac{3}{r}\frac{\partial\Phi}{\partial r} + \frac{\partial^2\Phi}{\partial z^2} = 0, \tag{11.3.19}$$

with the boundary condition

$$\frac{d\Phi}{ds} = 0 \quad \text{on } \Gamma. \tag{11.3.20}$$

A cone such as that shown in Fig. 11.5 is characterized by its constant cone angle, β, which can be expressed for this problem as

$$\cos\beta = \frac{z}{\sqrt{R^2 + z^2}}, \tag{11.3.21}$$

where $R = R(z)$ is the radius of the cone at a given cross section and z is the axial coordinate. In knowing that $\cos\beta$ is always constant, one also knows that the function of $\cos\beta$ will also be constant. Thus, the boundary condition is automatically satisfied in using this function. However, to also satisfy the governing equation, we find that

$$\Phi = K\left[\frac{z}{\sqrt{r^2 + z^2}} - \frac{1}{3}\frac{z^3}{(r^2 + z^2)^{3/2}}\right], \tag{11.3.22}$$

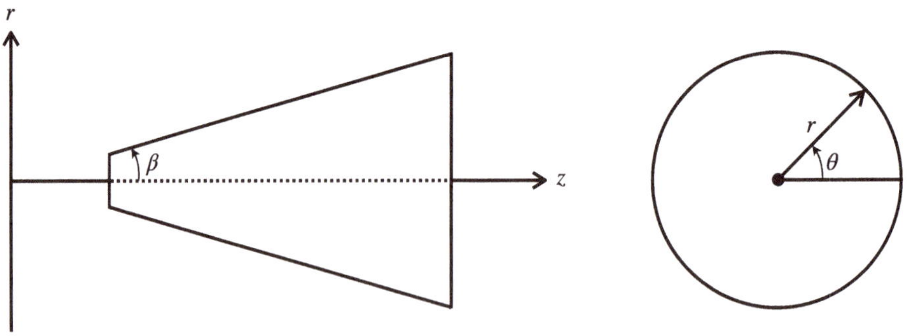

Figure 11.5: Schematic of a conical shaft.

where K is a constant which needs to be determined. Recall that the magnitude of the torque that induces $T_{\theta z}$ can be expressed (in cylindrical coordinates) as

$$M = \int_{\Omega} T_{\theta z} r^2 \, dA = 2\pi\mu \int_{0}^{R(z)} \frac{\partial \Phi}{\partial r} \, dr = 2\pi\mu \Phi \Big|_0^{R(z)}. \tag{11.3.23}$$

Noting that $\Phi(R(z), z) = K(\cos\beta - \frac{1}{3}\cos^3\beta)$ and $\Phi(0, z) = -\frac{2}{3}K$, one sees that

$$K = -\frac{M}{2\pi\mu(\frac{2}{3} - \cos\beta + \frac{1}{3}\cos^3\beta)}. \tag{11.3.24}$$

These results (through the appropriate differentiation of Eq. (11.3.22)) lead to shear stress components

$$T_{r\theta} = \frac{\mu}{r^2}\frac{\partial \Phi}{\partial z} = -\frac{K\mu r^2}{(r^2 + z^2)^{5/2}}, \tag{11.3.25a}$$

$$T_{\theta z} = -\frac{\mu}{r^2}\frac{\partial \Phi}{\partial r} = -\frac{K\mu r z}{(r^2 + z^2)^{5/2}}, \tag{11.3.25b}$$

where K is given by Eq. (11.3.24). Finally, the tangential displacement can be found through

$$u_\theta = \frac{1}{\mu}\int T_{\theta z} \, dz = -Kr \int \frac{z}{(r^2 + z^2)^{5/2}} \, dz \tag{11.3.26a}$$

$$= \frac{Kr}{3(r^2 + z^2)^{3/2}} + f(r), \tag{11.3.26b}$$

where $f(r)$ is an arbitrary function of r resulting from the integration. Using the strain-displacement relationships and the compatibility condition, it can be shown that $f(r) = \omega r$ such that

$$u_\theta = -\frac{Kr}{3(r^2 + z^2)^{3/2}} + \omega r, \tag{11.3.27}$$

where ωr defines the rigid body rotation about the z-axis and ω can be determined from the shaft rotation along z. Figure 11.6 shows the maximum stress as a function of z in comparison with the solution predicted from elementary mechanics of materials.

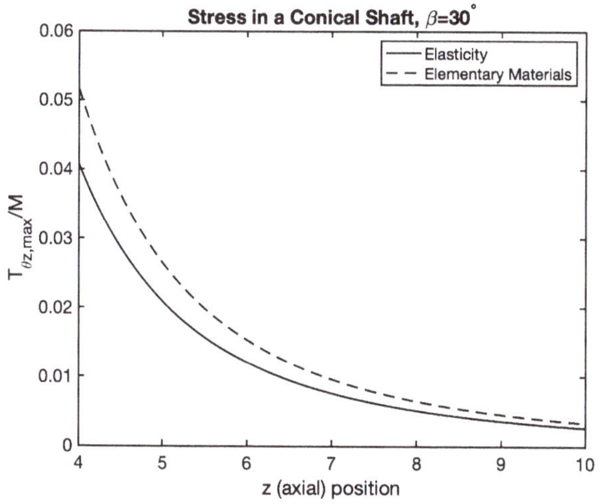

Figure 11.6: Maximum shear stress, $T_{\theta z}$, as a function of z for cone angle $\beta = 30°$.

References

[1] I. S. Sokolnikoff. Mathematical Theory of Elasticity. McGraw Hill, New York, 2nd edition, 1956.
[2] L. E. Malvern. Introduction to the Mechanics of a Continuous Medium, 1967.
[3] W. Jaunzemis. Continuum Mechanics. Macmillan, 1967.
[4] S. Timoshenko and J. N. Goodier. Theory of Elasticity. McGraw Hill, New York, 3rd edition.

12 Stress concentration

Holes and inclusions cause the stresses where they are located and in their neighborhood, to be more than what it would be at the same location, in their absence. The stress concentration factor is the ratio of the stress when a stress concentrator like a hole or inclusion is present and the stress in the absence of the stress concentrator. In this chapter, we analyze problems that elucidate this issue.

12.1 Circular hole in an infinite plate-uniaxial loading

We first observe that since all equations of linearized elasticity are linear, superposition can be applied. The method of solution is as follows:

(i) Solve the problem without the cavity: S_1.
(ii) Determine the traction that S_1 produces on the actual cavity.
(iii) Consider the original problem with no external loading but with the negative of the traction acting on the cavity.
(iv) Solve the problem: S_2.

The problem is illustrated in Fig. 12.1.

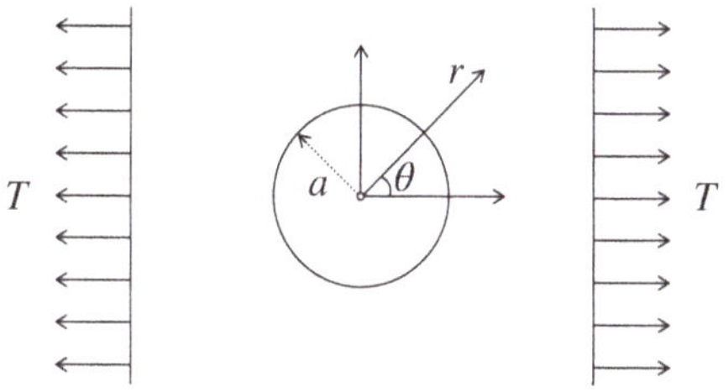

Figure 12.1: Infinite plate with a circular hole. Stress applied to lateral edges.

The solution to the problem is of the form

$$S = S_1 + S_2. \tag{12.1.1}$$

For the first step, we find S_1. The state of stress for this step corresponding the applied stress state is

$$T_{xx} = T, \quad T_{xy} = 0, \quad T_{yy} = 0. \tag{12.1.2}$$

Using polar coordinates yields

$$T_{rr} = T\cos^2\theta = \frac{T}{2}(1 + \cos 2\theta) \tag{12.1.3}$$

https://doi.org/10.1515/9783110789515-012

$$T_{\theta\theta} = T\sin^2\theta = \frac{T}{2}(1 - \cos 2\theta) \tag{12.1.4}$$

$$T_{r\theta} = -T\sin\theta\cos\theta = -\frac{T}{2}\sin 2\theta. \tag{12.1.5}$$

Evaluating the above at $r = a$ gives

$$T_{rr}^{(I)} = \frac{T}{2}(1 + \cos 2\theta) \tag{12.1.6}$$

$$T_{\theta\theta}^{(I)} = \frac{T}{2}(1 - \cos 2\theta) \tag{12.1.7}$$

$$T_{r\theta}^{(I)} = -\frac{T}{2}\sin 2\theta. \tag{12.1.8}$$

But on the cavity, for determining S_2, we have

$$T_{rr}^{(II)} = -\frac{T}{2}(1 + \cos 2\theta) \tag{12.1.9}$$

$$T_{\theta\theta}^{(II)} = -\frac{T}{2}(1 - \cos 2\theta) \tag{12.1.10}$$

$$T_{r\theta}^{(II)} = \frac{T}{2}\sin 2\theta. \tag{12.1.11}$$

We now proceed to find the solution S_2. Recall the Airy's stress function, ϕ. The stresses in polar coordinates are given in terms of the Airy stress function in the following manner:

$$T_{rr} = \frac{1}{r}\frac{\partial\phi}{\partial r} + \frac{1}{r^2}\frac{\partial^2\phi}{\partial\theta^2} \tag{12.1.12}$$

$$T_{\theta\theta} = \frac{\partial^2\phi}{\partial r^2} \tag{12.1.13}$$

$$T_{r\theta} = -\frac{1}{r}\left[\frac{\partial^2\phi}{\partial r\partial\theta} - \frac{1}{r}\frac{\partial\phi}{\partial\theta}\right]. \tag{12.1.14}$$

We want to use the Airy's stress function formulation to solve the problem. Recall

$$\nabla^4\phi = 0. \tag{12.1.15}$$

We will use the fact that a biharmonic function can be expressed in terms of three harmonic function as, $\phi = x\psi_1 + y\psi_2 + \psi_3$ where

$$\nabla^2\psi_{1,2,3} = 0. \tag{12.1.16}$$

Thus,

$$\phi = r\cos\theta\psi_1 + r\sin\theta\psi_2 + \psi_3. \tag{12.1.17}$$

Based on the form of the boundary condition, we seek ϕ of the form

$$\phi = F_2(r)\cos 2\theta + F_1(r). \tag{12.1.18}$$

Our aim is now to determine ψ_1, ψ_2, ψ_3. The general solution of the Laplace equation in polar coordinates is

$$\psi(r,\theta) = A_0\theta + B_0 + C_0 \ln r + D_0\theta \ln r + \cdots$$
$$\sum_{n=1}^{\infty}(A_n r^n + B_n r^{-n})(C_n \cos n\theta + D_n \sin n\theta). \quad (12.1.19)$$

Note, ϕ should have period 2π. This implies that

$$A_0 = 0, \quad D_0 = 0, \quad (12.1.20)$$

which applies for solutions corresponding to $\psi_i, i = 1, 2, 3$. In determining the solutions for $\psi_i, i = 1, 2, 3$ we appeal to the following trigonometric identities

$$\cos\theta \cos n\theta = \frac{1}{2}[\cos(n+1)\theta + \cos(n-1)\theta], \quad (12.1.21)$$
$$\sin\theta \sin n\theta = \frac{1}{2}[\cos(n-1)\theta - \cos(n+1)\theta], \quad (12.1.22)$$
$$\sin\theta \cos n\theta = \frac{1}{2}[\sin(n+1)\theta - \sin(n-1)\theta], \quad (12.1.23)$$
$$\cos\theta \sin n\theta = \frac{1}{2}[\sin(n+1)\theta + \sin(n-1)\theta]. \quad (12.1.24)$$

Determination of ψ_1

For ψ_1, we see that in order to meet (12.1.19) and for the stress to vanish at infinity

$$B_0 = 0, \quad C_0 = 0, \quad D_n = 0, \quad \text{for all } n \quad (12.1.25)$$
$$C_n = 0 \quad \text{for } n = 2, 3, \ldots, \quad (12.1.26)$$
$$B_n = 0 \quad \text{for } n = 2, 3, \ldots \quad (12.1.27)$$

We will also note that

$$A_n = 0, \quad \text{for all n} \quad (12.1.28)$$

since the stresses have to be zero at ∞. Thus,

$$\psi_1 = \frac{B_1 C_1}{r}\cos\theta = \frac{C_1'}{r}\cos\theta. \quad (12.1.29)$$

In the above equation prime does not mean a derivative, however when discussing the stress concentration due to a rigid inclusion, we do use prime to denote derivative. Whether prime, does or does not denote derivative ought to be clear from the context.

Determination of ψ_2

For ψ_2, we see that for us to meet (12.1.19) and the stress to vanish at infinity

$$B_0 = C_0 = 0, \quad C_n = 0, \quad \text{for all } n \quad (12.1.30)$$

$$D_n = 0 \quad \text{for } n = 2, 3, \ldots, \tag{12.1.31}$$

$$A_n = 0 \quad \text{for } n = 1, 2, \ldots \tag{12.1.32}$$

Thus,

$$\psi_2 = \frac{B_1 D_1}{r} \sin\theta = \frac{D_1'}{r} \sin\theta. \tag{12.1.33}$$

Determination of ψ_3

For ψ_3, we see that using the same approach as before, we need

$$D_n = 0 \quad \text{for all } n, \quad C_1 = 0 \tag{12.1.34}$$

$$C_n = 0 \quad \text{for } n = 3, 4, \ldots, \tag{12.1.35}$$

$$A_n = 0 \; \forall n \tag{12.1.36}$$

Thus,

$$\psi_3 = B_0 + C_0 \ln r + \frac{B_2 C_2}{r^2} \cos 2\theta \tag{12.1.37}$$

$$= B_0 + C_0 \ln r + \frac{B_2'}{r^2} \cos 2\theta. \tag{12.1.38}$$

Thus, in virtue of Eq. (12.1.17) and Eq. (12.1.29)-(12.1.38),

$$\phi(r,\theta) = C_1' \cos^2\theta + D_1' \sin^2\theta + B_0 + C_0 \ln r + \frac{B_2'}{r^2} \cos 2\theta \tag{12.1.39}$$

$$= \tilde{a} + b \ln r + c \cos 2\theta + \frac{d}{r^2} \cos 2\theta. \tag{12.1.40}$$

This expression for ϕ provides stresses which vanish at ∞ and satisfies the biharmonic equation. This is the solution that has to be used to determine S_2. Now, by Eq. (12.1.12) and Eq. (12.1.14),

$$T_{rr}^{(II)} = \frac{b}{r^2} - \left(\frac{4c}{r^2} + \frac{6d}{r^4}\right) \cos 2\theta \tag{12.1.41}$$

$$T_{r\theta}^{(II)} = -\left(\frac{2c}{r^2} + \frac{6d}{r^4}\right) \sin 2\theta. \tag{12.1.42}$$

Recall from Eq. (12.1.9)-(12.1.11) that

$$T_{rr}^{(II)}(a,\theta) = -\frac{T}{2}(1 + \cos 2\theta) \tag{12.1.43}$$

$$T_{r\theta}^{(II)}(a,\theta) = \frac{T}{2} \sin 2\theta. \tag{12.1.44}$$

Equating T_{rr} yields

$$T_{rr}^{(II)}(a,\theta) = \frac{b}{a^2} - \left(\frac{4c}{a^2} + \frac{6d}{a^4}\right)\cos 2\theta = -\frac{T}{2}(1+\cos 2\theta), \tag{12.1.45}$$

which implies that

$$\frac{b}{a^2} = -\frac{T}{2}, \quad \frac{4c}{a^2} + \frac{6d}{a^4} = \frac{T}{2}. \tag{12.1.46}$$

Next, from the $T_{r\theta}$ condition, we get

$$-\left(\frac{2c}{a^2} + \frac{6d}{a^4}\right) = \frac{T}{2}. \tag{12.1.47}$$

Hence,

$$b = -\frac{Ta^2}{2}, \quad c = \frac{Ta^2}{2}, \quad d = -\frac{Ta^4}{4}. \tag{12.1.48}$$

So, the solution S_2 for the stresses is

$$T_{rr}^{(II)} = -\frac{Ta^2}{2r^2} - \frac{Ta^2}{r^2}\left(2 - \frac{3}{2}\frac{a^2}{r^2}\right)\cos 2\theta \tag{12.1.49}$$

$$T_{r\theta}^{(II)} = -\frac{Ta^2}{r^2}\left(1 - \frac{3}{2}\frac{a^2}{r^2}\right)\sin 2\theta. \tag{12.1.50}$$

$$T_{\theta\theta}^{(II)} = \frac{Ta^2}{2r^2} - \frac{3}{2}\frac{Ta^4}{r^4}\cos 2\theta. \tag{12.1.51}$$

Finally, the complete solution, $S = S_1 + S_2$, becomes

$$T_{rr} = T_{rr}^{(I)} + T_{rr}^{(II)} = \frac{T}{2}\left(1 - \frac{a^2}{r^2}\right) + \frac{T}{2}\left(1 - \frac{4a^2}{r^2} + \frac{3a^4}{r^4}\right)\cos 2\theta \tag{12.1.52}$$

$$T_{r\theta} = T_{r\theta}^{(I)} + T_{r\theta}^{(II)} = -\frac{T}{2}\left(1 + \frac{2a^2}{r^2} - \frac{3a^4}{r^4}\right)\sin 2\theta \tag{12.1.53}$$

$$T_{\theta\theta} = T_{\theta\theta}^{(I)} + T_{\theta\theta}^{(II)} = \frac{T}{2}\left(1 + \frac{a^2}{r^2}\right) - \frac{T}{2}\left(1 + \frac{3a^4}{r^4}\right)\cos 2\theta. \tag{12.1.54}$$

At $r = a$, we have

$$T_{rr} = T_{r\theta} = 0 \tag{12.1.55}$$

$$T_{\theta\theta} = T - 2T\cos 2\theta = T(1 - 2\cos 2\theta). \tag{12.1.56}$$

At $\theta = \frac{\pi}{2}, \frac{3\pi}{2}$,

$$T_{\theta\theta} = 3T, \tag{12.1.57}$$

which is the maximum stress. Thus, the stress concentration factor for this particular problem is 3.

Also at $\theta = \frac{\pi}{2}$, $T_{\theta\theta} \longrightarrow T$ as $r \longrightarrow \infty$.

12.2 Circular hole in an infinite plate-biaxial loading

Consider the biaxial problem shown in Fig. 12.2. We use the superposition approach again. The solution due to T is known. The solution due to $\hat{T}$ is obtained by merely changing in (12.1.52)-(12.1.54) θ to $(\theta + 90)$ and thus:

$$\hat{T}_{rr} = \frac{\hat{T}}{2}\left(1 - \frac{a^2}{r^2}\right) - \frac{\hat{T}}{2}\left(1 - \frac{4a^2}{r^2} + \frac{3a^4}{r^4}\right)\cos 2\theta \tag{12.2.1}$$

$$\hat{T}_{r\theta} = \frac{\hat{T}}{2}\left(1 + \frac{2a^2}{r^2} - \frac{3a^4}{r^4}\right)\sin 2\theta \tag{12.2.2}$$

$$\hat{T}_{\theta\theta} = \frac{\hat{T}}{2}\left(1 + \frac{a^2}{r^2}\right) + \frac{\hat{T}}{2}\left(1 + \frac{3a^4}{r^4}\right)\cos 2\theta. \tag{12.2.3}$$

Add to the previous solution for T, then

$$T_{rr} = T\left(1 - \frac{a^2}{r^2}\right) \tag{12.2.4}$$

$$T_{\theta\theta} = T\left(1 + \frac{a^2}{r^2}\right) \tag{12.2.5}$$

$$T_{r\theta} = 0. \tag{12.2.6}$$

At $r = a$,

$$T_{\theta\theta}(a,\theta) = 2T. \tag{12.2.7}$$

Thus, the stress concentration factor is 2. Recall that we obtained this value as a limiting case when we discussed pressure vessels subject to uniform pressure.

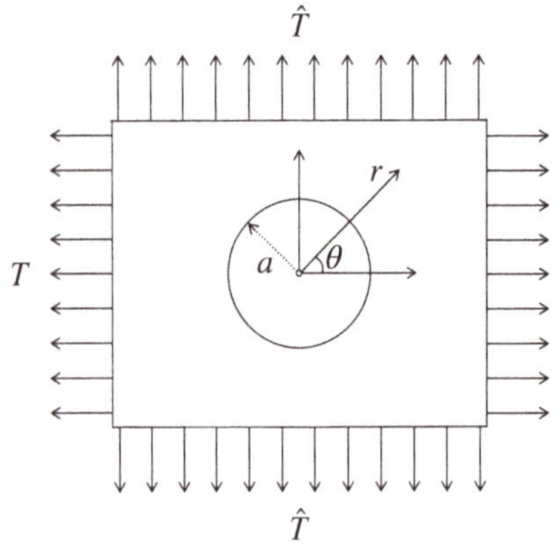

Figure 12.2: Infinite plate with hole at the center under a biaxial state of stress.

12.3 Rigid circular inclusion

Consider the problem of the infinite plate with a rigid circular inclusion as shown in Fig. 12.3. We will note the boundary condition on the interface of the inclusion: $\mathbf{u}(a,\theta) = \mathbf{0}$, (i. e., displacement should be zero if rigidly bonded). The solution to S_1 is the following:

$$T_{xx} = T, \quad T_{yy} = 0, \quad T_{xy} = 0. \tag{12.3.1}$$

Since the boundary condition at the interface between the rigid inclusion and the elastic material is a consequence of the rigid inclusion being bonded perfectly, we have to enforce that the displacement at the interface is continuous. As a rigid body cannot deform, this implies that the displacement is zero at $r = a$, for the problem as a whole. For the solution S_1, we need to find the displacement at $r = a$ by assuming that there is no inclusion. In virtue of Eq. (6.3.8),

$$\varepsilon_{xx} = \frac{T}{E}, \quad \varepsilon_{yy} = -\nu\frac{T}{E}, \quad \varepsilon_{xy} = 0. \tag{12.3.2}$$

It follows from the definition of ε that

$$u_x = \frac{Tx}{E} + f(y), \tag{12.3.3}$$

and

$$u_y = -\frac{\nu Ty}{E} + g(x), \tag{12.3.4}$$

where u_x and u_y are the displacements in the x and y directions, respectively. Hence, as consequence of Eq. (12.3.3) and Eq. (12.3.4) and the definition of ε_{xy}

$$f'(y) + g'(x) = 0 \tag{12.3.5}$$

$$f'(y) = -g'(x) = c \tag{12.3.6}$$

$$f(y) = cy + D \tag{12.3.7}$$

$$g(x) = -cx + E, \tag{12.3.8}$$

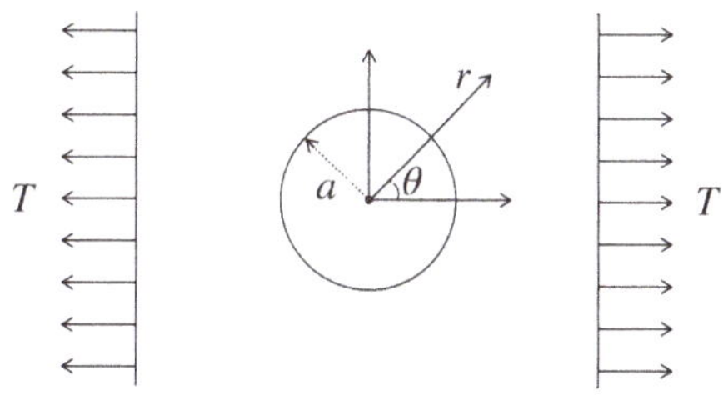

Figure 12.3: Infinite plate with rigid inclusion at the center.

where the prime denotes the differentiation with respect to the argument and the expression for $f(y)$ and $g(x)$ is just rigid body motion that is a consequence of a rotation plus a translation. However, u_x cannot depend on y or u_y on x. Thus, if rigid motion is to be ignored,

$$f(y) = g(x) = 0. \tag{12.3.9}$$

Hence,

$$u_x = \frac{\mathrm{Tr}\cos\theta}{E} \tag{12.3.10}$$

$$u_y = -\frac{\nu\,\mathrm{Tr}\sin\theta}{E}. \tag{12.3.11}$$

Recall,

$$u_r = u_x\cos\theta + u_y\sin\theta \tag{12.3.12}$$

$$u_\theta = -u_x\sin\theta + u_y\cos\theta. \tag{12.3.13}$$

Thus, the solution for the displacement associated with S_1 is

$$u_r^{(I)} = \frac{\mathrm{Tr}}{E}(\cos^2\theta - \nu\sin^2\theta) \tag{12.3.14}$$

$$u_\theta^{(I)} = -\frac{\mathrm{Tr}}{E}[(1+\nu)\sin\theta\cos\theta]. \tag{12.3.15}$$

For the field S_2, we have to specify at $r = a$,

$$u_r^{(II)}(a,\theta) = -\frac{Ta}{2E}[(1-\nu) + (1+\nu)\cos 2\theta] \tag{12.3.16}$$

$$u_\theta^{(II)}(a,\theta) = \frac{Ta}{2E}[(1+\nu)\sin 2\theta]. \tag{12.3.17}$$

We also have to satisfy $T_{rr}^{(II)}, T_{\theta\theta}^{(II)}, T_{r\theta}^{(II)} \longrightarrow 0$ as $r \longrightarrow \infty$. Following the same procedure as before, we find that the Airy's stress function is given by

$$\phi = \tilde{a} + b\ln r + c\cos 2\theta + \frac{d}{r^2}\cos 2\theta. \tag{12.3.18}$$

Thus,

$$T_{rr}^{(II)} = \frac{b}{r^2} - \left(\frac{4c}{r^2} + \frac{6d}{r^4}\right)\cos 2\theta \tag{12.3.19}$$

$$T_{\theta\theta}^{(II)} = -\frac{b}{r^2} + \frac{6d}{r^4}\cos 2\theta \tag{12.3.20}$$

$$T_{r\theta}^{(II)} = -\left(\frac{2c}{r^2} + \frac{6d}{r^4}\right)\sin 2\theta. \tag{12.3.21}$$

Next, recall

$$E\frac{\partial u_r}{\partial r} = T_{rr} - \nu T_{\theta\theta} = \frac{b}{r^2} - \left(\frac{4c}{r^2} + \frac{6d}{r^4}\right)\cos 2\theta - \nu\left(-\frac{b}{r^2} + \frac{6d}{r^4}\cos 2\theta\right). \tag{12.3.22}$$

Integrating yields

$$Eu_r = -\frac{b}{r}(1+\nu) + \frac{4c}{r}\cos 2\theta + \frac{2d}{r^3}(1+\nu)\cos 2\theta + f'(\theta). \tag{12.3.23}$$

Next,

$$E\left[\frac{\partial u_\theta}{\partial \theta} + u_r\right] = r[T_{\theta\theta} - \nu T_{rr}] \tag{12.3.24}$$

$$E\frac{\partial u_\theta}{\partial \theta} = r[T_{\theta\theta} - \nu T_{rr}] - Eu_r. \tag{12.3.25}$$

On substituting Eq. (12.3.23) into Eq. (12.3.25) and integrating we obtain

$$Eu_\theta = -\frac{2c}{r}(1-\nu)\sin 2\theta + \frac{2d}{r^3}(1+\nu)\sin 2\theta - f(\theta) + g(r). \tag{12.3.26}$$

Recall from Eq. (11.0.16a) that

$$\frac{ET_{r\theta}}{\mu} = E\left[\frac{1}{r}\frac{\partial u_r}{\partial \theta} + \frac{\partial u_\theta}{\partial r} - \frac{u_\theta}{r}\right] \tag{12.3.27}$$

$$= -2(1+\nu)\left(\frac{2c}{r^2} + \frac{6d}{r^4}\right)\sin 2\theta. \tag{12.3.28}$$

Thus, it follows from Eq. (12.3.25)-(12.3.28) that

$$f''(\theta) + f(\theta) + rg'(r) - g(r) = 0. \tag{12.3.29}$$

It immediately follows that

$$f''(\theta) + f(\theta) = -[rg'(r) - g(r)] = D = \text{const.} \tag{12.3.30}$$

Hence,

$$f(\theta) = D + a_1\cos\theta + a_2\sin\theta \tag{12.3.31}$$

$$g(r) = D + a_3 r. \tag{12.3.32}$$

However, $u_r^{(II)}, u_\theta^{(II)} \longrightarrow 0$ as $r \longrightarrow \infty$. This implies that

$$a_1 = a_2 = a_3 = 0. \tag{12.3.33}$$

Therefore,

$$f = D, \quad g = D, \quad f' = 0. \tag{12.3.34}$$

Boundary conditions for $u_r^{(II)}$ imply that

$$-\frac{b}{a}(1+\nu) = -\frac{Ta}{2}(1-\nu) \tag{12.3.35}$$

$$\frac{4c}{a} + \frac{2d}{a^3}(1+\nu) = -\frac{Ta}{2}(1+\nu). \tag{12.3.36}$$

The boundary conditions for $u_\theta^{(II)}$ implies that

$$-\frac{2c}{a}(1-\nu) + \frac{2d}{a^3}(1+\nu) = \frac{Ta}{2}(1+\nu). \tag{12.3.37}$$

Solving Eq. (12.3.35), (12.3.36), and (12.3.37) for b, c, and d yields

$$b = \frac{Ta^2(1-\nu)}{2(1+\nu)}, \tag{12.3.38}$$

$$c = -\frac{Ta^2(1+\nu)}{2(3-\nu)}, \tag{12.3.39}$$

$$d = \frac{Ta^4(1+\nu)}{4(3-\nu)}. \tag{12.3.40}$$

It follows that the total solution S is given by

$$T_{rr} = T_{rr}^{(I)} + T_{rr}^{(II)} \tag{12.3.41}$$

$$= \frac{T}{2}\left\{1 + \frac{(1-\nu)}{(1+\nu)}\frac{a^2}{r^2} + \left[1 + \frac{4(1+\nu)}{(3-\nu)}\frac{a^2}{r^2} - \frac{3(1+\nu)}{(3-\nu)}\frac{a^4}{r^4}\right]\cos 2\theta\right\} \tag{12.3.42}$$

$$T_{\theta\theta} = T_{\theta\theta}^{(I)} + T_{\theta\theta}^{(II)} \tag{12.3.43}$$

$$= \frac{T}{2}\left\{1 - \frac{(1-\nu)}{(1+\nu)}\frac{a^2}{r^2} + \left[-1 + \frac{3(1+\nu)}{(3-\nu)}\frac{a^4}{r^4}\right]\cos 2\theta\right\} \tag{12.3.44}$$

$$T_{r\theta} = T_{r\theta}^{(I)} + T_{r\theta}^{(II)} \tag{12.3.45}$$

$$= \frac{T}{2}\left\{-1 + \frac{2(1+\nu)}{(3-\nu)}\frac{a^2}{r^2} - \frac{3(1+\nu)}{(3-\nu)}\frac{a^4}{r^4}\right\}\sin 2\theta. \tag{12.3.46}$$

Note that

$$T_{rr}(r,0) = \frac{T}{2}\left\{2 + \left(\frac{(1-\nu)}{(1+\nu)} + \frac{4(1+\nu)}{3(-\nu)}\right)\frac{a^2}{r^2} - \frac{3(1+\nu)}{(3-\nu)}\frac{a^4}{r^4}\right\}. \tag{12.3.47}$$

We now pose the question: where is $T_{rr}(r,0)$ a maximum?. A necessary condition for an extremum is that

$$\frac{dT_{rr}(r,0)}{dr} = -T\left[\left(\frac{(1-\nu)}{(1+\nu)} + \frac{4(1+\nu)}{(3-\nu)}\right)\frac{a^2}{r^3} - \frac{6(1+\nu)}{(3-\nu)}\frac{a^4}{r^5}\right] = 0. \tag{12.3.48}$$

Thus at $r = \bar{r}$, we have

$$\bar{r} = a\left\{\frac{6 + 12\nu + 6\nu^2}{7 + 4\nu + 5\nu^2}\right\}^{1/2}. \tag{12.3.49}$$

The maximum occurs at $\bar{r} > a$ if $6+12\nu+6\nu^2 > 7+4\nu+5\nu^2$. This implies that if $\nu > 0.1231$, the maximum radial stress does not occur at the interface but in front of it. However, if $\nu < 0.1231$, the maximum radial stress occurs at $r = a$.

12.4 State of stress and strain due to an elliptic hole in a plane

Let us write Cartesian coordinates x and y in terms of elliptic coordinates ξ and θ as

$$x = D\cosh\xi.\cos\theta \tag{12.4.1}$$

$$y = D\sinh\xi.\sin\theta. \tag{12.4.2}$$

Let $\xi = \xi_0$ be fixed, then the equation for the ellipse which is shown in the figure 12.5 given by

$$\frac{x^2}{D^2\cosh^2\xi_0} + \frac{y^2}{D^2\sinh^2\xi_0} = 1 \tag{12.4.3}$$

where $a = D\cosh\xi_0$ and $b = D\sinh\xi_0$. Therefore, $D = \sqrt{a^2 - b^2}$ and $\tanh\xi_0 = \frac{b}{a}$. Let $\theta = \theta_0$, $0 < \theta_0 < \frac{\pi}{2}$. Then, the equation of hyperbola as shown in the figure 12.6 is given by

$$\frac{x^2}{D^2\cos^2\theta_0} - \frac{y^2}{D^2\sin^2\theta_0} = 1 \tag{12.4.4}$$

The system is a con-focal system of orthogonal ellipses and hyperbolas (see Figure 12.7).

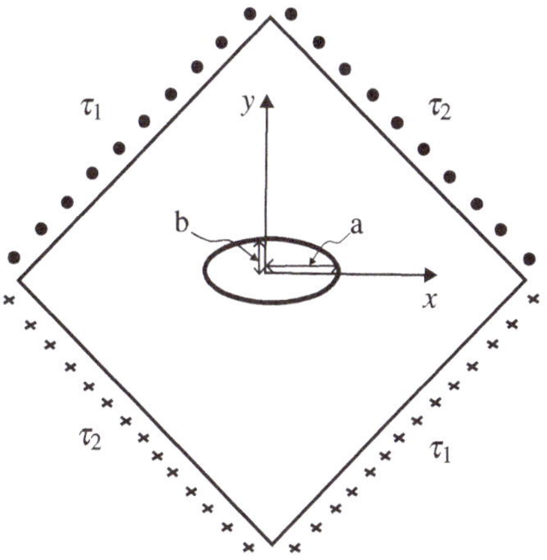

Figure 12.4: Infinite plate with a hole at the center with shear stress applied at infinity.

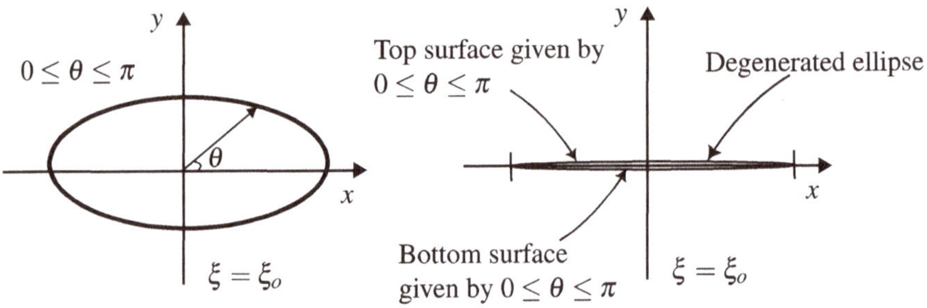

Figure 12.5: Typical representation of an ellipse belonging to the con-focal system of ellipses and hyperbolas and a degenerate ellipse.

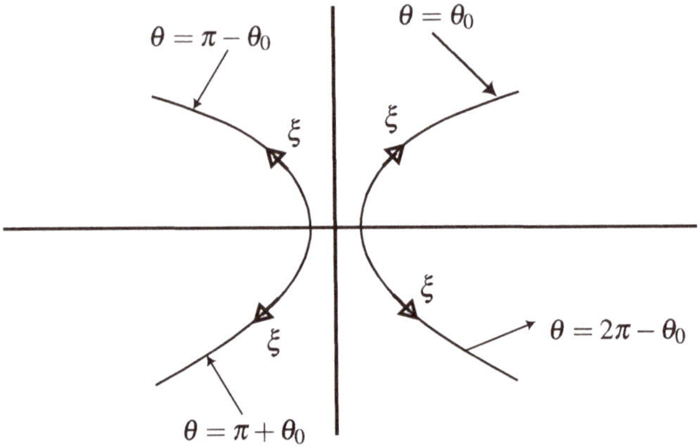

Figure 12.6: A typical hyperbola belonging to the con-focal system of ellipses and hyperbolas.

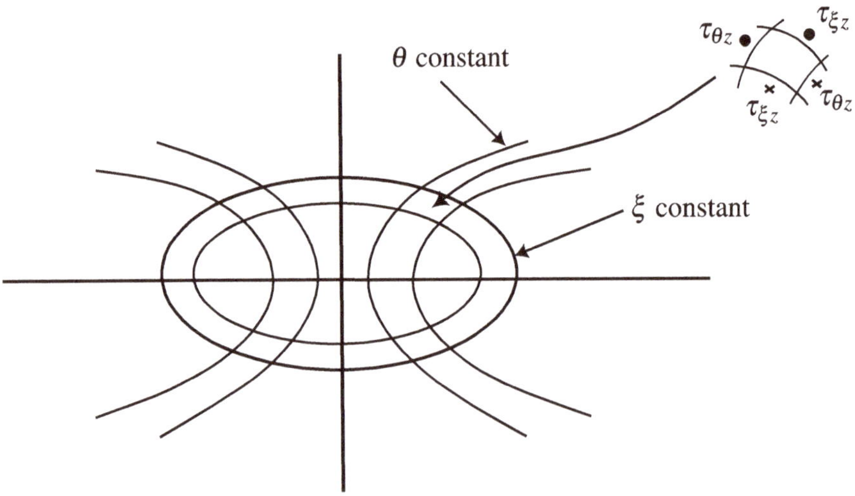

Figure 12.7: Con-focal system of orthogonal ellipses and hyperbolas.

$T_{\xi z}$ is the shear stress in the z direction acting on the surface ξ = constant and acts in the + z direction when the θ axis is pointing from the material. $T_{\theta z}$ is the shear stress in the z direction acting on the surface θ = constant.

$$T_{\xi z} = 2\mu(\epsilon_{\xi z}) \tag{12.4.5}$$

$$T_{\theta z} = 2\mu(\epsilon_{\theta z}) \tag{12.4.6}$$

$\epsilon_{\xi z}$ is the decrease of right angle between two fibres originally along θ = constant and the z-axis, where $\epsilon_{\theta z}$ is the decrease of right angle between two fibres originally along ξ = constant and the z-axis

12.4.1 Transformation of stress in elliptic coordinates

As shown in the figure 12.8, the stress $T_{\xi z}$ can be written in the components T_{xz} and T_{yz} as

$$T_{\xi z} = T_{xz}\cos\alpha + T_{yz}\sin\alpha \tag{12.4.7}$$

The Cartesian coordinates x and y in terms of elliptic coordinates ξ and θ are given by

$$x = D\cosh\xi\cos\theta \tag{12.4.8}$$

$$y = D\sinh\xi\sin\theta. \tag{12.4.9}$$

$$\frac{\partial x}{\partial \xi} = D\sinh\xi\cos\theta, \quad \frac{\partial y}{\partial \xi} = D\cosh\xi\sin\theta. \tag{12.4.10}$$

Therefore, $\cos\alpha$ and $\sin\alpha$ are given by (Refer to figure 12.8)

$$\cos\alpha = \frac{\frac{\partial x}{\partial \xi}}{\sqrt{(\frac{\partial x}{\partial \xi})^2 + (\frac{\partial y}{\partial \xi})^2}}, \tag{12.4.11}$$

$$\sin\alpha = \frac{\frac{\partial y}{\partial \xi}}{\sqrt{(\frac{\partial x}{\partial \xi})^2 + (\frac{\partial y}{\partial \xi})^2}}. \tag{12.4.12}$$

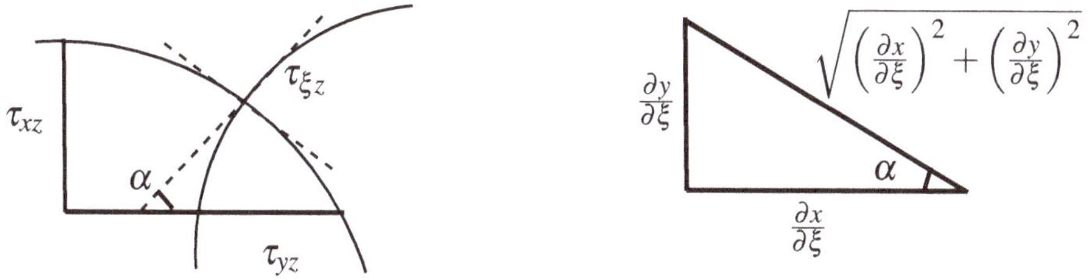

Figure 12.8: Components of $T_{\xi z}$.

$$T_{\xi z} = \frac{T_{xz}\frac{\partial x}{\partial \xi} + T_{yz}\frac{\partial y}{\partial \xi}}{\sqrt{(\frac{\partial x}{\partial \xi})^2 + (\frac{\partial y}{\partial \xi})^2}} \tag{12.4.13}$$

$$T_{\theta z} = \frac{T_{xz}\frac{\partial x}{\partial \theta} + T_{yz}\frac{\partial y}{\partial \theta}}{\sqrt{(\frac{\partial x}{\partial \theta})^2 + (\frac{\partial y}{\partial \theta})^2}}. \tag{12.4.14}$$

From the definition of shear strain in terms of the displacement w, the above equations can be written as

$$T_{\xi z} = \frac{\mu\frac{\partial w}{\partial \xi}}{\sqrt{(\frac{\partial x}{\partial \xi})^2 + (\frac{\partial y}{\partial \xi})^2}}, \tag{12.4.15}$$

$$T_{\theta z} = \frac{\mu\frac{\partial w}{\partial \theta}}{\sqrt{(\frac{\partial x}{\partial \theta})^2 + (\frac{\partial y}{\partial \theta})^2}}. \tag{12.4.16}$$

Substituting equations (12.4.10) in (12.4.15) and (12.4.16), we get

$$T_{\xi z} = \frac{\mu}{D}\frac{\frac{\partial w}{\partial \xi}}{\sqrt{\frac{1}{2}(\cosh 2\xi - \cos 2\theta)}}, \tag{12.4.17}$$

$$T_{\theta z} = \frac{\mu}{D}\frac{\frac{\partial w}{\partial \theta}}{\sqrt{\frac{1}{2}(\cosh 2\xi - \cos 2\theta)}}. \tag{12.4.18}$$

The final governing equation in terms of w after invoking the equilibrium equations, are

$$\frac{\partial^2 w}{\partial \xi^2} + \frac{\partial^2 w}{\partial \theta^2} = 0. \tag{12.4.19}$$

Now applying the boundary conditions on $\xi = \xi_0$, namely $t_z = 0$, and recognizing that

$$t_z = \mu \nabla w.\mathbf{n}, \quad \text{where } \mathbf{n}.\nabla w \text{ is evaluated at } \xi = \xi_0 \tag{12.4.20}$$

we get

$$\frac{\partial w}{\partial \xi}|_{\xi=\xi_0} = 0 \tag{12.4.21}$$

Similarly, at infinity, let β be the angle between the x-axis on which T_{yz} acts and the surface on which τ_1 acts. The angle between the y-axis on which T_{xz} acts and the surface on which τ_2 acts is also β. The stress components T_{yz} and T_{xz} are given by

$$T_{yz} = \tau_1 \cos\beta + \tau_2 \sin\beta \tag{12.4.22}$$

$$T_{xz} = \tau_2 \cos\beta - \tau_1 \sin\beta \tag{12.4.23}$$

As $x, y \to \infty$, then $\frac{\partial w}{\partial x}$ and $\frac{\partial w}{\partial y}$ are given by

$$\frac{\partial w}{\partial x} = \frac{1}{\mu}(\tau_2 \cos\beta - \tau_1 \sin\beta) \tag{12.4.24}$$

$$\frac{\partial w}{\partial y} = \frac{1}{\mu}(\tau_1 \cos\beta + \tau_2 \sin\beta) \tag{12.4.25}$$

The field I solution w_1 for the governing equation $\nabla^2 w_1 = 0$ is given by

$$w_1 = \frac{1}{\mu}(\tau_2 \cos\beta - \tau_1 \sin\beta)x + \frac{1}{\mu}(\tau_2 \sin\beta + \tau_1 \cos\beta)y \tag{12.4.26}$$

Using equations (12.4.8), the above equation becomes

$$\begin{aligned}\mu\frac{w_1}{D} &= \tau_2(\cosh\xi \cos\theta \cos\beta + \sinh\xi \sin\theta \sin\beta) \\ &\quad + \tau_1(-\cosh\xi \cos\theta \sin\beta + \sinh\xi \sin\theta \cos\beta)\end{aligned} \tag{12.4.27}$$

The stress component $T^{(1)}_{\xi z}$ is

$$T^{(1)}_{\xi z} = \frac{\mu\frac{\partial w_1}{\partial \xi}}{D\sqrt{\frac{1}{2}(\cosh 2\xi - \cos 2\theta)}}. \tag{12.4.28}$$

Using the first field solution (12.4.27) in (12.4.28) and computing $T^{(1)}_{\xi z}$ at $\xi = \xi_0$, we get

$$\begin{aligned}T^{(1)}_{\xi z}(\xi_0, \theta) &= \frac{\tau_1(-\sinh\xi_0 \cos\theta \sin\beta + \cosh\xi_0 \sin\theta \cos\beta)}{\sqrt{\frac{1}{2}(\cosh 2\xi_0 - \cos 2\theta)}} \\ &\quad + \frac{\tau_2(\sinh\xi_0 \cos\theta \cos\beta + \cosh\xi_0 \sin\theta \sin\beta)}{\sqrt{\frac{1}{2}(\cosh 2\xi_0 - \cos 2\theta)}}.\end{aligned} \tag{12.4.29}$$

Similarly, the field w_2 solution for the governing equation $\nabla^2 w_2 = 0$ is obtained by trying at $\xi = \xi_0$, a trial solution in the form of

$$w_2 = f(\xi)\cos\theta + g(\xi)\sin\theta. \tag{12.4.30}$$

Substituting in $\nabla^2 w_2 = 0$, we get

$$f'' - f = 0, \tag{12.4.31}$$

$$g'' - g = 0. \tag{12.4.32}$$

The solution for above equations is given by

$$f = A\exp(-\xi) + C\exp(\xi), \tag{12.4.33}$$

$$g = B\exp(-\xi) + E\exp(\xi). \tag{12.4.34}$$

Since f and g must vanish at $\xi = \infty$, the possible solution for w_2 is

$$w_2 = A\exp(-\xi)\cos\theta + B\exp(-\xi)\sin\theta. \tag{12.4.35}$$

Computing $\frac{\partial w}{\partial x}$ at $\xi = \xi_0$ we obtain

$$\frac{\partial w_2}{\partial \xi}\Big|_{\xi=\xi_0} = -A\exp(-\xi_0)\cos\theta - B\exp(-\xi_0)\sin\theta. \tag{12.4.36}$$

Comparing with equation (12.4.29), the constants A and B are given by

$$A = \frac{D}{\mu}\exp(\xi_0)(\tau_2\sinh\xi_0\cos\beta - \tau_1\sinh\xi_0\sin\beta) \tag{12.4.37}$$

$$B = \frac{D}{\mu}\exp(\xi_0)(\tau_2\cosh\xi_0\sinh\beta + \tau_1\cosh\xi_0\cos\beta) \tag{12.4.38}$$

Then, final solution for w is

$$w = w_1 + w_2 \tag{12.4.39}$$

$$\begin{aligned} w = \frac{D}{\mu}\{&\tau_2[(\cosh\xi + \exp(-(\xi-\xi_0))\sinh\xi_0)\cos\beta\cos\theta \\ &+(\sinh\xi + \exp(-(\xi-\xi_0))\cosh\xi_0)\sin\beta\sin\theta] \\ &+\tau_1[-(\cosh\xi + \exp(-(\xi-\xi_0))\sinh\xi_0)\sin\beta\cos\theta \\ &+(\sinh\xi + \exp(-(\xi-\xi_0))\cosh\xi_0)\cos\beta\sin\theta]\}. \end{aligned} \tag{12.4.40}$$

Evaluating at ξ_0, we find that w is given by

$$w(\xi_0,\theta) = \frac{D}{\mu}\exp(\xi_0)[\tau_2\cos(\theta-\beta) + \tau_1\sin(\theta-\beta)] \tag{12.4.41}$$

Using the equation (12.4.27), the stress component $\tau_{\theta z}$ at $\xi = \xi_0$ is expressed by

$$T_{\theta z}|_{\xi=\xi_0} = \frac{\exp(\xi_0)[\tau_1\cos(\theta-\beta) + \tau_2\sin(\theta-\beta)]}{\sqrt{\frac{1}{2}(\cosh 2\xi_0 - \cos 2\theta)}} \tag{12.4.42}$$

where $T_{\theta z}$ is the shear stress at the boundary of the elliptical hole acting on the surfaces perpendicular to the boundary.

Next we consider the crack case at $\xi_o = 0$. We can take $\beta = 0$ without loss of generality and omit τ_2. For this case w becomes

$$w = \frac{D}{\mu}\tau_1(\sinh\xi + \exp(-\xi))\sin\theta. \tag{12.4.43}$$

Simplifying the above equation further, we obtain

$$w = \frac{D}{\mu}\tau_1 \cosh\xi \sin\theta \tag{12.4.44}$$

Then, the stress components $T_{\xi z}$ and $T_{\theta z}$ become

$$T_{\xi z} = \frac{\tau_1 \sinh\xi \sin\theta}{\sqrt{\frac{1}{2}(\cosh 2\xi - \cos 2\theta)}}, \tag{12.4.45}$$

$$T_{\theta z} = \frac{\tau_1 \cosh\xi \cos\theta}{\sqrt{\frac{1}{2}(\cosh 2\xi - \cos 2\theta)}}. \tag{12.4.46}$$

12.4.2 Analysis in the vicinity of a crack tip

Let us analyse the stresses in the vicinity of the crack tip and to do the same, lets define coordinates x and y as (see Figure 12.9)

$$x = D + r\cos\psi, \tag{12.4.47}$$

$$y = r\sin\psi. \tag{12.4.48}$$

Then the equation of a circle becomes

$$(x - D)^2 + y^2 = r^2 = D^2(\cosh\xi - \cos\theta)^2 \tag{12.4.49}$$

Therefore, radius r is given by

$$r = D(\cosh\xi - \cos\theta) \tag{12.4.50}$$

From the figure, $\tan\psi$ is given as

$$\tan\psi = \frac{\sinh\xi \sin\theta}{\cosh\xi \cos\theta - 1}. \tag{12.4.51}$$

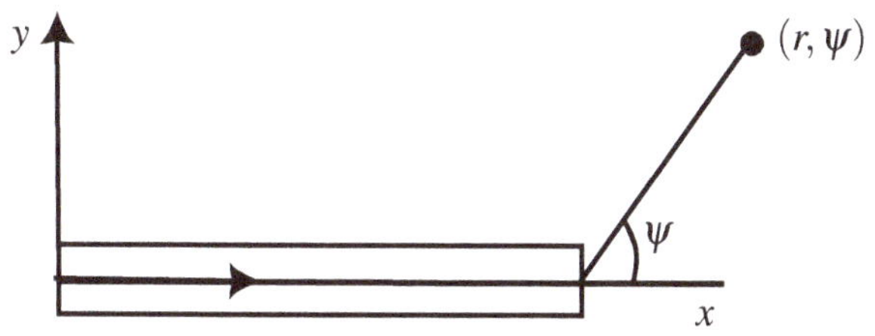

Figure 12.9: Vicinity of crack tip described in terms of a Polar-coordinate system located at a tip of the crack.

Expanding equations (12.4.50) and (12.4.51) using exponential form of $\cosh\xi$ and $\cos\theta$, we will get approximate expression for r and $\tan\psi$ as

$$r \approx \frac{D}{2}(\xi^2 + \theta^2) \tag{12.4.52}$$

$$\tan\xi \approx \frac{2\xi\theta}{\xi^2 - \theta^2}, \tag{12.4.53}$$

where $\xi = \sqrt{\frac{2r}{D}}\cos\frac{\psi}{2}$ and $\theta = \sqrt{\frac{2r}{D}}\sin\frac{\psi}{2}$. Finally, the value of the stresses τ_{xz} and τ_{yz} at the vicinity of the crack tip are computed as

$$T_{xz} = -\tau_1\sqrt{\frac{D}{2r}}\sin\frac{\psi}{2}, \tag{12.4.54}$$

$$T_{yz} = -\tau_1\sqrt{\frac{D}{2r}}\cos\frac{\psi}{2}. \tag{12.4.55}$$

Above values agrees with the result arrived by the complex variable method.

Let us return to the case where $\xi = \xi_0$. For convenience take $\tau_2 = \beta = 0$, then the final solution w becomes

$$w = \frac{D}{\mu}\tau_1(\sinh\xi + \exp(-(\xi - \xi_o))\cosh\xi_o)\sin\theta \tag{12.4.56}$$

$$w(\xi_o, \theta) = \frac{D}{\mu}\tau_1\exp(\xi_o)\sin\theta \tag{12.4.57}$$

Then, the stresses at $\xi = \xi_0$ are

$$T_{\theta z} = \frac{\tau_1\exp(\xi_o)\cos\theta}{\sqrt{\frac{1}{2}(\cosh 2\xi_o - \cos 2\theta)}}. \tag{12.4.58}$$

Maximum value of stresses in terms of Cartesian coordinates occur at $\theta = 0$ and $\theta = \pi$ and are

$$T_{xz} = \frac{-\tau_1\exp(\xi_o)\cosh\xi_o\sin 2\theta}{\cosh 2\xi_o - \cos 2\theta}, \tag{12.4.59}$$

$$T_{yz} = \frac{2\tau_1\exp(\xi_o)\sinh\xi_o\cos^2 2\theta}{\cosh 2\xi_o - \cos 2\theta}. \tag{12.4.60}$$

Now if $\theta \approx 0$, $T_{xz} \approx 0$ and

$$T_{yz} = \frac{\tau_1\exp(\xi_o)}{\sinh\xi_o} \quad \text{and as } \xi_o \to 0,\ T_{yz} \to \infty, \tag{12.4.61}$$

as in the crack problem. Let ρ denote the radius of curvature of tip, then

$$\rho = \frac{D\sinh^2\xi_0}{\cosh\xi_0}. \tag{12.4.62}$$

Then $\theta \to D\sinh^2 \xi_0$ as $\xi_0 \to 0$. Hence,

$$T_{yz} \to \sqrt{\frac{D}{\rho}}\tau_1 \quad \text{as } \xi_o \to 0. \tag{12.4.63}$$

12.5 Stresses around cracks

We will consider one of many possible problems involving cracks in a linearized elastic half-plane. The problem that we shall consider is anti-plane loading of a half plane containing a crack. Other important problems are the crack in a half-plane subject to uni-axial and bi-axial loading, and also the same body subject to shear. A combination of such loadings can be obtained by just adding the individual solutions in view of the linearity of the governing equations.

12.5.1 Anti-plane strain: Mode III loading

For this problem, we note that

$$T_{xz} \longrightarrow \tau_2 \quad \text{as } x \longrightarrow \infty \tag{12.5.1}$$

$$T_{yz} \longrightarrow \tau_1 \quad \text{as } y \longrightarrow \infty. \tag{12.5.2}$$

The problem is illustrated in Fig. 12.10. We observe that (by considering Figure 12.11 and assuming that the surface on which τ_0 acts is of unit length and unit thickness along the z-direction) that

$$\tau_0(1) = \tau_2 \cos\alpha + \tau_1 \sin\alpha \tag{12.5.3}$$

$$\tau_1 \cos\alpha = \tau_2 \sin\alpha \tag{12.5.4}$$

$$\tau_0^2 = \tau_1^2 + \tau_2^2. \tag{12.5.5}$$

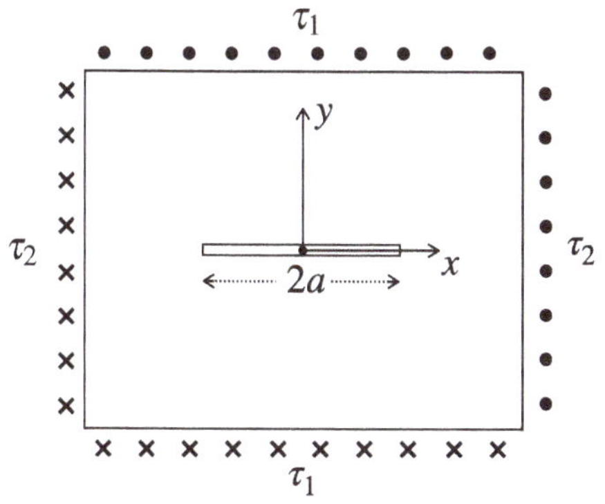

Figure 12.10: Infinite plate with shear stress applied and crack in the center.

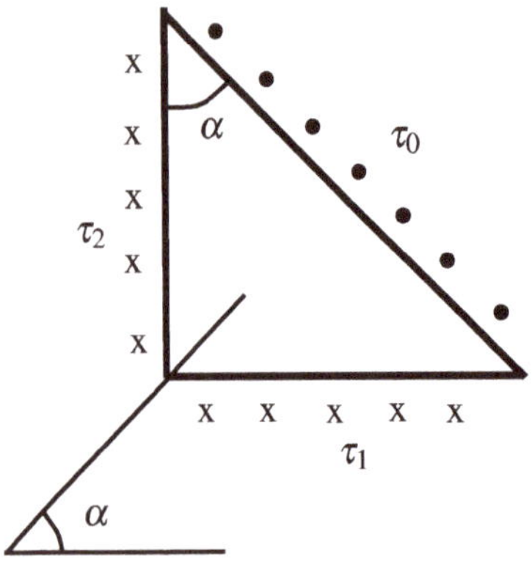

Figure 12.11: Stresses acting on a element in equilibrium.

We will use the notation τ_{xy} and τ_{yz} for the shear stresses.

For solution S_1, we have

$$T_{xz}^{(I)} = \tau_2 \tag{12.5.6}$$

$$T_{yz}^{(I)} = \tau_1. \tag{12.5.7}$$

The boundary condition at the crack is

$$T_{yz}(x,0) = 0, \quad |x| < a. \tag{12.5.8}$$

The solution S_1 does not satisfy the boundary condition for the full problem unless $\tau_1 = 0$. In that case for any τ_2, the boundary condition is satisfied. The loading τ_2 has no effect on the crack. Since τ_2 has no effect on the presence of the crack, we can drop this loading in our analysis.

For the solution S_1, we thus ignore (12.5.6) and the negative of the condition (12.5.7) is enforced as the boundary condition for the solution S_2. Thus, for the solution S_2, the boundary condition is (see Figure 12.12)

$$T_{yz}^{(II)}(x,0^{\pm}) = -\tau_1, \quad |x| < a, \tag{12.5.9}$$

and $T_{yz}^{(II)}, T_{xz}^{(II)} \longrightarrow 0$ at ∞.

We will denote a function $W^{(II)}$ that is the displacement in the z direction and is an odd function of y. From the symmetry of the problem,

$$W^{(II)}(x,0) = 0, \quad |x| > a. \tag{12.5.10}$$

The equations of equilibrium are

$$\nabla^2 W^{(II)} = 0. \tag{12.5.11}$$

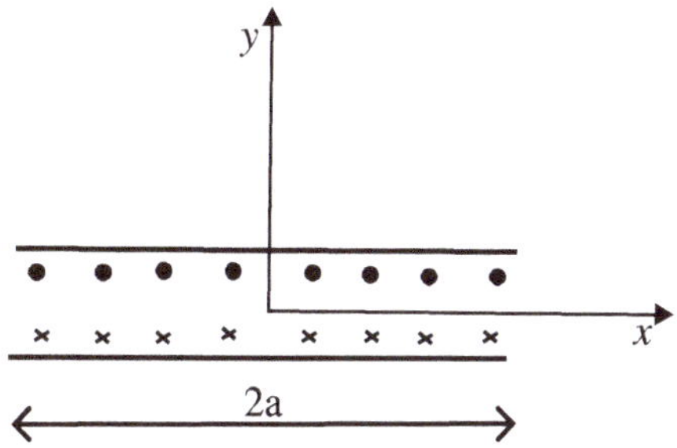

Figure 12.12: Loading for field (II).

From the stress and strain relationships, we have

$$T_{xz}^{(II)} = \mu \frac{\partial W}{\partial x} \tag{12.5.12}$$

$$T_{yz}^{(II)} = \mu \frac{\partial W}{\partial y}. \tag{12.5.13}$$

Thus, we have the boundary condition

$$\frac{\partial W^{(II)}(x, 0^{\pm})}{\partial y} = -\frac{\tau_1}{\mu}, \quad |x| < a, \tag{12.5.14}$$

and

$$W^{(II)}(x, 0) = 0, \quad |x| > a, \tag{12.5.15}$$

and

$$|\nabla W^{(II)}| = \text{grad } W^{(II)} \longrightarrow 0 \quad \text{at } \infty. \tag{12.5.16}$$

The solution to the problem is (see Chapter 14 where the use of complex variables to analyze problems in elasticity).

$$W^{(II)}(x, y) = \text{Re}\left[\frac{i\tau_1}{\mu}(z - \sqrt{z^2 - a^2})\right], \tag{12.5.17}$$

where Re stands for the real part of the complex number $z = x + iy$. Note,

$$W^{(I)}(x, y) = -\text{Re}\left(\frac{i\tau_1}{\mu} z\right) = \frac{\tau_1}{\mu} y. \tag{12.5.18}$$

Because we are setting $\tau_2 = 0$,

$$\mu \frac{\partial W^{(I)}}{\partial x} = 0. \tag{12.5.19}$$

Also, we have

$$\mu \frac{\partial W^{(I)}}{\partial y} = \tau_1. \tag{12.5.20}$$

Thus,

$$W = W^{(I)} + W^{(II)} = \frac{\tau_1}{\mu}\mathrm{Re}(-i\sqrt{z^2 - a^2}) = \frac{\tau_1}{\mu}\mathrm{Im}(\sqrt{z^2 - a^2}). \tag{12.5.21}$$

Now, we seek to verify the function satisfies the boundary conditions. Let $|z| \longrightarrow \infty$,

$$W \longrightarrow \frac{\tau_1 y}{\mu} = W^{(I)}. \tag{12.5.22}$$

At $y = 0$,

$$W = \frac{\tau_1}{\mu}\mathrm{Im}(\sqrt{x^2 - a^2}) = 0, \quad |x| > a. \tag{12.5.23}$$

Also note that

$$\mu \frac{\partial W}{\partial y} = \mu \frac{\partial w}{\partial z}\frac{\partial z}{\partial y} = \tau_1 \mathrm{Im}\left[\frac{iz}{\sqrt{z^2 - a^2}}\right] = \tau_{yz}. \tag{12.5.24}$$

If we let $y = 0$,

$$T_{yz} = \tau_1 \frac{x}{\sqrt{x^2 - a^2}}, \quad |x| > a, \tag{12.5.25}$$

and

$$T_{yz} = 0, \quad x < a. \tag{12.5.26}$$

Therefore we have for the analysis,

$$T_{xz} = \mu \frac{\partial W}{\partial x} = \tau_1 \frac{y}{\sqrt{x^2 - a^2}}, \quad |x| > a, \tag{12.5.27}$$

$$T_{yz} = \tau_1 \frac{x}{\sqrt{x^2 - a^2}}, \quad |x| > a. \tag{12.5.28}$$

Consider the crack tip at $x = a$. In polar coordinates, we have

$$x = a + r\cos\theta, \quad y = r\sin\theta. \tag{12.5.29}$$

Thus (see Figure 12.13),

$$z = a + r(\cos\theta + i\sin\theta) = a + re^{i\theta}, \tag{12.5.30}$$

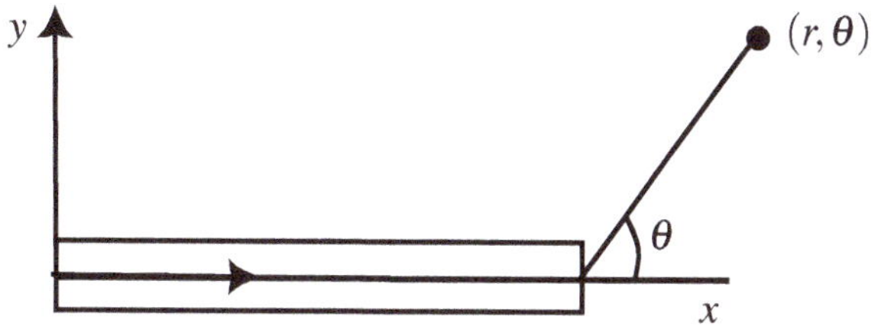

Figure 12.13: Vicinity of crack tip described in terms of a Polar-coordinate system located at a tip of the crack.

where we shall take r to be small. Noting that

$$z - a = re^{i\theta}, \tag{12.5.31}$$

$$z + a = 2a + re^{i\theta}, \tag{12.5.32}$$

implies that

$$z^2 - a^2 = 2are^{i\theta}\left(1 + \frac{r}{2a}e^{i\theta}\right). \tag{12.5.33}$$

Taking the square root yields

$$\sqrt{z^2 - a^2} = \sqrt{2ar}e^{\frac{i\theta}{2}}\left(1 + \frac{r}{2a}e^{i\theta}\right)^{1/2} \approx \sqrt{2ar}e^{\frac{i\theta}{2}}\left(1 + \frac{r}{4a}e^{i\theta}\right). \tag{12.5.34}$$

This implies that

$$\frac{z}{\sqrt{z^2 - a^2}} \approx \frac{a(1 + \frac{r}{a}e^{i\theta})}{\sqrt{2ar}e^{\frac{i\theta}{2}}(1 + \frac{r}{4a}e^{i\theta})}, \tag{12.5.35}$$

$$\approx \sqrt{\frac{a}{2r}}e^{-\frac{i\theta}{2}}\left(1 + \frac{r}{a}e^{i\theta}\right)\left(1 - \frac{r}{4a}e^{i\theta}\right) \tag{12.5.36}$$

$$\approx \sqrt{\frac{a}{2r}}e^{-\frac{i\theta}{2}}\left(1 + \frac{3r}{4a}e^{i\theta}\right). \tag{12.5.37}$$

Further simplification yields

$$\frac{z}{\sqrt{z^2 - a^2}} \approx \sqrt{\frac{a}{2r}}e^{-\frac{i\theta}{2}} + \frac{3}{4}\sqrt{\frac{r}{2a}}e^{\frac{i\theta}{2}} \tag{12.5.38}$$

$$= \left(\sqrt{\frac{a}{2r}} + \frac{3}{4}\sqrt{\frac{r}{2a}}\right)\cos\frac{\theta}{2} - i\left(\sqrt{\frac{a}{2r}} - \frac{3}{4}\sqrt{\frac{r}{2a}}\right)\sin\frac{\theta}{2}. \tag{12.5.39}$$

Therefore,

$$T_{yz} = \tau_1\left(\sqrt{\frac{a}{2r}} + \frac{3}{4}\sqrt{\frac{r}{2a}}\right)\cos\frac{\theta}{2} + \mathcal{O}\left[\left(\frac{r}{a}\right)^{3/2}\right] \tag{12.5.40}$$

$$T_{xz} = -\tau_1\left(\sqrt{\frac{a}{2r}} - \frac{3}{4}\sqrt{\frac{r}{2a}}\right)\sin\frac{\theta}{2} + \mathcal{O}\left[\left(\frac{r}{a}\right)^{3/2}\right]. \tag{12.5.41}$$

Notice that at $\theta = \pm\pi$, $T_{yz} = 0$ (i. e., we recover the boundary condition).

Thus, stresses are singular near the crack tip in the order of $r^{-1/2}$. For real materials, this implies either yielding or in the case of brittle crack propagation, fracture. However, a real crack does not have infinite curvature at its edges. There is some finite curvature which will keep the stress finite.

12.5.2 Plane strain or plane stress

Let us consider an infinite plate with normal and shear stress applied and crack in the center as shown in the figure 12.14. For this problem, we note that

$$T_{xx} \longrightarrow T \tag{12.5.42}$$

$$T_{yy} \longrightarrow S \tag{12.5.43}$$

$$T_{xy} \longrightarrow \tau \tag{12.5.44}$$

as x and y tend to ∞. The boundary condition at the crack is

$$T_{yy}(x,0) = 0, \quad |x| < a, \tag{12.5.45}$$

$$T_{yx}(x,0) = 0, \quad |x| < a. \tag{12.5.46}$$

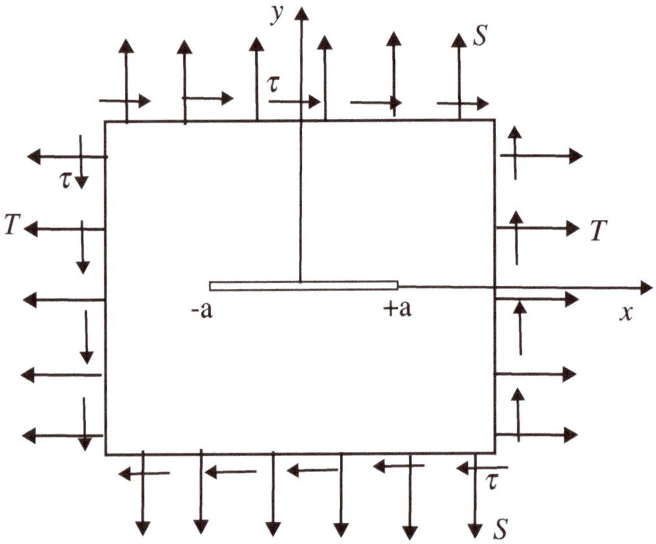

Figure 12.14: Infinite plate with normal and shear stress applied and crack in the center.

For solution S_1, we have

$$T_{xx}^{(I)} = T, \tag{12.5.47}$$

$$T_{yy}^{(I)} = S, \tag{12.5.48}$$

$$T_{xy}^{(I)} = \tau, \tag{12.5.49}$$

where T loading is non interactive with the crack. Hence its action is not of interest and we shall take $T = 0$. For solution S_2, we have

$$T_{yy}^{(II)}(x,0) = -S, \quad \text{for } |x| < a, \tag{12.5.50}$$

$$T_{yx}^{(II)}(x,0) = -\tau, \quad \text{for } |x| < a. \tag{12.5.51}$$

We will split the problem in two parts. The first part is $\tau = 0$ and $S \neq 0$.

Part:1 solution is

$$T_{yy}^{(II)}(x,0) = -S, \quad \text{for } |x| < a, \tag{12.5.52}$$

$$T_{yx}^{(II)}(x,0) = 0, \quad \text{for } |x| < a. \tag{12.5.53}$$

Above loading corresponds to Mode I loading and the stresses tend to 0 at ∞. Let u and v be the displacement in the x and y direction, respectively. The displacement u is even in y and v is odd in y, and

$$T_{xy} = \mu\left(\frac{\partial v}{\partial x} + \frac{\partial u}{\partial y}\right). \tag{12.5.54}$$

Therefore, $T_{xy}^{(II)} = 0$ on $y = 0$ for $|x| < a$. Hence, $T_{yx}(x,0) = 0$ for all x. Thus we can use the Westergaard potential.

$$T_{xx}^{(II)} = y\frac{\partial\psi}{\partial y} + \psi + \hat{\hat{a}}\frac{y^2}{2} + cy + d \tag{12.5.55}$$

$$T_{yy}^{(II)} = -y\frac{\partial\psi}{\partial y} + \psi - \hat{\hat{a}}\frac{x^2}{2} + ex + b \tag{12.5.56}$$

$$T_{xy}^{(II)} = -y\frac{\partial\psi}{\partial x} \tag{12.5.57}$$

From the boundary condition that stresses tend to 0 at ∞, the value of the constants $\hat{\hat{a}}$, e, c in the equations from (12.5.55) to (12.5.57) take the value of 0. The other constants b and d will be found from the analysis. The final solution of ψ that satisfies the governing equation $\nabla^2\psi = 0$ is of the form (once again we appeal to complex analysis to find the solution)

$$\psi = S\mathrm{Re}\left[\frac{Az}{\sqrt{z^2 - a^2}} - 1\right], \tag{12.5.58}$$

where A is real constant.

We want to verify this solution and pin down values for A, b and d in the process. Let's find T_{yy} at $y = 0$, then

$$T_{yy}|_{y=0} = S\left[\frac{Ax}{\sqrt{x^2 - a^2}}\right] - S + b \tag{12.5.59}$$

but $T_{yy}^{(II)}(x, 0) = -S$, for $|x| < a$, therefore $b = 0$. So, finally

$$\frac{\partial \psi}{\partial x} = S\mathrm{Re}\left\{A\left[\frac{1}{\sqrt{z^2 - a^2}} - \frac{z^2}{(z^2 - a^2)^{3/2}}\right]\right\}, \tag{12.5.60}$$

$$\frac{\partial \psi}{\partial y} = S\mathrm{Re}\left\{Ai\left[\frac{1}{\sqrt{z^2 - a^2}} - \frac{z^2}{(z^2 - a^2)^{3/2}}\right]\right\}. \tag{12.5.61}$$

So the stresses are

$$T_{xx}^{(II)} = S\mathrm{Re}A\left[\frac{iy}{\sqrt{z^2 - a^2}} - \frac{iyz^2}{(z^2 - a^2)^{3/2}} + \frac{z}{\sqrt{z^2 - a^2}}\right] - S + d \tag{12.5.62}$$

$$T_{yy}^{(II)} = S\mathrm{Re}A\left[\frac{-iy}{\sqrt{z^2 - a^2}} + \frac{iyz^2}{(z^2 - a^2)^{3/2}} + \frac{z}{\sqrt{z^2 - a^2}}\right] - S, \tag{12.5.63}$$

$$T_{xy}^{(II)} = -S\mathrm{Re}A\left[\frac{y}{\sqrt{z^2 - a^2}} - \frac{yz^2}{(z^2 - a^2)^{3/2}}\right]. \tag{12.5.64}$$

Let $|z| \to \infty$, then $\sigma_x^{(II)}$ and $\sigma_y^{(II)}$ tend to 0. The equations (12.5.62) and (12.5.63) become

$$S(A - 1) + d = 0 \tag{12.5.65}$$

$$S(A - 1) = 0 \tag{12.5.66}$$

Solving (12.5.65) and (12.5.66), we get $A = 1$ and $d = 0$.

Also, $T_{xy}^{(II)} = 0$.

Now, the total solution is

$$T_{xx} = -S\mathrm{Re}\left[\frac{iya^2}{(z^2 - a^2)^{3/2}} - \frac{z}{\sqrt{z^2 - a^2}} + 1\right] \tag{12.5.67}$$

$$T_{yy} = -S\mathrm{Re}\left[\frac{iya^2}{(z^2 - a^2)^{3/2}} + \frac{z}{\sqrt{z^2 - a^2}}\right] \tag{12.5.68}$$

$$T_{xy} = S\mathrm{Re}\left[\frac{ya^2}{(z^2 - a^2)^{3/2}}\right] \tag{12.5.69}$$

Analysis in the vicinity of crack tip

Let us analyse the stresses in the vicinity of the crack tip. Consider $x = a$ and $y = 0$ (refer figure 12.15) and define the coordinates as

$$x = a + r\cos\theta, \tag{12.5.70}$$

$$y = r\sin\theta, \tag{12.5.71}$$

$$z = a + re^{i\theta}. \tag{12.5.72}$$

$$z^2 - a^2 = 2are^{i\theta}\left[1 + \frac{r}{2a}e^{i\theta}\right]. \tag{12.5.73}$$

The equation (12.5.73) approximated as

$$\left(z^2 - a^2\right)^{-1/2} \approx (2ar)^{(-1/2)}e^{\frac{-i\theta}{2}}\left[1 - \frac{r}{4a}e^{i\theta}\right]. \tag{12.5.74}$$

Substituting (12.5.73) in (12.5.67), we get T_{xx} as

$$T_{xx} \approx -S\mathrm{Re}\left[\frac{ia^2 r\sin\theta}{(2ar)^{3/2}}e^{\frac{-3i\theta}{2}}\left(1 - \frac{3r}{4a}e^{i\theta}\right) - \frac{a(1 + \frac{r}{a}e^{i\theta})}{\sqrt{2ar}}e^{\frac{-i\theta}{2}}\left(1 - \frac{r}{4a}e^{i\theta}\right) + 1\right]. \tag{12.5.75}$$

Use $\sin\theta = \frac{1}{2i}(e^{i\theta} - e^{-i\theta})$, then the equation becomes

$$T_{xx} \approx -S - S\mathrm{Re}\left[\frac{1}{4}\sqrt{\frac{a}{2r}}(e^{\frac{-i\theta}{2}} - e^{\frac{-5i\theta}{2}})\left(1 - \frac{3r}{4a}e^{i\theta}\right) - \sqrt{\frac{a}{2r}}e^{\frac{-i\theta}{2}} - \frac{3}{4}\sqrt{\frac{r}{2a}}e^{\frac{i\theta}{2}}\right]. \tag{12.5.76}$$

Final expressions for T_{xx}, T_{yy} and T_{xy} are

$$T_{xx} \approx \frac{1}{4}\sqrt{\frac{a}{2r}}S\left(3\cos\frac{\theta}{2} + \cos\frac{5\theta}{2}\right) - S + \frac{1}{16}\sqrt{\frac{r}{2a}}S\left(15\cos\frac{\theta}{2} - 3\cos\frac{3\theta}{2}\right), \tag{12.5.77}$$

$$T_{yy} \approx \frac{1}{4}\sqrt{\frac{a}{2r}}S\left(5\cos\frac{\theta}{2} - \cos\frac{5\theta}{2}\right) + \frac{1}{16}\sqrt{\frac{r}{2a}}S\left(9\cos\frac{\theta}{2} + 3\cos\frac{3\theta}{2}\right), \tag{12.5.78}$$

$$T_{xy} \approx -\frac{1}{4}\sqrt{\frac{a}{2r}}S\left(\sin\frac{\theta}{2} - \sin\frac{5\theta}{2}\right) - \frac{3}{16}\sqrt{\frac{r}{2a}}S\left(\sin\frac{\theta}{2} + 3\sin\frac{3\theta}{2}\right). \tag{12.5.79}$$

It follows from the above equations that

1. Stresses exhibit $r^{-\frac{1}{2}}$ singularity at the crack tip.
2. At $\theta = \pm\pi$, $T_{yy} = 0$ and $T_{xy} = 0$.

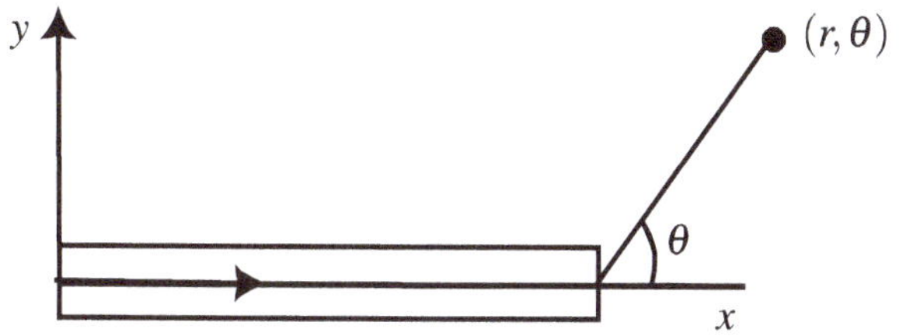

Figure 12.15: Vicinity of crack tip described in terms of a Polar-coordinate system located at a tip of the crack.

3. At $\theta = 0$, $T_{xy} = 0$. Hence, T_{xx} and T_{yy} are principal stresses along this line.

$$T_{xx} \longrightarrow \sqrt{\frac{a}{2r}}S \quad \text{for } y = 0, x > a, \tag{12.5.80}$$

$$T_{yy} \longrightarrow \sqrt{\frac{a}{2r}}S \quad \text{for } y = 0, x > a. \tag{12.5.81}$$

4. At $\theta = \frac{\pi}{2}$, $T_{xy} = 0$. Here,

$$T_{yy} = \frac{3\sqrt{3}}{4}\sqrt{\frac{a}{2r}}S = T_{yy}|_{\max}, \tag{12.5.82}$$

$$T_{xx} = \frac{\sqrt{3}}{4}\sqrt{\frac{a}{2r}}S. \tag{12.5.83}$$

12.6 Problems

12.1 Show that the general solution to Eq. (12.1.16) (Laplace equation in polar coordinates) is

$$\begin{aligned}\psi(r,\theta) &= A_0\theta + B_0 + C_0 \ln r + D_0\theta \ln r + \cdots \\ &\sum_{n=1}^{\infty}(A_n r^n + B_n r^{-n})(c_n \cos n\theta + D_n \sin n\theta).\end{aligned} \tag{12.6.1}$$

12.2 Show that for the case of a rigid circular inclusion, ϕ is given by

$$\phi = \tilde{a} + b \ln r + c \cos 2\theta + \frac{d}{r^2}\cos 2\theta. \tag{12.6.2}$$

12.3 Determine the total stress field in an infinite plate with a circular hole when uniform shear traction is applied at infinity, such as that shown in Fig. 12.16.

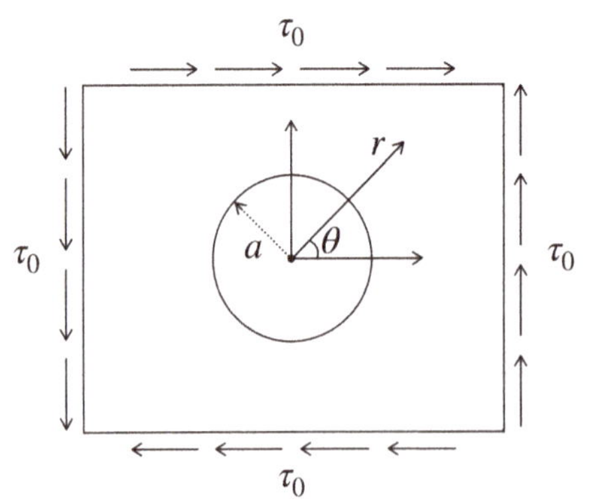

Figure 12.16: Infinite plate with hole subjected to shear stress at infinity.

13 Concentrated loads

By a concentrated load we mean a finite load that acts at a point. This gives rise to a singularity in the stresses as we have a finite force acting at a point which supposedly has zero area measure. The problem can be traced to the notion of a point, one of the most complex notions in mathematics. While it is taken for granted that it is a self-evident notion, it is far from being so, a rigorous definition of which has taxed some of the greatest minds. We cannot get into a discussion of the issues as it is beyond the scope of this book (see [8] for a discussion of the same). Here, as earlier in the lecture notes we will take for granted that we comprehend what is meant by a point. Our study is not mathematically rigorous (this would require knowledge of the theory of distributions), but it can be made to be so. The linearized theory of elasticity is an useful approximation to analyze problems wherein the displacement gradient, and hence the strain are sufficiently small. Thus, using such as theory to study problems wherein the strains are large is unsound. However, such studies might yet throw some light on problems where though strains are locally large at some locations, they are appropriately small in a significant part of the domain of the body. Examples of such problems are elastic bodies wherein there are cracks, elastic bodies subject to concentrated forces, doublets, etc., wherein the analysis of the problem presuming the body to be a linearized elastic body leads to singularities in the strain. This is a consequence of the theory being linear, with singularity in the stress leading automatically to singularity in the strain. The justification of such studies usually is that the results seem to apply in most of the body except for a small neighborhood of the singularity. Such logic however does not seem to be on sound ground as the region of interest, say with regard to some sort of "failure" of the material, is at the location of the singularity. As an example, if one is interested in the progression of a "crack", it is precisely the values of the stresses and strains the crack tip that one is concerned with, and if the theory erroneously predicts that this would be infinite, as does the linearized theory of elasticity, one is then left to appeal to adhoc fixes to come up with a criterion for the propagation of the crack. If the material is ductile, then the body responds inelastically near the crack tip, and the response of the body is described in a small neighborhood of the crack tip with a constitutive relation for inelastic response. However, one cannot resort to such a fix in the case of a brittle body. Such shortcomings notwithstanding, studies concerning problems wherein singularities arise provide us with some rudimentary understanding of these problems. Moreover, the study of such problems have a historical context. In the early days of the development of the field, one could at best solve only idealized problems for which certain mathematical tools such as Green's functions, etc., were available. The interplay between the solution of these idealized problems and the progress in the development of mathematical tools to study partial differential equations and integral equations cannot be over emphasized. In view of this, we shall study some problems wherein singularities manifest themselves.

https://doi.org/10.1515/9783110789515-013

13.1 Some general remarks

While the following results are stated as theorems and proofs in books that are mathematically inclined, here we shall merely observe them as remarks relevant to the study governing the equation of equilibrium of a linearized elastic body. Recall that the equation of equilibrium for a linearized elastic body is

$$\nabla^2 \mathbf{u} + \frac{1}{1-2\nu}\nabla(\operatorname{div}\mathbf{u}) + \frac{1}{\mu}\mathbf{f} = 0, \quad \text{in } \kappa_\tau(\mathcal{B}) \tag{13.1.1}$$

13.1.1 Boussinesq-Papkovitch-Neuber Solution

Suppose ϕ and ψ are sufficiently smooth functions and further suppose that they satisfy

$$\Delta\phi = -\frac{\mathbf{x}.\mathbf{f}}{2(1-\nu)}, \quad \Delta\psi = \frac{\mathbf{f}}{2(1-\nu)} \tag{13.1.2}$$

Then, a displacement field $\mathbf{u}$ such that

$$\mathbf{u} = \frac{1}{2\mu}[\operatorname{grad}(\phi + \mathbf{x}.\psi) - 4(1-\nu)\psi], \tag{13.1.3}$$

is a solution to (13.1.1).

The stress and linearized strain are given by

$$T_{ij} = \phi - 2\mu\delta_{ij}\psi_{k,k} - (1-2\nu)(\psi_{i,j} + \psi_{j,i}) + x_k\psi_{k,ij} \tag{13.1.4}$$

$$\epsilon_{ij} = 2\mu[\phi, ij - (1-2\nu)(\psi_{i,j} + \psi_{j,i}) + x_k\psi_{k,ij}] \tag{13.1.5}$$

The above result is known as the Boussinesq-Papkovitch-Neuber solution [1, 7, 6].

13.1.2 Completeness of the Boussinesq-Paplovitch-Neuber solutoin

If the domain is bounded and the boundary is "regular", and if the body force is sufficiently smooth, then the representation is complete, that is all the displacement fields allow such a representation.

13.1.3 Solution for sufficiently smooth body force

If the body force is sufficiently smooth then the equations of equilibrium admits a solution, where (see Figure 13.1)

$$\phi(\mathbf{x}) = \frac{1}{8\pi(1-\nu)}\int_{k_\tau(\mathcal{B})} \frac{\boldsymbol{\xi}.\mathbf{f}(\boldsymbol{\xi})}{R(\mathbf{x}.\boldsymbol{\xi})} dv(\boldsymbol{\xi}) \tag{13.1.6}$$

$$\psi(\mathbf{x}) = -\frac{1}{8\pi(1-\nu)}\int_{k_\tau(\mathcal{B})} \frac{\mathbf{f}(\boldsymbol{\xi})}{R(\mathbf{x}.\boldsymbol{\xi})} dv(\boldsymbol{\xi}) \tag{13.1.7}$$

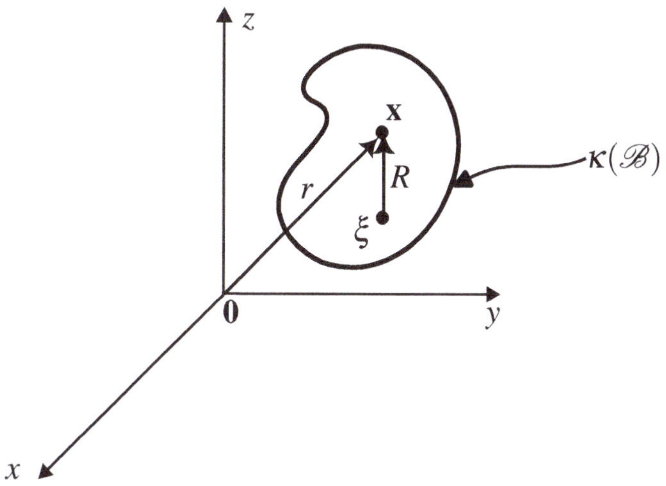

Figure 13.1: Domain for the application of concentrated force.

While one can develop the solutions for concentrated loads, etc., with a certain amount of rigor using the above tools (the interested reader can find a discussion of the relevant issues in Gurtin [3]) we shall adopt a much more heuristic analysis of the problem. The problem of a concentrated load acting on an infinite body was first studied by [4]. The problem of a concentrated load acting on a half space was first studied by [1] (see [2]).

13.2 Concentrated force

Suppose we have a smooth distribution of traction acting over a finite area on a body. Let

$$\mathbf{P} = \iint_A \mathbf{T} dA \tag{13.2.1}$$

be the resultant force. Also, let S_d (solution for the distributed load) be the solution due to this loading. Then, we define the solution for the *concentrated force* in the direction of $\mathbf{P}$ as

$$\lim_{A \to 0, \|\mathbf{T}\| \to \infty} S_d \equiv S_c, \tag{13.2.2}$$

such that $\iint_A \mathbf{T} dA = \mathbf{P}$. Here, S_c denotes the solution for the concentrated load.

The solution for the concentrated load is defined to be the limit of the *unique* solution for the distributed load as the area shrinks to zero but the amplitude of the distribution grows without bounds such that the resultant force remains fixed. We must have the distribution smooth to guarantee a unique solution. One might ask the following question: is S_c equal to the solution of the same problem with a prescribed concentrated force **P**? Generally the answer is no, since there is no unique solution for singular loadings. However, if we prescribe the concentrated force by a Dirac delta function,[1] the solution usually will be S_c (i. e., the physically consistent solution).

A motivation for the "generalized function" that we have in mind is portrayed in Figure 13.2. Here, we see a sequence of functions whose area under the curve remains constants as n increases. In the limit, as n tends to infinity we envision the result to be the function that we seek. Unfortunately there is no such function, that is the limiting process does not lead to a function with the properties that we desire. We shall however not worry about the mathematical rigor of our analysis but see how far we can go with our intuition. To do this properly, we need to work within the context of distribution theory, but then as we mentioned at the beginning of the chapter, the problem to start with is not a sensible problem within the context of linearized elasticity in that the strains

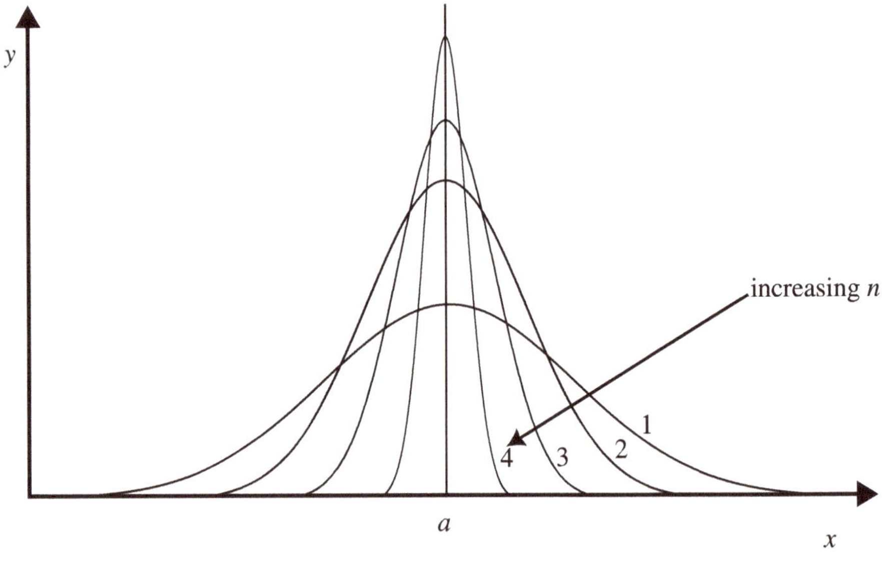

Figure 13.2: Example of distributed force leading in the limit to an equivalent concentrated force. The area under the curve is a constant and as $n \to \infty$, the height tends to infinity, but the area remains a constant. It is very simple to imagine it within the context of non-smooth functions. Imagine a sequence of rectangles of width w_n and height $\frac{1}{w_n}$. The area is unity. However, suppose that as $n \to \infty$, $w_n \to 0$. The area however remains constant as the height $\frac{1}{w_n} \to \infty$.

1 It is incorrect to call this a function. While one usually uses the terminology "generalized function", it is what is termed a "measure" in mathematics. It is also referred to as the Dirac distribution. We shall be somewhat cavalier about it and refer to it as "delta function" and carry out a study, which from the rigorous point is not justified, it yet leads to a physically meaningful solution.

that are engendered will not meet the requirement that it be small for the linearization to hold. Now, let us introduce the delta function through

$$f(x) = \delta(x-a) \begin{cases} 0 & x \neq a \\ \text{undefined} & x = a, \end{cases} \tag{13.2.3}$$

with the properties

$$\int_{-\infty}^{\infty} \delta(x-a)\,dx = 1, \tag{13.2.4a}$$

$$\int_{x_1}^{x_2} \delta(x-a)\,dx = \begin{cases} 1 & \text{if } x_1 < a < x_2 \\ 0 & \text{if } a \text{ does not lie between } x_1 \text{ and } x_2 \end{cases}, \tag{13.2.4b}$$

$$\int_{x_1}^{x_2} g(x)\delta(x-a)\,dx = \begin{cases} g(a) & \text{if } x_1 < a < x_2 \\ 0 & \text{if } a \text{ does not lie between } x_1 \text{ and } x_2 \end{cases}. \tag{13.2.4c}$$

To solve problems involving loads given in terms of the delta function, we can use transform methods.

The Fourier transform pairs are given by

$$\bar{f}(\xi) = \int_{-\infty}^{\infty} f(x)e^{-i\xi x}\,dx, \quad f(x) = \frac{1}{2\pi}\int_{-\infty}^{\infty} \bar{f}(\xi)e^{i\xi x}\,dx, \tag{13.2.5a}$$

where the first is the Fourier transform and the second is the inverse Fourier transform. We will also note the transform of derivatives are given by

$$\int_{-\infty}^{\infty} \frac{df}{dx}e^{-i\xi x}\,dx = f(x)e^{-i\xi x}\big|_{-\infty}^{\infty} + i\xi\int_{-\infty}^{\infty} fe^{-i\xi x}\,dx, \tag{13.2.6}$$

where we will assume that $f \to 0$ as $x \to \pm\infty$. Thus, if F denotes the Fourier transform, then

$$F\{f'(x)\} = i\xi\bar{f}(\xi), \tag{13.2.7}$$

and

$$F\{f''(x)\} = i\xi F\{f'(x)\} = -\xi^2\bar{f}. \tag{13.2.8}$$

In general

$$F\{f^n(x)\} = (i\xi)^n\bar{f}. \tag{13.2.9}$$

Now, consider

$$F\{\delta(x-a)\} = \int_{-\infty}^{\infty} \delta(x-a)e^{-i\xi x}\,dx = e^{-i\xi a}, \tag{13.2.10}$$

such that

$$\delta(x-a) = \frac{1}{2\pi}\int_{-\infty}^{\infty} e^{-i\xi a}e^{i\xi x}\,dx. \tag{13.2.11}$$

So, for the generalized function, we have

$$\delta(x-a) = \frac{1}{2\pi}\int_{-\infty}^{\infty} e^{i\xi(x-a)}\,dx. \tag{13.2.12}$$

Consider a concentrated force at

$$x = a, \quad y = b, \quad z = c, \tag{13.2.13}$$

in an arbitrary direction (Fig. 13.3)

$$\mathbf{P} = P_x\mathbf{e}_x + P_y\mathbf{e}_y + P_z\mathbf{e}_z. \tag{13.2.14}$$

We can represent the force by

$$\mathbf{F} = \mathbf{P}\delta(x-a)\delta(y-b)\delta(z-c), \tag{13.2.15}$$

where **F** has units of force per unit volume because $\delta(x-a)$ has units of inverse length, we are thus treating **P** as a body force. Now, consider a concentrated force acting in an infinite solid [5]. Take the z-axis along the vector **P** and choose the point at which **P** acts as the origin, there is no loss of generality since the solid is infinite. Recall that the

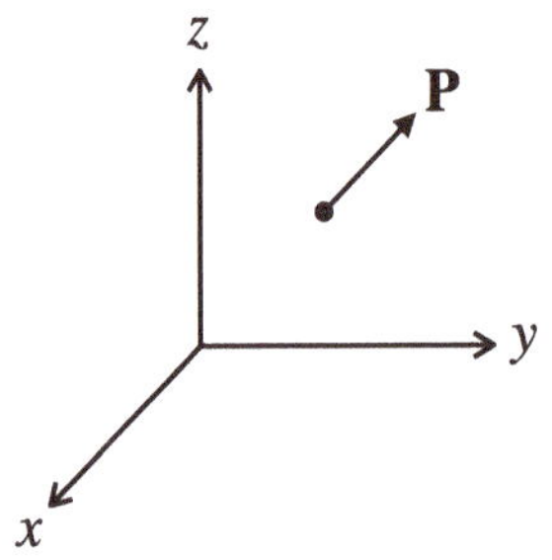

Figure 13.3: Arbitrary force, **P**, located at (a, b, c).

equation of equilibrium for a linearized elastic solid can be written as

$$\nabla^2\mathbf{u} + \frac{1}{1-2\nu}\nabla(\operatorname{div}\mathbf{u}) + \frac{1}{\mu}\mathbf{f} = 0, \tag{13.2.16}$$

where $\mathbf{u} = u\mathbf{e}_x + v\mathbf{e}_y + z\mathbf{e}_z$. We do not bother with compatibility as the equation is for the displacement and we do not have to verify if a given or assumed system of strains are compatible. We have three unknowns and we have a three-dimensional vector equation. Now, let us define the Fourier transform of the displacement in the x direction as

$$\bar{u}(\xi, y, z) = \int_{-\infty}^{\infty} u(x, y, z)e^{-i\xi x}\, dx, \tag{13.2.17}$$

and similarly in the y and z directions through

$$\bar{\bar{u}}(\xi, \eta, z) = \int_{-\infty}^{\infty} u(\xi, y, z)e^{-i\eta y}\, dy, \tag{13.2.18a}$$

$$\bar{\bar{\bar{u}}}(\xi, \eta, \gamma) = \int_{-\infty}^{\infty} u(\xi, \eta, z)e^{-i\gamma z}\, dz. \tag{13.2.18b}$$

We will also note that if $\mathrm{e} = \operatorname{div}\mathbf{u}$

$$F\{\nabla \mathrm{e}\} = \iiint_{-\infty}^{\infty} (\nabla \mathrm{e})e^{-i\xi x - i\eta y - i\gamma z}\, dxdydz, \tag{13.2.19}$$

where

$$\nabla \mathrm{e} = \frac{\partial \mathrm{e}}{\partial x}\mathbf{e}_x + \frac{\partial \mathrm{e}}{\partial y}\mathbf{e}_y + \frac{\partial \mathrm{e}}{\partial z}\mathbf{e}_z. \tag{13.2.20}$$

Thus, we can write (assuming appropriate conditions at ∞)

$$F\{\nabla \mathrm{e}\} = i(\xi\mathbf{e}_x + \eta\mathbf{e}_y + \gamma\mathbf{e}_z)\bar{\bar{\bar{\mathrm{e}}}}. \tag{13.2.21}$$

Now, let us define $\boldsymbol{\xi} = \xi\mathbf{e}_x + \eta\mathbf{e}_y + \gamma\mathbf{e}_z$. This allows us to write

$$\bar{\bar{\bar{\mathrm{e}}}} = \iiint_{-\infty}^{\infty} \left(\frac{\partial u}{\partial x} + \frac{\partial v}{\partial y} + \frac{\partial w}{\partial z}\right)e^{-i(\xi x + \eta y + \gamma z)}\, dxdydz, \tag{13.2.22}$$

or simply,

$$\bar{\bar{\bar{\mathrm{e}}}} = i\xi\bar{\bar{\bar{u}}} + i\eta\bar{\bar{\bar{v}}} + i\gamma\bar{\bar{\bar{w}}} = i(\boldsymbol{\xi} \cdot \bar{\bar{\bar{\mathbf{u}}}}). \tag{13.2.23}$$

This finally implies that

$$F\{\nabla e\} = -\boldsymbol{\xi}(\boldsymbol{\xi}\cdot\bar{\bar{\bar{\mathbf{u}}}}). \tag{13.2.24}$$

Now, consider the first term in Eq. (13.2.16). Performing the Fourier transform yields

$$F\{\nabla^2\mathbf{u}\} = \iiint\limits_{-\infty}^{\infty}\left(\frac{\partial^2\mathbf{u}}{\partial x^2}+\frac{\partial^2\mathbf{u}}{\partial y^2}+\frac{\partial^2\mathbf{u}}{\partial z^2}\right)e^{-i(\xi x+\eta y+\gamma z)}\,dxdydz, \tag{13.2.25}$$

or simply

$$F\{\nabla^2\mathbf{u}\} = -(\xi^2+\eta^2+\gamma^2)\bar{\bar{\bar{\mathbf{u}}}} = -(\boldsymbol{\xi}\cdot\boldsymbol{\xi})\bar{\bar{\bar{\mathbf{u}}}}. \tag{13.2.26}$$

It is also important to note that

$$\iiint\limits_{-\infty}^{\infty}\delta(x)\delta(y)\delta(z)e^{-i(\xi x+\eta y+\gamma z)}\,dxdydz = 1. \tag{13.2.27}$$

Finally, the Fourier transform of the equations of equilibrium (Eq. (13.2.16)) becomes

$$-(\boldsymbol{\xi}\cdot\boldsymbol{\xi})\bar{\bar{\bar{\mathbf{u}}}} + \left(\frac{-1}{1-2\nu}\right)(\boldsymbol{\xi}\cdot\bar{\bar{\bar{\mathbf{u}}}})\boldsymbol{\xi} + \frac{P}{\mu}\mathbf{e}_z = 0. \tag{13.2.28}$$

If we form the scalar product of Eq. (13.2.28) with $\boldsymbol{\xi}$, we see that

$$(\bar{\bar{\bar{\mathbf{u}}}}\cdot\boldsymbol{\xi}) = \frac{1-2\nu}{(2-2\nu)}\frac{P\gamma}{\mu(\boldsymbol{\xi}\cdot\boldsymbol{\xi})}. \tag{13.2.29}$$

Substituting Eq. (13.2.29) into Eq. (13.2.28) yields

$$\bar{\bar{\bar{\mathbf{u}}}} = -\frac{1}{(2-2\nu)}\frac{P\gamma}{\mu(\boldsymbol{\xi}\cdot\boldsymbol{\xi})^2}\boldsymbol{\xi} + \frac{P}{\mu(\boldsymbol{\xi}\cdot\boldsymbol{\xi})}\mathbf{e}_z, \tag{13.2.30}$$

or, in scalar form,

$$\bar{\bar{\bar{u}}} = -\frac{1}{(2-2\nu)}\frac{P}{\mu}\frac{\gamma\xi}{(\xi^2+\eta^2+\gamma^2)^2} = \frac{1}{(2-2\nu)}\frac{P}{\mu}\frac{(i\gamma)(i\xi)}{(\xi^2+\eta^2+\gamma^2)^2} \tag{13.2.31a}$$

$$\bar{\bar{\bar{v}}} = -\frac{1}{(2-2\nu)}\frac{P}{\mu}\frac{\gamma\eta}{(\xi^2+\eta^2+\gamma^2)^2} = \frac{1}{(2-2\nu)}\frac{P}{\mu}\frac{(i\gamma)(i\eta)}{(\xi^2+\eta^2+\gamma^2)^2} \tag{13.2.31b}$$

$$\bar{\bar{\bar{w}}} = \left[\frac{1}{(2-2\nu)}\frac{(i\gamma)^2}{(\xi^2+\eta^2+\gamma^2)^2} + \frac{1}{(\xi^2+\eta^2+\gamma^2)^2}\right]\frac{P}{\mu}. \tag{13.2.31c}$$

Now, in order to find the displacement we have to obtain the inverse Fourier transform of Equations (13.2.31) for the inverse transforms, and doing so we find that

$$\mathbf{u}(x,y,z) = \frac{1}{(2\pi)^3}\iiint\limits_{-\infty}^{\infty}\bar{\bar{\bar{\mathbf{u}}}}(\xi,\eta,\gamma)e^{i(\xi x+\eta y+\gamma z)}\,d\xi d\eta d\gamma. \tag{13.2.32}$$

Thus, taking the inverse Fourier transform of Eq. (13.2.31) yields

$$u = \frac{P}{16\mu(1-\nu)\pi^3}\frac{\partial^2}{\partial x\partial z}\iiint_{-\infty}^{\infty}\frac{e^{i(\boldsymbol{\xi}\cdot\mathbf{r})}}{(\boldsymbol{\xi}\cdot\boldsymbol{\xi})^2}d\xi d\eta d\gamma, \tag{13.2.33a}$$

$$v = \frac{P}{16\mu(1-\nu)\pi^3}\frac{\partial^2}{\partial y\partial z}\iiint_{-\infty}^{\infty}\frac{e^{i(\boldsymbol{\xi}\cdot\mathbf{r})}}{(\boldsymbol{\xi}\cdot\boldsymbol{\xi})^2}d\xi d\eta d\gamma, \tag{13.2.33b}$$

$$w = \frac{P}{16\mu(1-\nu)\pi^3}\frac{\partial^2}{\partial z^2}\iiint_{-\infty}^{\infty}\frac{e^{i(\boldsymbol{\xi}\cdot\mathbf{r})}}{(\boldsymbol{\xi}\cdot\boldsymbol{\xi})^2}d\xi d\eta d\gamma + \frac{P}{8\mu\pi^3}\iiint_{-\infty}^{\infty}\frac{e^{i(\boldsymbol{\xi}\cdot\mathbf{r})}}{(\boldsymbol{\xi}\cdot\boldsymbol{\xi})}d\xi d\eta d\gamma \tag{13.2.33c}$$

We must therefore evaluate

$$I_1 = \frac{1}{8\pi^3}\iiint_{-\infty}^{\infty}\frac{e^{i(\boldsymbol{\xi}\cdot\mathbf{r})}}{(\boldsymbol{\xi}\cdot\boldsymbol{\xi})}d\xi d\eta d\gamma, \quad I_2 = \frac{1}{8\pi^3}\iiint_{-\infty}^{\infty}\frac{e^{i(\boldsymbol{\xi}\cdot\mathbf{r})}}{(\boldsymbol{\xi}\cdot\boldsymbol{\xi})^2}d\xi d\eta d\gamma. \tag{13.2.34}$$

To do this, let us shift to spherical polar coordinates in ξ, η, γ as depicted in Fig. 13.4. Noting that

$$\boldsymbol{\xi}\cdot\mathbf{r} = |\xi| r\cos\beta, \quad |\xi| = \sqrt{\xi^2+\eta^2+\gamma^2}, \quad \boldsymbol{\xi}\cdot\boldsymbol{\xi} = |\xi|^2, \tag{13.2.35}$$

the infinitesimal volume element in spherical-polar coordinates can be expressed as

$$d\xi d\eta d\gamma = |\xi|^2 \sin\beta d\beta d\xi d\theta. \tag{13.2.36}$$

Thus, I_1 can be computed as

$$I_1 = \frac{1}{8\pi^3}\int_0^{2\pi}\int_0^{\infty}\int_0^{\pi} e^{ir|\xi|\cos\beta}\sin\beta d\beta d\xi d\theta. \tag{13.2.37}$$

Noticing that

$$\int_0^{\pi} e^{i|\xi|r\cos\beta}\sin\beta d\beta = -\frac{1}{i|\xi|r}e^{i|\xi|r\cos\beta}\Big|_0^{\pi} = \frac{2}{|\xi|r}\sin|\xi|r, \tag{13.2.38}$$

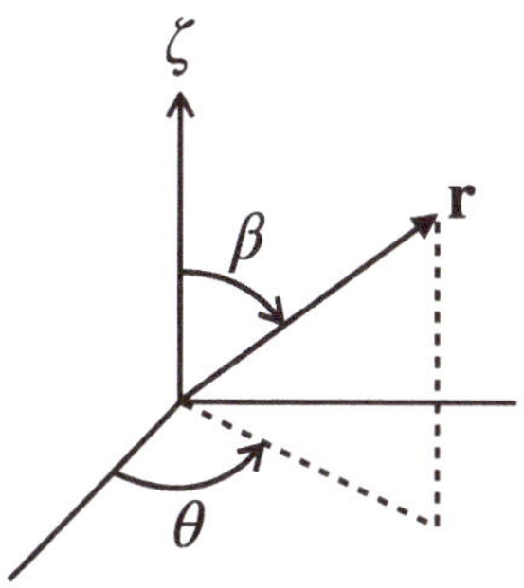

Figure 13.4: Depiction of spherical coordinate frame.

and letting $\lambda = |\xi| r$, the integral I_1 becomes

$$I_1 = \frac{1}{2\pi^2 r}\int_0^\infty \frac{\sin\lambda}{\lambda}\,d\lambda = \frac{1}{4\pi r}. \tag{13.2.39}$$

For the second integral, I_2, we have

$$I_2 = \frac{1}{8\pi^3}\int_0^{2\pi}\int_0^\infty\int_0^\pi \frac{e^{i(\boldsymbol{\xi}\cdot\mathbf{r})}}{|\xi|^2}\sin\beta\, d\beta\, d\xi\, d\theta \tag{13.2.40a}$$

$$= \frac{1}{2\pi^2 r}\int_0^\infty \frac{\sin|\xi| r}{|\xi|^3}\,d\xi \tag{13.2.40b}$$

$$\equiv \frac{1}{2\pi^2 r}\lim_{\epsilon\to 0}\int_0^\infty \frac{\sin|\xi| r}{|\xi|(|\xi|^2+\epsilon^2)}\,d\xi \tag{13.2.40c}$$

$$= \frac{1}{4\pi r}\lim_{\epsilon\to 0}\frac{1}{\epsilon^2}\left(1-e^{-\epsilon r}\right) \tag{13.2.40d}$$

$$= \frac{1}{4\pi}\lim_{\epsilon\to 0}\left(\frac{1}{\epsilon}-\frac{r}{2}\right). \tag{13.2.40e}$$

Now, the term $1/\epsilon$ goes to (an infinite) constant as ϵ approaches zero but the displacements involve derivatives of I_2 and hence for the purpose of generating displacements, we will take $I_2 \equiv -r/8\pi$, thus yielding

$$u = \frac{Cxz}{r^3}, \quad v = \frac{Cyz}{r^3}, \quad w = -\frac{C(x^2+y^2)}{r^3} + \frac{P}{4\pi\mu r}, \tag{13.2.41}$$

where $C = P/[16\pi\mu(1-\nu)]$. Note that Eq. (13.2.41) only depends on $R = \sqrt{x^2+y^2}$ and z. So, switching to cylindrical coordinates yields (Fig. 13.5)

$$u = C\frac{Rz\cos\theta}{(R^2+z^2)^{3/2}}, \tag{13.2.42a}$$

$$v = C\frac{Rz\sin\theta}{(R^2+z^2)^{3/2}}, \tag{13.2.42b}$$

$$w = -C\frac{R^2}{(R^2+z^2)^{3/2}} + \frac{P}{4\pi\mu(R^2+z^2)^{1/2}}. \tag{13.2.42c}$$

It may seem as if the displacement is not axi-symmetric due to the dependence on θ. However, these are the Cartesian displacements. So, if we substitute Eq. (13.2.42) into

$$u_R = u\cos\theta + v\sin\theta, \quad u_\theta = -u\sin\theta + v\cos\theta, \tag{13.2.43}$$

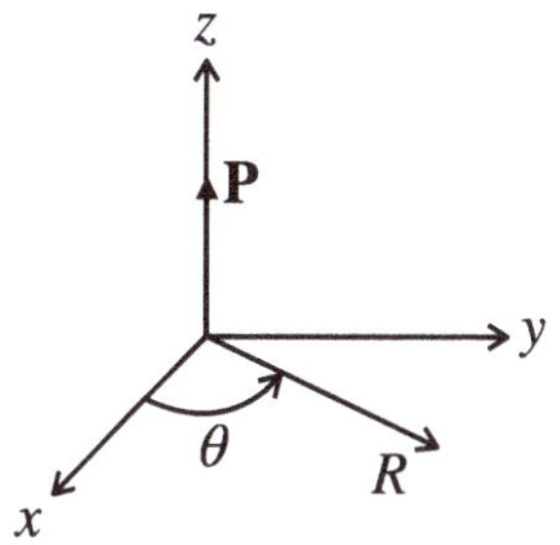

Figure 13.5: Force P directed along z in cylindrical coordinates.

we see that

$$u_R = \frac{CRz}{(R^2 + z^2)^{3/2}}, \qquad u_\theta = 0, \tag{13.2.44}$$

and w is the same as in Eq. (13.2.42). Therefore, the problem is axi-symmetric. The linearized strains in this case (in cylindrical coordinates) are

$$\varepsilon_{RR} = \frac{\partial u_R}{\partial R} = C\frac{z(z^2 - 2R^2)}{(R^2 + z^2)^{5/2}}, \tag{13.2.45a}$$

$$\varepsilon_{\theta\theta} = \frac{1}{R}\frac{\partial u_\theta}{\partial \theta} + \frac{u_R}{R} = \frac{Cz}{(R^2 + z^2)^{3/2}}, \tag{13.2.45b}$$

$$\varepsilon_{zz} = \frac{\partial w}{\partial z} = 3C\frac{R^2 z}{(R^2 + z^2)^{5/2}} - \frac{Pz}{4\pi G(R^2 + z^2)^{3/2}}, \tag{13.2.45c}$$

$$\varepsilon_{\theta z} = \varepsilon_{\theta R} = 0, \tag{13.2.45d}$$

$$\varepsilon_{Rz} = \frac{1}{2}\left(\frac{\partial w}{\partial R} + \frac{\partial u_R}{\partial z}\right) = \frac{3}{2}C\frac{R^3}{(R^2 + z^2)^{5/2}} - C\frac{R}{(R^2 + z^2)^{3/2}}. \tag{13.2.45e}$$

Example 13.2.1. Show that the stresses for the case of a concentrated force become

$$\sigma_{RR} = 2Cz\left[(1 - 2\nu)\frac{1}{r^3} - 3\frac{R^2}{r^5}\right] \tag{13.2.46a}$$

$$\sigma_{\theta\theta} = 2C\mu(1 - 2\nu)\frac{z}{r^3}, \tag{13.2.46b}$$

$$\sigma_{zz} = -2\mu Cz\left[(1 - 2\nu)\frac{1}{r^3} + \frac{3z^2}{r^5}\right], \tag{13.2.46c}$$

$$\sigma_{Rz} = 2\mu CR\left[(1 - 2\nu)\frac{1}{r^3} + \frac{3z^2}{r^5}\right]. \tag{13.2.46d}$$

13.3 Doublet

Let $S(x,y,z)$ be the Kelvin solution for a unit force along z as shown in the Fig. 13.6. The previous solution now leads to, on expressing the solution to the doublet as $PS^z(x,y,z) - PS^z(x,y,z+\epsilon)$, and furthermore re-writing the same as

$$\frac{\epsilon P[S^z(x,y,z) - S^z(x,y,z+\epsilon)]}{\epsilon}, \tag{13.3.1}$$

and taking the limit as $\epsilon \to 0$, the solution for the doublet to be

$$-Q\frac{\partial S(x,y,z)}{\partial z}, \tag{13.3.2}$$

where the limit of ϵP tends to some value Q (recall, we have a delta function associated with the concentrated force). The superscript z on S denotes the concentrated force is acting along the z-direction. The above development is far from rigorous. As mentioned earlier the Dirac delta is not a function and it is a measure. The result obtained can be established rigorously.

The displacements corresponding to the doublet along z is then given by

$$u = -\frac{Qx}{16\pi\mu(1-\nu)}\left[\frac{1}{r^3} - \frac{3z^2}{r^5}\right] \tag{13.3.3}$$

$$v = -\frac{Qy}{16\pi\mu(1-\nu)}\left[\frac{1}{r^3} - \frac{3z^2}{r^5}\right] \tag{13.3.4}$$

$$w = -\frac{3QR^2z}{16\pi\mu(1-\nu)r^5} + \frac{Qz}{4\pi G r^3} \tag{13.3.5}$$

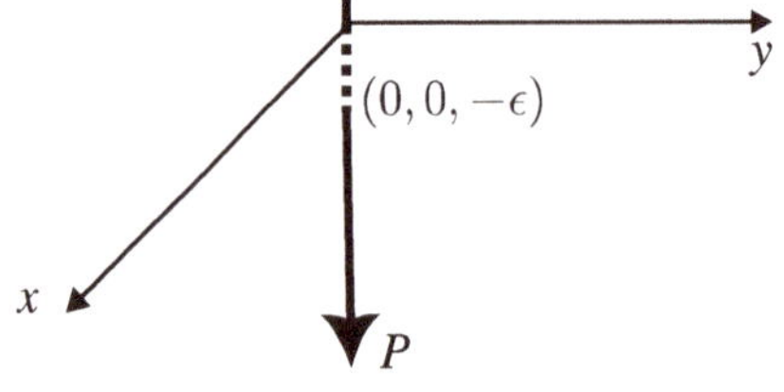

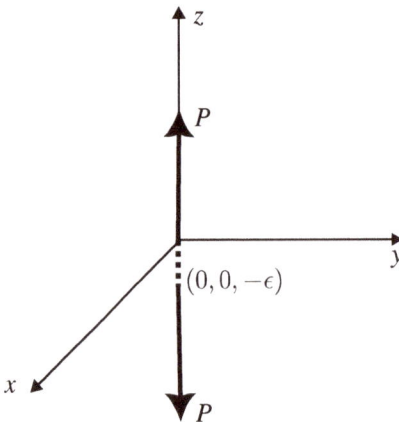

Figure 13.6: Concentrated forces acting along the z-axis.

13.4 Couplets

Let us consider a couple about x due to forces acting along the z-axis as shown in the Figure 13.7 (assuming right hand rule for positive couples). The Couplet about the x-axis has the solution $P[S^z(x,y,z) - S^z(x,y+\epsilon,z)]$, and expressing this as $\frac{\epsilon P[S^z(x,y,z)-S^z(x,y+\epsilon,z)]}{\epsilon}$, $\epsilon \to 0$, defining the product ϵP as $\epsilon \to 0$ as M, we arrive at the solution for a couplet to be $-M\frac{\partial S^z}{\partial y}$, where the superscript z expresses the fact that the Couplet about the x-axis.

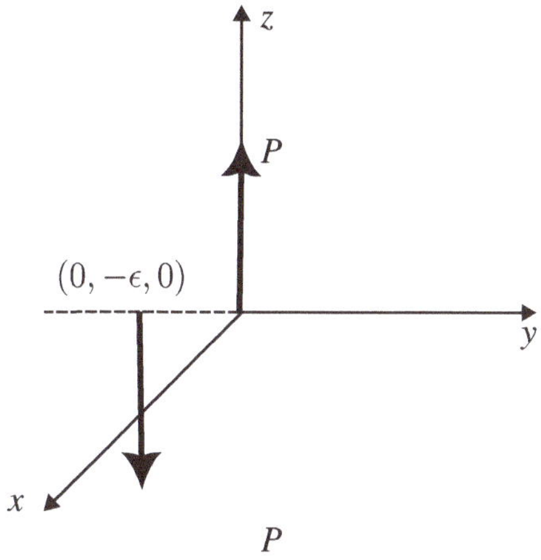

Figure 13.7: Couplet due to forces acting at a distance ϵ apart along the y-axis, directed along the z-axis.

13.4.1 Alternate arrangement

If we let $S^y(x,y,z)$ be the Kelvin solution for the unit load along the y-axis as shown in the Figure 13.8, then the alternate solution is

$$\frac{\epsilon P[S^{(y)}(x,y,z) - S^{(y)}(x,y,z-\epsilon)]}{\epsilon}, \tag{13.4.1}$$

which in the limit of $\epsilon \to 0$ leads to

$$= M\frac{\partial S^{(y)}}{\partial z} \tag{13.4.2}$$

13.5 Center of dilatation

An arrangement of three mutually orthogonal doublets of like strength, Q, is known as a centre of dilatation. It is an approximation for a small spherical cavity subjected to uniform constant pressure as shown in the Fig. 13.9. It becomes an exact solution as the

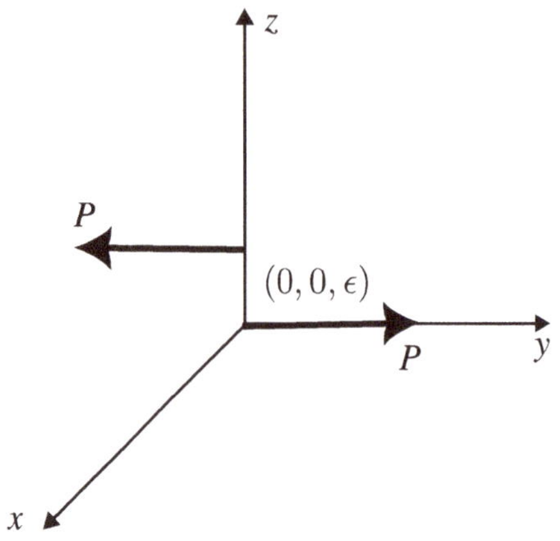

Figure 13.8: Couplet due to forces acting at a distance ϵ apart along the z-axis, directed along the y-axis.

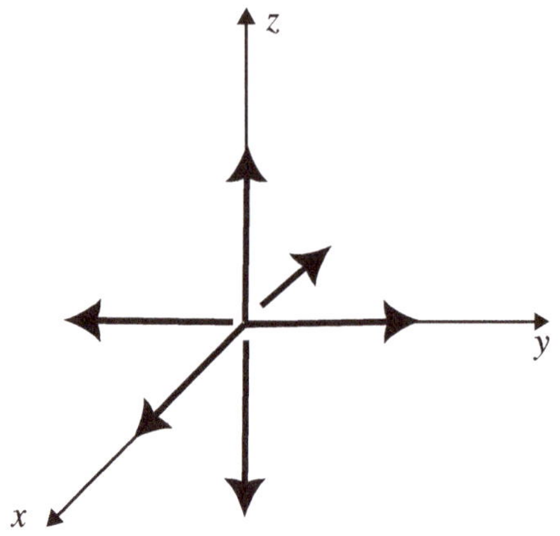

Figure 13.9: Center of dilation.

cavity shrinks to a point and the pressure increases to infinity such that the product pressure and cavity volume remains fixed.

The determination of the solution for a center of dilation is given as a problem below.

13.6 Problems

13.1 Determine rigorously whether or not

$$\frac{\partial S^{(z)}}{\partial y} = -\frac{\partial S^{(y)}}{\partial z} \tag{13.6.1}$$

13.2 Show that for a center of dilatation $u = A\frac{x}{r^3}$, $v = A\frac{y}{r^3}$, $w = A\frac{z}{r^3}$

$$A = \frac{(1-2\nu)}{(2-2\nu)}\frac{Q}{4\pi\mu} \tag{13.6.2}$$

By combining the results for the three orthogonal doublets.

References

[1] J. Boussinesq. Application des potentiels à l'étude de l'équilibre et du mouvement des solides élastiques: principalement au calcul des deformations et des pressions que produisent, dans ces solides, des efforts quelconques exercés sur und petite partie de leur surface ou de leur intérieur; memoire suivi de notes étendues sur divers points de physique mathématique et d'analyse. Gauthier-Villars, 1885.

[2] R. A. Eubanks and E. Sternberg. On the completeness of the Boussinesq-Papkovich stress functions. Journal of Rational Mechanics and Analysis, 5(5):735–746, 1956.

[3] M. E. Gurtin. The Linear Theory of Elasticity. In: Truesdell, C., Ed., Handbuch der Physik, Vol. VIa/2, Springer-Verlag, Berlin, 1–296.

[4] W. Thomson (Lord Kelvin). Displacement due to a point load in an indefinitely extended solid. Mathematical and Physical papers (London), 1, 1848.

[5] W. Thomson (Lord Kelvin). Note on the Integration of the Equations of Equilibrium of an Elastic Solid. Cambridge & Dublin Mathematical Journal, 1848.

[6] H. v Neuber. Ein neuer ansatz zur lösung räumlicher probleme der elastizitätstheorie. der hohlkegel unter einzellast als beispiel. ZAMM-Journal of Applied Mathematics and Mechanics/Zeitschrift für Angewandte Mathematik und Mechanik, 14(4):203–212, 1934.

[7] P. F. Papkovich. Solution generale des equations differentielles fondamentales d'elasticite, experimee par trois fonctions harmoniques. CR Acad Sci Paris, 195:513–515, 1932.

[8] K. R. Rajagopal. Particle-free bodies and point-free spaces. International Journal of Engineering Science, 72:155–176, 2013.

14 Complex variable methods

The material discussed in this chapter, unlike a chapter in a text book is very much in the spirit of lecture notes in that the student is expected, even more than in the other chapters, to consult the books [3, 4] and carry out a great deal of computation to fill in all the details of the analysis. A lot of information concerning complex analysis, as well as the lengthy algebra that is necessary to flesh out the details, is missing. Analysis using complex variables is particularly well-suited for studying two dimensional problems in elasticity. While they were very popular in the past, in view of their utility being limited to two dimensional problems and with the advent of powerful computational methods, their use has diminished considerably. However, it is instructive to add this arsenal to the repertoire of the student, both in view of its elegance and its usability in being a check against computational results, and hence we will discuss the use of complex analysis to problems in elasticity, albeit briefly. Since plane stress and plane strain problems lead to two dimensional problems, such problems are ideal to illustrate the usefulness of the method. The student is expected to have a nodding acquaintance with complex analysis (see [1] and [2] for an introduction to the subject matter; for a detailed treatment of the use of complex variables to analyze two dimensional problems in elasticity look at [4] and [3]).

14.1 Muskhelishvili's technique

Recall that plane stress and plane strain problems, on the introduction of the Airy's stress function, in the absence of body forces, are governed by the two dimensional biharmonic equation,

$$\nabla^4 \phi = 0, \tag{14.1.1}$$

namely

$$\frac{\partial^4 \phi}{\partial x^4} + 2\frac{\partial^4 \phi}{\partial x^2 \partial y^2} + \frac{\partial^4 \phi}{\partial y^4} = 0. \tag{14.1.2}$$

We now proceed to show that the biharmonic function of two variables can be expressed in terms of two functions of a complex variable (see [4] for details of the same). Now, a complex number z and its conjugate $\bar{z}$ can be expressed as

$$z = x + iy, \quad \bar{z} = x - iy, \tag{14.1.3}$$

where the overbar will denote the complex conjugate, x is the real part and y is the complex part of the complex number, with i being defined through $i = \sqrt{-1}$. We will also note that in terms of the complex variable z and its conjugate, (14.1.2) reduces to

$$16\frac{\partial^4 \phi}{\partial z^2 \partial \bar{z}^2} = 0. \tag{14.1.4}$$

https://doi.org/10.1515/9783110789515-014

Let us integrate (14.1.4) with respect to z and let us call the integrated quantity as shown below:

$$\frac{\partial^3 \phi}{\partial z \partial \bar{z}^2} = \overline{f_1''(z)}. \tag{14.1.5}$$

On repeatedly integrating the above thrice we obtain ϕ

$$\frac{\partial^2 \phi}{\partial \bar{z}^2} = z\overline{f_1''(z)} + \overline{f_2''(z)} \tag{14.1.6}$$

$$\frac{\partial \phi}{\partial \bar{z}} = z\overline{f_1'(z)} + \overline{f_2'(z)} + f_3(z) \tag{14.1.7}$$

$$\phi = z\overline{f_1(z)} + \overline{f_2(z)} + \bar{z}f_3(z) + f_4(z) \tag{14.1.8}$$

Thus the conjugate of ϕ is given by

$$\bar{\phi} = \bar{z}f_1(z) + f_2(z) + z\overline{f_3(z)} + \overline{f_4(z)}, \tag{14.1.9}$$

where $f_i(z), i = 1, 2, 3, 4$ are arbitrary complex functions.

Now, we require that

$$\phi - \bar{\phi} = 0, \tag{14.1.10}$$

that is, we require ϕ to be real. Thus,

$$\phi - \bar{\phi} = z[\overline{f_1(z)} - \overline{f_3(z)}] + \bar{z}[f_3(z) - f_1(z)] + [\overline{f_2(z)} - \overline{f_4(z)}] + [f_4(z) - f_2(z)] = 0. \tag{14.1.11}$$

The stress function is expressed without loss of generality as

$$\phi = z\overline{f_1(z)} + \bar{z}f_1(z) + f_2(z) + \overline{f_2(z)} = \mathrm{Re}[\bar{z}f_1(z) + f_2(z)]. \tag{14.1.12}$$

Now, let $G(z)$ be a function of a complex variable. It can be expressed as

$$G(z) = p(x, y) + iq(x, y) \tag{14.1.13}$$

$$G(\bar{z}) = p(x, -y) + iq(x, -y) \tag{14.1.14}$$

$$\bar{G}(z) = p(x, -y) - iq(x, -y), \tag{14.1.15}$$

where we have changed y to $-y$ and i to $-i$. Thus, we have

$$\bar{G}(\bar{z}) \equiv \overline{G(z)} = p(x, y) - iq(x, y), \tag{14.1.16}$$

where z is changed to $\bar{z}$ and i to $-i$. Note, $\overline{G(z)}$ is the complex conjugate of $G(z)$.

We note that if $f(z)$ is a regular function, then $f(z)$ depends only on z (i. e., $\partial f / \partial \bar{z} = 0$). Thus, the stress function can be expressed as

$$\phi = \mathrm{Re}[\bar{z}f(z) + \tilde{g}(z)] - 4\frac{\partial^2\phi}{\partial z\partial\bar{z}} = 4\mathrm{Re}[f'(z)], \tag{14.1.17}$$

and thus

$$T_{xx} + T_{yy} = 4\mathrm{Re}[f'(z)]. \tag{14.1.18}$$

The above implies that, using the stress function,

$$T_{yy} - T_{xx} + 2iT_{xy} = \phi_{,xx} - \phi_{,yy} - 2i\phi_{,xy}, \tag{14.1.19}$$

Eq. (14.1.19) can be written as

$$T_{yy} - T_{xx} + 2iT_{xy} = 4\phi_{,zz}(z,\bar{z}). \tag{14.1.20}$$

Therefore, from the above relations, we arrive at

$$T_{yy} - T_{xx} = 2\mathrm{Re}[\bar{z}f''(z) + \tilde{g}''(z)], \tag{14.1.21a}$$

$$T_{yy} + T_{xx} = 4\mathrm{Re}[f'(z)], \tag{14.1.21b}$$

$$T_{xy} = \mathrm{Im}[\bar{z}f''(z) + \tilde{g}''(z)], \tag{14.1.21c}$$

where

$$\phi = \mathrm{Re}[\bar{z}f(z) + \tilde{g}(z)]. \tag{14.1.22}$$

Now we seek to determine the strains. The strains for plane stress are

$$\varepsilon_{xx} = \frac{1}{2\mu}\left[T_{xx} - \frac{\nu}{1+\nu}(T_{xx} + T_{yy})\right], \tag{14.1.23a}$$

$$\varepsilon_{yy} = \frac{1}{2\mu}\left[T_{yy} - \frac{\nu}{1+\nu}(T_{xx} + T_{yy})\right], \tag{14.1.23b}$$

and in the case of plane strain they are given by

$$\varepsilon_{xx} = \frac{1}{2\mu}[T_{xx} - \nu(T_{xx} + T_{yy})], \tag{14.1.24a}$$

$$\varepsilon_{yy} = \frac{1}{2\mu}[T_{yy} - \nu(T_{xx} + T_{yy})]. \tag{14.1.24b}$$

In terms of the displacements we can express (14.1.23) as

$$\frac{\partial v}{\partial y} - \frac{\partial u}{\partial x} = \frac{1}{2\mu}(T_{yy} - T_{xx}) \tag{14.1.25}$$

$$\frac{\partial v}{\partial x} + \frac{\partial u}{\partial y} = \frac{1}{\mu}(T_{xy}), \tag{14.1.26}$$

and by multiplying (14.1.26) by i and adding it to (14.1.25) we arrive at

$$\left(\frac{\partial}{\partial y} + i\frac{\partial}{\partial x}\right)v - \left(\frac{\partial}{\partial x} - i\frac{\partial}{\partial y}\right)u = \frac{1}{2\mu}(T_{yy} - T_{xx} + 2iT_{xy}). \tag{14.1.27}$$

Therefore, in virtue of Eq. (14.1.3) and Eq. (14.1.21), Eq. (14.1.27) can be expressed as

$$-2u'(z) + 2iv'(z) = \frac{1}{\mu}[\bar{z}f''(z) + \tilde{g}''(z)], \tag{14.1.28}$$

whereby integration yields

$$u - iv = -\frac{1}{2\mu}[\bar{z}f'(z) + \tilde{g}'(z) + h(\bar{z})], \text{where } h(\bar{z}) \text{ is an arbitrary function.} \tag{14.1.29}$$

Now, by virtue of Eq. (14.1.24), the sum of the strains is

$$\varepsilon_{xx} + \varepsilon_{yy} = \begin{cases} \frac{1}{2\mu}(1-2\nu)(T_{xx} + T_{yy}) & \text{in the case of plane strain} \\ \frac{1}{2\mu}(\frac{1-\nu}{1+\nu})(T_{xx} + T_{yy}) & \text{in the case of plane stress} \\ \frac{1}{2\mu}\eta(T_{xx} + T_{yy}) & \text{in both the cases but } \eta \text{ standing} \\ & \text{for different constants.} \end{cases} \tag{14.1.30}$$

From the above relations and the definition of strain, the derivative of the displacement u with respect x can be expressed as

$$\frac{\partial u}{\partial x} = \frac{1}{2\mu}\text{Re}[\bar{z}f''(z) + \tilde{g}''(z) + f'(z) + h'(\bar{z})], \tag{14.1.31}$$

and likewise for the derivative of v with respect to y we have

$$\frac{\partial v}{\partial y} = \frac{1}{2\mu}\frac{\partial}{\partial y}\text{Im}[\bar{z}f'(z) + \tilde{g}'(z) + h(\bar{z})], \tag{14.1.32a}$$

$$= \frac{i}{2\mu}\text{Im}[\bar{z}f''(z) + \tilde{g}''(z) - f'(z) - h'(\bar{z})], \tag{14.1.32b}$$

$$= \frac{1}{2\mu}\text{Re}[\bar{z}f''(z) + \tilde{g}''(z) - f'(z) - h'(\bar{z})]. \tag{14.1.32c}$$

By summing Eq. (14.1.31) and Eq. (14.1.32) together, we see that

$$\frac{\partial u}{\partial x} + \frac{\partial v}{\partial y} = -\frac{1}{\mu}\text{Re}[f'(z) + h'(\bar{z})] = \frac{2}{\mu}\eta\text{Re}[f'(z)], \tag{14.1.33}$$

implying that $h(\bar{z})$ is given by

$$h(\bar{z}) = -(2\eta + 1)f(z). \tag{14.1.34}$$

Thus, by revisiting Eq. (14.1.29) and substituting for $h(\bar{z})$, we see that

$$u - iv = \frac{1}{2\mu}\left[K\overline{f(z)} - \bar{z}f'(z) - \tilde{g}'(z)\right], \tag{14.1.35}$$

where

$$K = \begin{cases} 3 - 4\nu & \text{for Plane strain} \\ \frac{3-\nu}{1+\nu} & \text{for Plane stress.} \end{cases} \tag{14.1.36}$$

Now, without loss of generality we rename $\tilde{g}'(z) = g(z)$ so that

$$T_{yy} + T_{xx} = 4\mathrm{Re}[f'(z)], \tag{14.1.37a}$$

$$T_{yy} - T_{xx} = 2\mathrm{Re}[\bar{z}f''(z) + g'(z)], \tag{14.1.37b}$$

$$T_{xy} = \mathrm{Im}[\bar{z}f''(z) + g'(z)], \tag{14.1.37c}$$

and

$$u - iv = \frac{1}{2\mu}\left[zf(z) - \bar{z}f'(z) - g(z)\right]. \tag{14.1.38}$$

Let us now inspect various relevant boundary conditions. First, consider the traction prescribed on the boundary $\partial\Omega$ where

$$t_x = T_{xx}n_x + T_{xy}n_y = \phi_{,yy}n_x - \phi_{,xy}n_y, \tag{14.1.39a}$$

$$t_y = T_{xy}n_x + T_{yy}n_y = -\phi_{,xy}n_x + \phi_{,xx}n_y \tag{14.1.39b}$$

where n_x and n_y are the usual components of the normal given through

$$n_x = \frac{dy}{dS}, \quad n_y = \frac{dx}{dS}. \tag{14.1.40}$$

Using Eq. (14.1.40), Eq. (14.1.39) can be rewritten as

$$t_x = \phi_{,yy}\frac{dy}{dS} + \phi_{,xy}\frac{dx}{dS} = \frac{d\phi_{,y}}{dS}, \tag{14.1.41a}$$

$$t_y = -\phi_{,xy}\frac{dy}{dS} - \phi_{,xx}\frac{dx}{dS} = -\frac{d\phi_{,x}}{dS}, \tag{14.1.41b}$$

on $\partial\Omega$, where now

$$t_x = t_x(S), \quad t_y = t_y(S). \tag{14.1.42}$$

We can integrate the tractions given in Eq. (14.1.41) to arrive at

$$\phi_{,x} = -\int_S t_y(S')dS' + c_1, \tag{14.1.43a}$$

$$\phi_{,y} = \int_S t_x(S')dS' + c_2 \tag{14.1.43b}$$

which after multiplying (14.1.43b) by i and adding it to (14.1.41) yields

$$\phi_{,x} + i\phi_{,y} = i\int_S [t_x(S') + it_y(S')]\, dS' + c, \tag{14.1.44}$$

or, using the definition of z,

$$2\phi_{,\bar{z}} = i\int_S t_x(S') + it_y(S')\, dS' + c. \tag{14.1.45}$$

We also know, however, that

$$2\phi_{,\bar{z}} = f(z) + z\overline{f'(z)} + \overline{g(z)}, \tag{14.1.46}$$

thus implying that

$$f(z) + z\overline{f'(z)} + \overline{g(z)} = i\int_S (t_x + it_y)\, dS + c. \tag{14.1.47}$$

The state of stress is not altered if we were to replace f by $f + az + \gamma$, and g by $g + \gamma_1$.
Thus,

$$az + \gamma_2 + z\bar{a} + \overline{\gamma_1} = c. \tag{14.1.48}$$

So,

$$z(a + \bar{a}) + \gamma_2 + \overline{\gamma_1} = c \tag{14.1.49}$$

and

$$a + \bar{a} = 0 \implies \gamma_2 + \overline{\gamma_1} = c. \tag{14.1.50}$$

Here, we will note that (see [3])

$$a = ib, \tag{14.1.51}$$

where b is real. Therefore,

$$f'(z) = ib, \quad g'(z) = 0, \quad \mathrm{Re}[f'(z)] = 0, \tag{14.1.52}$$

implying that arbitrary displacement fields are

$$u - iv = \frac{1}{2\mu}\left[K(-ib\bar{z} + \overline{\gamma_2}) - ib\bar{z} - \gamma_1\right], \tag{14.1.53}$$

$$= -\frac{1}{2\mu}\left[(K\overline{\gamma_2} - \gamma_1) - ib(K+1)\bar{z}\right], \tag{14.1.54}$$

where $K\overline{\gamma_2} - \gamma_1$ represents rigid body translation while $b(K+1)\bar{z}$ represents rigid body rotation. Modulo constants

$$u = -b(K+1)y \tag{14.1.55}$$

$$v = b(K+1)x \tag{14.1.56}$$

$$\frac{1}{2}\left(\frac{\partial v}{\partial x} - \frac{\partial u}{\partial y}\right) = b(K+1) \tag{14.1.57}$$

Let us now consider another boundary condition, namely a displacement prescribed on S given by

$$2\mu u(x,y)|_S = g_1(S), \quad 2\mu v(x,y)|_S = g_2(S), \tag{14.1.58}$$

which implies that

$$K\overline{f(z)} - \bar{z}f'(z) - g(z) = g_1 - ig_2. \tag{14.1.59}$$

Therefore

$$K\overline{\gamma_2} - \gamma_1 = 0, \quad b = 0, \quad \gamma_1 = K\overline{\gamma_2} \tag{14.1.60}$$

All stresses and displacements are determined completely even though f and g can contain one unknown constant between them.

We will now consider the problem in polar coordinates. Recall the complex number z can be expressed as,

$$z = re^{i\theta}. \tag{14.1.61}$$

Also, let us recall that

$$T_{rr} + T_{\theta\theta} = T_{xx} + T_{yy} = 4\mathrm{Re}\left[f'(z)\right]. \tag{14.1.62}$$

$$T_{rr} = \frac{1}{2}(T_{xx} + T_{yy}) + \frac{1}{2}(T_{yy} - T_{xx})\cos 2\theta + T_{xy}\sin 2\theta, \tag{14.1.63a}$$

$$T_{\theta\theta} = \frac{1}{2}(T_{xx} + T_{yy}) - \frac{1}{2}(T_{yy} - T_{xx})\cos 2\theta - T_{xy}\sin 2\theta, \tag{14.1.63b}$$

$$T_{r\theta} = -\frac{1}{2}(T_{yy} - T_{xx})\sin 2\theta + T_{xy}\cos 2\theta \tag{14.1.63c}$$

The above relations imply that

$$T_{\theta\theta} - T_{rr} + 2iT_{r\theta} = -(T_{yy} - T_{xx})e^{2i\theta} + 2iT_{xy}e^{2i\theta}. \tag{14.1.64}$$

Therefore,

$$T_{\theta\theta} - T_{rr} + 2iT_{r\theta} = 2e^{2i\theta}\left[\bar{z}f''(z) + g'(z)\right]. \tag{14.1.65}$$

Now, the displacements in polar coordinates are given by

$$u_r = u\cos\theta + v\sin\theta, \tag{14.1.66a}$$

$$u_\theta = -u\sin\theta + v\cos\theta \tag{14.1.66b}$$

whereby yielding the relation

$$u_r - iu_\theta = (u - iv)e^{i\theta} = \frac{e^{i\theta}}{2\mu}\left[K\overline{f(z)} - \bar{z}f'(z) - g(z)\right]. \tag{14.1.67}$$

So, we now have the relations

$$T_{\theta\theta} + T_{rr} = 4\mathrm{Re}[f'(z)], \tag{14.1.68a}$$

$$T_{\theta\theta} - T_{rr} + 2iT_{r\theta} = 2e^{2i\theta}\left[\bar{z}f''(z) + g'(z)\right] \tag{14.1.68b}$$

We will now implement a Fourier representation for f and g. Consider

$$f(z) + z\overline{f'(z)} + \overline{g(z)} = i\int\limits^{\theta}[t_x + it_y]\,d\theta + c, \tag{14.1.69}$$

where $f(z)$ and $g(z)$ are given by

$$f(z) = \sum_{n=0}^{\infty} a_n z^n, \quad g(z) = \sum_{n=0}^{\infty} b_n z^n, \tag{14.1.70}$$

and thus

$$f'(z) = \sum_{n=0}^{\infty} na_n z^{n-1}. \tag{14.1.71}$$

By letting $z = e^{i\theta}$ such that $|z| = 1$, Eq. (14.1.69) and Eq. (14.1.70) yield

$$\sum_{n=0}^{\infty}\left[a_n e^{in\theta} + e^{i\theta}xn\bar{a}_n e^{-i(n-1)\theta} + \bar{b}_n e^{-in\theta}\right] = i\int\limits^{\theta}(t_x + it_y)\,d\theta + c. \tag{14.1.72}$$

Now, let

$$t_1(\theta) = -\int\limits^{\theta} t_y\,d\theta, \quad t_2(\theta) = \int\limits^{\theta} t_x\,d\theta \tag{14.1.73}$$

such that

$$\sum_{n=0}^{\infty} a_n e^{in\theta} + n\bar{a}_n e^{-i(n-2)\theta} + \bar{b}_n e^{-n\theta} = t_1 + it_2 + c \tag{14.1.74}$$

or, more concisely,

$$t_1 + it_2 = \sum_{n=-\infty}^{\infty} A_n e^{in\theta}, \tag{14.1.75}$$

where

$$A_n = \frac{1}{2\pi} \int_0^{2\pi} e^{-in\theta}(t_1 + it_2)\, d\theta. \tag{14.1.76}$$

In terms of A_n, we now have

$$\sum_{n=0}^{\infty} [a_n e^{in\theta} + n\bar{a}_n e^{-i(n-2)\theta} + \overline{b_n} e^{-in\theta}] = \sum_{n=1}^{\infty} A_n e^{in\theta} + A_{(-n)} e^{-in\theta} + A_0 + c, \tag{14.1.77}$$

where the coefficient of e^0, $e^{i\theta}$ and $e^{2i\theta}$ are, respectively,

$$A_0 + c = a_0 + 2\bar{a}_2 + \bar{b}_0, \tag{14.1.78a}$$

$$A_1 = a_1 + \bar{a}_1, \tag{14.1.78b}$$

$$A_n = a_2, \; n \geq 2, \tag{14.1.78c}$$

and

$$(n+2)\bar{a}_{n+2} + \bar{b}_n = A_{(-n)}, \quad n \leq 1. \tag{14.1.79}$$

Thus, for the constants, we have

$$a_0 + \bar{b}_0 - c = A_0 - 2\bar{a}_2, \tag{14.1.80}$$

where $A_0 - 2\bar{a}_2$ is known. Let $a_0 = c$ and $\bar{b}_0 = A_0 - 2\bar{a}_2$. The fact that $a_1 + \bar{a}_1$ is real implies that A_1 should be real.

Therefore, for $f(z)$, we have

$$f(z) = a_0 + i\mathrm{Im}(a_1)z + \mathrm{Re}(a_1)z + \sum_{n=2}^{\infty} a_n z^n \tag{14.1.81}$$

Here, $a_0 + i\mathrm{Im}(a_1)$ is arbitrary and $\mathrm{Re}(a_1)$ is known. Now, consider A_1 which is given by

$$A_1 = \frac{1}{2\pi} \int_0^{2\pi} (t_1 + it_2)e^{-i\theta}\, d\theta = \frac{1}{2\pi} \int_0^{2\pi} [t_1 \cos\theta + t_2 \sin\theta + i(t_2 \cos\theta - t_1 \sin\theta)]\, d\theta, \tag{14.1.82}$$

where

$$\frac{1}{2\pi}\int_0^{2\pi}(t_2\cos\theta - t_1\sin\theta)\,d\theta = \frac{1}{2\pi}(t_2\sin\theta + t_1\cos\theta)|_0^{2\pi} - \frac{1}{2\pi}\int_0^{2\pi} t_2'\sin\theta + t_1'\cos\theta\,d\theta, \tag{14.1.83}$$

$$= \frac{1}{2\pi}[t_1(2\pi) - t_1(0)] + \frac{1}{2\pi}\int_0^{2\pi}[t_y\cos\theta - t_x\sin\theta]\,d\theta, \tag{14.1.84}$$

$$= -\frac{1}{2\pi}\int_0^{2\pi} t_y\,d\theta + \frac{1}{2\pi}\int_0^{2\pi}[t_y\cos\theta - t_x\sin\theta]\,d\theta = 0. \tag{14.1.85}$$

For equilibrium, we have the conditions that

$$\int_0^{2\pi} dM_0 = 0, \quad \int_0^{2\pi} t_y\,d\theta = 0. \tag{14.1.86}$$

Therefore, A_1 is real.

Let us now consider a displacement prescribed on the boundary of the unit circle, where

$$z = e^{i\theta}, \tag{14.1.87}$$

yielding us

$$K\overline{f(z)} - \bar{z}f'(z) - g(z) = g_1(\theta) - ig_2(\theta), \tag{14.1.88}$$

or

$$-Kf(z) + z\overline{f'(z)} + \overline{g(z)} = -(g_1(\theta) + ig_2(\theta)). \tag{14.1.89}$$

Again, let $f(z)$ and $g(z)$ be represented by

$$f(z) = \sum_{n=0}^{\infty} a_n z^n, \quad g(z) = \sum_{n=0}^{\infty} b_n z^n, \tag{14.1.90}$$

such that

$$\sum_{n=0}^{\infty}[Ka_n e^{in\theta} + n\bar{a}_n e^{i(n-2)\theta} + \bar{b}_n e^{-in\theta}] = \sum_{n=1}^{\infty} B_n e^{in\theta} + B_{(-n)}e^{-in\theta} + B_0 \tag{14.1.91}$$

where $-(g_1 + ig_2)$ is given as a summation. The coefficient B_n may be expressed as

$$B_n = -\frac{1}{2\pi}\int_0^{2\pi}(g_1 + ig_2)e^{-in\theta}\,d\theta. \tag{14.1.92}$$

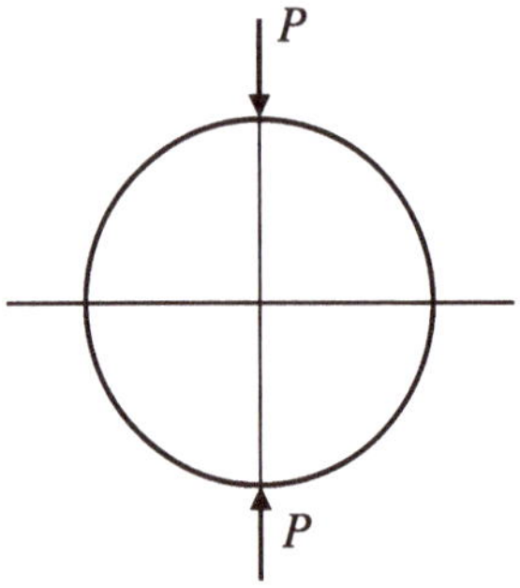

Figure 14.1: Example.

Here, for the coefficients of e^0, $e^{i\theta}$, $e^{in\theta}$, we have, respectively,

$$B_0 = -Ka_0 + 2\bar{a}_2 + \bar{b}_0, \tag{14.1.93a}$$
$$B_1 = -Ka_1 + \bar{a}_1, \tag{14.1.93b}$$
$$B_n = -Ka_n, \quad n \geq 2, \tag{14.1.93c}$$
$$B_n = (n+2)\bar{a}_n + 2 + \bar{b}_n, \tag{14.1.93d}$$

and

$$-Ka_0 + \bar{b}_0 = B_0 - 2\bar{a}_2, \tag{14.1.94}$$

where

$$\bar{b}_0 = B_0 - 2\bar{a}_2 + ua_0, \tag{14.1.95}$$

and a_0 is arbitrary. Note, a_1 is not arbitrary.

Example 14.1.1. Let us consider an example where the traction given by (see Figure 14.1)

$$t_x = 0, \quad t_y = -P\delta\left(\theta - \frac{\pi}{2}\right) + P\delta\left(\theta - \frac{3\pi}{2}\right), \tag{14.1.96}$$

where δ is the dirac measure (generalized function) implying that

$$t_1 = P\int_0^\theta \left[\delta\left(\theta' - \frac{\pi}{2}\right) - \delta\left(\theta' - \frac{3\pi}{2}\right)\right] d\theta' = P\begin{cases} 0 & 0 < \theta < \pi/2 \\ 1 & \pi/2 < \theta < 3\pi/2 \\ 0 & 3\pi/2 < \theta < 2\pi \end{cases}, \tag{14.1.97}$$

and

$$t_2 = 0. \tag{14.1.98}$$

For this problem, the coefficient A_n can be expressed as

$$A_n = \frac{P}{2\pi}\int_{\pi/2}^{3\pi/2} e^{-in\theta}\,d\theta = \frac{iP}{2\pi i}e^{-in\theta}\Big|_{\pi/2}^{3\pi/2}, \quad n \neq 0, \tag{14.1.99}$$

and A_0 becomes

$$A_0 = \frac{P}{2\pi}\int_{\pi/2}^{3\pi/2} d\theta = \frac{P}{2}. \tag{14.1.100}$$

After integration of Eq. (14.1.99), we see that

$$A_n = \frac{iPe^{-in\pi/2}}{2\pi n}\left[(-1)^n - 1\right] = \frac{Pi^{n+1}}{2\pi n}\left[1 - (-1)^n\right], \quad n \neq 0. \tag{14.1.101}$$

Therefore, the final coefficients are

$$A_0 = \frac{P}{2}, \quad A_{2n} = 0, \quad A_{2n+1} = \frac{(-1)^{n+1}P}{(2n+1)\pi}. \tag{14.1.102}$$

The interested student will find a detailed discussion of the material in this chapter in chapter 15 of [4].

References

[1] L. V. Ahlfors. Complex Analysis: An Introduction to the Theory of Analytic Functions of One Complex Variable. McGraw Hill Co., New York, 1976.

[2] R. V. Churchill. Complex Variables With Applications. McGraw-Hill, New York, 2nd edition, 1960.

[3] A. H. England. Complex Variable Methods in Elasticity. Dover Publications, Mineola, New York, 1971.

[4] N. I. Muskhelishvili. Some Basic Problems of the Mathematical Theory of Elasticity. P. Nootfhoff limited, Groningen-Holland, 1953.

15 Approximations

15.1 Elementary beam theory

By a beam we mean a body which is essentially much longer along one direction than the other two directions, and furthermore the deformation to which it is subject to is in a state of plane stress. We shall consider the beam of unit thickness as portrayed in Fig. 15.1. A state of plane stress reduces the equilibrium equations to

$$\frac{\partial T_{xx}}{\partial x} + \frac{\partial T_{xy}}{\partial y} = 0 \tag{15.1.1}$$

$$\frac{\partial T_{yx}}{\partial x} + \frac{\partial T_{yy}}{\partial y} = 0. \tag{15.1.2}$$

Recall the definition of the bending moment and shear force:

$$M(x) = \int_{-h/2}^{h/2} y T_{xx}(x, y)\, \mathrm{d}y, \tag{15.1.3}$$

$$V(x) = \int_{-h/2}^{h/2} T_{xy}(x, y)\, \mathrm{d}y. \tag{15.1.4}$$

Note, we are assuming unit length along the z-direction. Multiply Eq. (15.1.1) by y and integrate between $-h/2$ and $h/2$ (averaging). Then,

$$\int_{-h/2}^{h/2} y \frac{\partial T_{xx}}{\partial x}\, \mathrm{d}y + \int_{-h/2}^{h/2} y \frac{\partial T_{xy}}{\partial y}\, \mathrm{d}y = \frac{\mathrm{d}M}{\mathrm{d}x} + y T_{xy}\Big|_{-h/2}^{h/2} - \int_{-h/2}^{h/2} T_{xy}\, \mathrm{d}y \tag{15.1.5}$$

$$= \frac{\mathrm{d}M}{\mathrm{d}x} - V(x) \quad (\because\ T_{xy}(x, \pm h/2) = 0) \tag{15.1.6}$$

$$= 0. \tag{15.1.7}$$

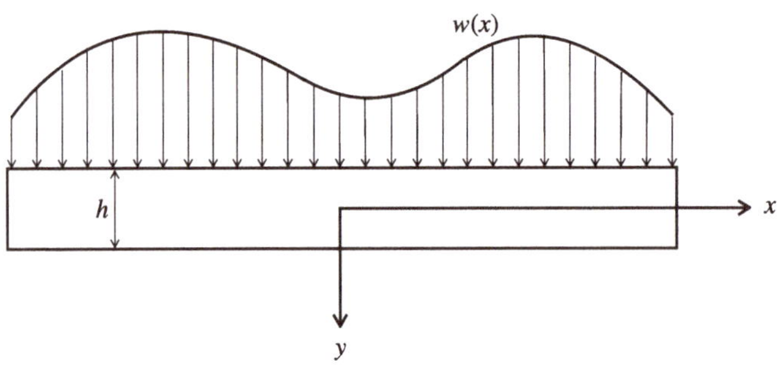

Figure 15.1: Beam carrying a distributed load.

https://doi.org/10.1515/9783110789515-015

Hence,

$$\frac{\mathrm{d}M}{\mathrm{d}x} - V(x) = 0. \tag{15.1.8}$$

Next, let us integrate Eq. (15.1.2) from $-h/2$ to $h/2$ (averaging):

$$\int_{-h/2}^{h/2} \frac{\partial T_{xy}}{\partial x}\,\mathrm{d}y + \int_{-h/2}^{h/2} \frac{\partial T_{yy}}{\partial y}\,\mathrm{d}y = \frac{\mathrm{d}V}{\mathrm{d}x} + T_{yy}(x,y)|_{-h/2}^{h/2} = 0. \tag{15.1.9}$$

Now,

$$T_{yy}\left(x, \frac{h}{2}\right) - T_{yy}\left(x, -\frac{h}{2}\right) = w(x), \tag{15.1.10}$$

where $w(x)$ is the load per unit length. So

$$\frac{\mathrm{d}V}{\mathrm{d}x} - w(x) = 0. \tag{15.1.11}$$

By Hooke's Law, we know that

$$\varepsilon = \frac{1}{E}\big[(1+\nu)\mathbf{T} - \nu(\mathrm{tr}\mathbf{T})\mathbf{I}\big]. \tag{15.1.12}$$

It then follows that

$$T_{xx} = \frac{E}{1-\nu^2}(\varepsilon_{xx} + \nu\varepsilon_{yy}), \tag{15.1.13}$$

$$T_{yy} = \frac{E}{1-\nu^2}(\varepsilon_{yy} + \nu\varepsilon_{xx}), \tag{15.1.14}$$

$$T_{xy} = \frac{E}{2(1+\nu)}\varepsilon_{xy} = \mu\varepsilon_{xy}. \tag{15.1.15}$$

Now we will make assumptions on the displacements. The displacements in the x and y directions, respectively, can be expressed as

$$u(x,y) = u(x,0) + \frac{\partial u}{\partial y}(x,0)y + \text{H.O.T. in } y, \tag{15.1.16}$$

$$v(x,y) = v(x,0) + \frac{\partial v}{\partial y}(x,0)y + \frac{\partial^2 v}{\partial y^2}(x,0)\frac{y^2}{2} + \text{H.O.T. in } y. \tag{15.1.17}$$

We will assume that the neutral axis will have no displacement in the x-direction. Thus, $u(x,0) = 0$. We will also assume that $\frac{\partial v}{\partial y}(x,0) = 0$. So

$$u(x,y) = y\beta(x), \tag{15.1.18}$$

where

$$\beta(x) \equiv \frac{\partial u}{\partial y}(x,0), \tag{15.1.19}$$

and

$$v(x,y) = v_0(x) + v_2(x)\frac{y^2}{2}, \tag{15.1.20}$$

where

$$v_0(x) \equiv v(x,0), \quad v_2(x) \equiv \frac{\partial^2 v}{\partial y^2}(x,0). \tag{15.1.21}$$

Thus,

$$\varepsilon_{xx} = y\beta'(x) \tag{15.1.22}$$

$$\varepsilon_{yy} = yv_2(x) \tag{15.1.23}$$

$$\varepsilon_{xy} = \frac{1}{2}\left[\beta(x) + v_0'(x) + v_2'(x)\frac{y^2}{2}\right] \approx [\beta(x) + v_0(x)], \tag{15.1.24}$$

where it is important to note that there are many ad hoc assumptions. Next, using Eq. (15.1.13), (15.1.22), and (15.1.23), we see that

$$M(x) = \int_{-h/2}^{h/2} yT_{xx}\,\mathrm{d}y = \frac{E}{1-\nu^2}\int_{-h/2}^{h/2} y^2[\beta'(x) + \nu v_2(x)]\,\mathrm{d}y \tag{15.1.25}$$

$$= \frac{EI}{1-\nu^2}[\beta'(x) + \nu v_2(x)], \tag{15.1.26}$$

where I is the moment of inertia of the cross section. Similarly,

$$\int_{-h/2}^{h/2} yT_{yy}\,\mathrm{d}y = \frac{EI}{1-\nu^2}[v_2(x) + \nu\beta'(x)] = 0, \tag{15.1.27}$$

which implies that

$$v_2 = -\nu\beta'. \tag{15.1.28}$$

On substituting Eq. (15.1.28) into Eq. (15.1.26) we find

$$M(x) = \frac{EI}{1-\nu^2}[\beta'(x)(1-\nu^2)] = EI\beta'(x). \tag{15.1.29}$$

Next,

$$\int_{-h/2}^{h/2} T_{xy}\,\mathrm{d}y = V(x) = \mu[\beta(x) + v_0'(x)] \int_{-h/2}^{h/2} \mathrm{d}y. \tag{15.1.30}$$

Hence,

$$V(x) = \mu A[\beta(x) + v_0'(x)], \tag{15.1.31}$$

where A is the cross-sectional area. To improve accuracy, the above is modified as

$$V(x) = kA\mu[\beta(x) + v_0'(x)], \tag{15.1.32}$$

where k is referred to as the Timoshenko shape factor. In summary,

$$M'(x) = V(x), \tag{15.1.33}$$
$$V'(x) = w(x), \tag{15.1.34}$$
$$M(x) = EI\beta'(x), \tag{15.1.35}$$
$$V(x) = kA\mu[\beta(x) + v_0'(x)]. \tag{15.1.36}$$

From the above,

$$\beta(x) = \frac{V}{kA\mu} - v_0'(x). \tag{15.1.37}$$

So,

$$\beta'(x) = \frac{V'}{kA\mu} - v_0''(x) = \frac{M}{EI}. \tag{15.1.38}$$

This implies that

$$\frac{w(x)}{kA\mu} - v_0''(x) = \frac{M}{EI} \implies \frac{w''(x)}{kA\mu} - v_0''''(x) = \frac{w(x)}{EI}. \tag{15.1.39}$$

Thus,

$$EI\frac{\mathrm{d}^4 v_0}{\mathrm{d}x^4} = -w(x) + \frac{EI}{kA\mu}\left(\frac{\mathrm{d}^2 w}{\mathrm{d}x^2}\right). \tag{15.1.40}$$

In Euler-Bernoulli beam theory (EBT), one does not have the term involving $\frac{\mathrm{d}^2 w}{\mathrm{d}x^2}$. This term incorporates the effect of *shear*. Another way to think about it is that in the EBT theory, $\mu = \infty$. In EBT,

$$\beta(x) = -v_0'(x), \tag{15.1.41}$$
$$M(x) = -EIv_0''(x), \tag{15.1.42}$$

$$V(x) = -EIv_0'''(x). \tag{15.1.43}$$

Below is a list of possible boundary conditions for beam theory:
1. Pin or roller: $v_0 = 0$, $M = 0$, $\beta' = 0$.
2. Free end: $M = 0$, $V = 0$, $\beta' = 0$, $\beta + v_0' = 0$.
3. Built-in section: $v_0 = 0$, $\beta = 0$.

One could develop approximations for rods, membranes, plates, shells, etc., but we shall not do so here. There are several books devoted to these subjects.

16 Elastodynamics

The balance of linear momentum in linearized elasticity for a homogeneous isotropic solid takes the form

$$(\lambda + \mu)\nabla_{\mathbf{x}}(\operatorname{div}_{\mathbf{x}}(\mathbf{u})) + \mu\Delta_{\mathbf{x}}\mathbf{u} + \rho\mathbf{b} = \rho\frac{d^2\mathbf{u}}{dt^2} \tag{16.0.1}$$

where $\nabla_{\mathbf{x}}$ and $\Delta_{\mathbf{x}}$ are the gradient and Laplacian with respect to $\mathbf{x}$. Since we are considering small displacement gradients, the above equation reduces to

$$(\lambda + \mu)\nabla_{\mathbf{x}}(\operatorname{div}_{\mathbf{x}}(\mathbf{u})) + \mu\Delta\mathbf{u} + \rho\mathbf{b} = \rho\frac{\partial^2\mathbf{u}}{\partial t^2}, \tag{16.0.2}$$

where now we are considering the partial time derivative of $\mathbf{u}$. The equation (16.0.2) has numerous important applications including those for the non-destructive testing of materials, for the study of earthquakes, vibration of strings, beams, membranes, plates and shells, etc. In fact, there have been so many studies within the context of elastodynamics that it would be impossible to do justice to the field in a short chapter. However, in order to bring the attention of the students to a very important class of problems concerning elastic bodies, we shall merely illustrate by means of few examples the study of wave propagation in elastic media using the above equation (the interested reader can find problems in elastodynamics discussed in detail in Love [4], Eringen and Suhubi [2], Achenbach [1] and Graff [3]). In general, the body is inhomogeneous and the Lame constants λ and μ will depend on $\mathbf{X}$, and as no distinction is made between $\mathbf{X}$ and $\mathbf{x}$ in linearized elasticity, the Lame' constants will depend on $\mathbf{x}$ and the governing equation becomes

$$\nabla_{\mathbf{x}}(\lambda(\mathbf{x})\operatorname{div}_{\mathbf{x}}\mathbf{u}) + \mu\operatorname{div}_{\mathbf{x}}[\mu(\mathbf{x})\{\Delta_{\mathbf{x}}\mathbf{u} + (\Delta_{\mathbf{x}}\mathbf{u})^T\}] + \rho\mathbf{b} = \rho\frac{\partial^2\mathbf{u}}{\partial t^2}. \tag{16.0.3}$$

If in addition, if the body is anisotropic, then one will have to obtain the appropriate equation taking the anisotropy of the body in to consideration. In this case, the equation will be given by

$$\operatorname{div}_{\mathbf{x}}[\mathbb{C}\{\Delta_{\mathbf{x}}\mathbf{u} + (\Delta_{\mathbf{x}}\mathbf{u})^T\}] + \rho\mathbf{b} = \rho\frac{\partial^2\mathbf{u}}{\partial t^2}, \tag{16.0.4}$$

where $\mathbb{C}$ is the elasticity tensor for the anisotropic body.

Let us suppose there are no body forces, in that case, the equation (16.0.1) for a homogeneous isotropic linearized elastic body reduces to

$$(\lambda + \mu)\nabla_{\mathbf{x}}(\operatorname{div}_{\mathbf{x}}(\mathbf{u})) + \mu\Delta_{\mathbf{x}}\mathbf{u} = \rho\frac{\partial^2\mathbf{u}}{\partial t^2}. \tag{16.0.5}$$

Let us further suppose the motion is such that the $\operatorname{div}_{\mathbf{x}}\mathbf{u} = 0$, then

https://doi.org/10.1515/9783110789515-016

$$\mu\Delta_{\mathbf{x}}\mathbf{u} = \rho\frac{\partial^2\mathbf{u}}{\partial t^2}. \tag{16.0.6}$$

Let us define

$$C_1 = \sqrt{\frac{\mu}{\rho}}. \tag{16.0.7}$$

Then, (16.0.6) can be rewritten as

$$C_1^2(\Delta_{\mathbf{x}}\mathbf{u}) = \frac{\partial^2\mathbf{u}}{\partial t^2}, \tag{16.0.8}$$

which is the wave equation, C_1 being the wave speed. The assumption that $\operatorname{div}_{\mathbf{x}} \mathbf{u} = 0$ implies that motion is isochoric. Then C_1 is called the isochoric wave speed.

Next, let us recall the vector identity

$$\Delta_{\mathbf{x}}\mathbf{u} = \nabla_{\mathbf{x}}(\operatorname{div}_{\mathbf{x}} \mathbf{u}) - \operatorname{curl}_{\mathbf{x}} \operatorname{curl}_{\mathbf{x}} \mathbf{u}. \tag{16.0.9}$$

Substituting (16.0.9) into (16.0.5), we obtain

$$(\lambda + \mu)\Delta_{\mathbf{x}}\mathbf{u} - (\lambda + \mu) \operatorname{curl}_{\mathbf{x}} \operatorname{curl}_{\mathbf{x}} \mathbf{u} = \rho\frac{\partial^2\mathbf{u}}{\partial t^2}. \tag{16.0.10}$$

If we now suppose that the motion is irrotational, then

$$(\lambda + 2\mu)\Delta_{\mathbf{x}}\mathbf{u} = \rho\frac{\partial^2\mathbf{u}}{\partial t^2}. \tag{16.0.11}$$

Now, on defining

$$C_2 = \sqrt{\frac{\lambda + 2\mu}{\rho}}. \tag{16.0.12}$$

Equation (16.0.11) reduces to the wave equation

$$C_2^2\Delta_{\mathbf{x}}\mathbf{u} = \frac{\partial^2\mathbf{u}}{\partial t^2}. \tag{16.0.13}$$

C_2 is called the irrotational wave speed.

Next, we notice that the larger the value of the density with other parameters being constant, the slower the wave speed. Also, note that

$$C_2^2 - C_1^2 = \frac{\lambda + \mu}{\rho}. \tag{16.0.14}$$

If ρ and $(\lambda + \mu)$ are positive, then

$$C_2^2 - C_1^2 > 0. \tag{16.0.15}$$

Thus, the irrotational wave speed is greater than the isochoric wave speed if $(\lambda + \mu)$ is positive. Next, let us take divergence of (16.0.11), then

$$\rho \frac{\partial^2}{\partial t^2}(\operatorname{div}_{\mathbf{x}} \mathbf{u}) = (\lambda + 2\mu)\Delta_{\mathbf{x}}(\operatorname{div}_{\mathbf{x}} \mathbf{u}), \tag{16.0.16}$$

that is, we obtain a wave equation for the divergence of the displacement, i. e.,

$$\frac{\partial^2}{\partial t^2}(\operatorname{div}_{\mathbf{x}} \mathbf{u}) = C_2^2 \Delta_{\mathbf{x}}(\operatorname{div}_{\mathbf{x}} \mathbf{u}). \tag{16.0.17}$$

Next, let us define two wave operators

$$\Box_\alpha \mathbf{f} = C_\alpha^2 \Delta_{\mathbf{x}} \mathbf{f} - \frac{\partial^2 \mathbf{f}}{\partial t^2}, \quad \alpha = 1, 2. \tag{16.0.18}$$

Now, if $\operatorname{div}_{\mathbf{x}} \mathbf{u} = 0$, equation (16.0.5) becomes,

$$C_1^2 \nabla_{\mathbf{x}} \mathbf{u} = \frac{\partial^2 \mathbf{u}}{\partial t^2}, \tag{16.0.19}$$

and thus

$$\Box_1 \mathbf{u} = \mathbf{0}. \tag{16.0.20}$$

If $\operatorname{curl}_{\mathbf{x}} \mathbf{u} = 0$, then (16.0.10) reduces to

$$\Box_2 \mathbf{u} = \mathbf{0}. \tag{16.0.21}$$

When $\mathbf{b} = 0$, it follows that

$$\Box_2 \operatorname{div}_{\mathbf{x}} \mathbf{u} = \mathbf{0}, \tag{16.0.22}$$

$$\Box_1 \operatorname{curl}_{\mathbf{x}} \mathbf{u} = \mathbf{0}. \tag{16.0.23}$$

It follows from the relation between λ, μ, Young's modulus and the Poisson's ratio that

$$C_1 = \sqrt{\frac{\mu}{\rho}} = \sqrt{\frac{E}{2\rho(1+\nu)}}, \tag{16.0.24}$$

and

$$C_2 = \sqrt{\frac{\lambda + 2\mu}{\rho}} = \sqrt{\frac{E(1-\nu)}{\rho(1+\nu)(1-2\nu)}}. \tag{16.0.25}$$

Numerous interesting and important elastodynamic problems such as vibration of elastic bodies, surface waves, shocks, etc., have been studied using the above equations.

We conclude this chapter with a discussion of waves in a one-dimensional body. Let us suppose that the body is homogeneous and let us ignore body forces. In this case, the constitutive relation reduces to

$$\epsilon(\mathbf{x}, t) = \frac{T(\mathbf{x}, t)}{E}. \tag{16.0.26}$$

When the equation (16.0.26) is substituted into the balance of linear momentum one obtains

$$\rho \frac{\partial^2 u}{\partial^2 t^2} = E \frac{\partial^2 u}{\partial^2 x}, \tag{16.0.27}$$

and hence

$$\frac{\partial^2 u}{\partial^2 t^2} = C^2 \frac{\partial^2 u}{\partial^2 x}, \tag{16.0.28}$$

where $C = \sqrt{\frac{E}{\rho}}$. The one dimensional wave equation has the general solution, namely the D'Alembert solution, wherein

$$u(x, t) = g_1(x - Ct) + g_2(x + Ct). \tag{16.0.29}$$

An example of such solutions are $g_1(x - Ct) = \sin(x - Ct)$. As an example, let us consider the wave

$$u = A \sin\big(a(x - Ct)\big). \tag{16.0.30}$$

The above solution satisfies the equation (16.0.28), A is called the amplitude of the wave, it is the largest displacement that is possible due to the propagation of the wave. Notice that at $t = 0$, the displacement at $x = 0$ is 0, and the displacement is 0 at $x = \frac{2\pi}{a}$. The quantity $\frac{2\pi}{a}$ is called the wavelength as the wave pattern is the same over the length $\frac{2\pi}{a}$, and it is denoted by λ. Let's define a parameter κ called the wave number, then

$$\lambda = \frac{2\pi}{\kappa}, \quad \kappa = \frac{2\pi}{\lambda}. \tag{16.0.31}$$

The time taken for one complete wave pattern to go through a specific point is called the period T, and it follows that

$$T = \frac{\lambda}{C}. \tag{16.0.32}$$

The inverse of the period is a measure of how the wave oscillates and is referred to as the frequency f, where

$$f = \frac{1}{T}. \tag{16.0.33}$$

There are many types of waves. A simple interesting wave pattern is the standing wave, which is also referred to as a stationary wave. In such waves, the location of maximum amplitude does not move in space. These waves are caused by two identical waves that move in opposite directions. The maximum amplitude remains at the exact same point in space. An example of such a standing wave is the composition of two waves, namely

$$u = A\sin(kx - \omega t) + B\sin(kx + \omega t). \tag{16.0.34}$$

References

[1] J. D. Achenbach. Wave Propagation in Elastic Solids. North Holland Publishing Company, Amsterdam-New York-Oxford, 1973.
[2] A. C. Eringen and E. S. Suhubi. Elastodynamics, Vol. II: Linear Theory. Academic, New York, 1975.
[3] K. F. Graff. Wave Motion in Elastic Solids. Dover Publications Inc., New York, 1991.
[4] A. E. H. Love. A Treatise on the Mathematical Theory of Elasticity. Dover Publications, New York, 1944.

17 Nonstandard approach to elasticity

17.1 Implicit theories for elasticity and their linearization

The Cauchy theory of elasticity presumes that the stress is a function of the density and the deformation gradient, and possibly on $\mathbf{X}$, and Green's theory assumes the existence of a stored energy function that depends on the density, the deformation gradient and $\mathbf{X}$, from which the stress can be derived. Since the applied traction is the cause (the resulting stress due to the applied traction the proximal cause), and deformation is the effect, the starting point for Cauchy elasticity (as also Green elasticity) has the cause being given in terms of the effect, thereby standing causality on its head. The linearization of constitutive relations for Cauchy elasticity bodies that leads to the expression for the stress in terms of the linearized strain also has the cause-effect relationship topsy-turvy. On the other hand, the approach of expressing the linearized strain in terms of the stress is consistent with the demands of causality as the linearized strain is the effect of the stress. As mentioned earlier, the class of solid bodies that are incapable of dissipation is far larger than the class of Cauchy or Green elastic bodies, if "elastic bodies" are to be understood as bodies being incapable of dissipation in any thermodynamic process that they undergo.

A simple example of a body that is elastic that cannot be described within the purview of Green elasticity or Cauchy elasticity is depicted in Figure 17.1, a system made up of a perfectly elastic spring and an inextensible string. The response of such a system cannot be described in terms of a force (correspondingly stress) as a function of the elongation (correspondingly stretch), but rather as the elongation in terms of the force. That is, the constitutive relation is explicit for a kinematical quantity in terms of the stress. Recognizing that there are bodies, described by implicit constitutive relations, that are incapable of dissipating energy, Rajagopal questioned the presumption that the stress or stored energy in elastic bodies be explicitly dependent only on the deformation gradient (see [6, 3, 4]). Rajagopal suggested a generalization wherein the stress and the deformation gradient are related implicitly, that includes Cauchy elastic and hence Green elastic bodies as special sub-classes. Necessary and sufficient conditions for such non-dissipative behavior, on the basis of thermodynamics, has been provided in [2]. A whole new horizon opens up with regard to the response of such non-dissipative bodies and it seems that such bodies might be able to provide a means for attacking problems that have been hitherto beyond the purview of classical Cauchy elasticity.

There are numerous reasons why the development of implicit constitutive relations to describe the elastic response of bodies is warranted. Here we shall merely discuss one of those reasons, namely the consequences of the linearization of such constitutive relations when the displacement gradients are small, as it provides a rational explanation for experimentally observed data. It is possible within the purview of implicit theory, linearization can lead to approximations wherein the strain can be limited to be as small as desired, a'priori, and yet the relationship between the linearized strain

https://doi.org/10.1515/9783110789515-017

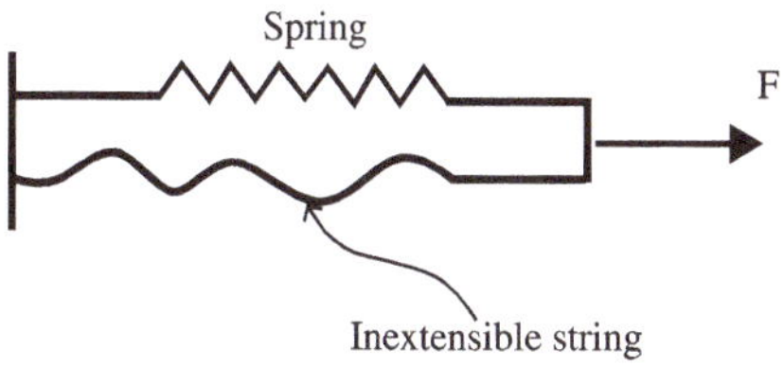

a: Unstretched spring and inextensible string.

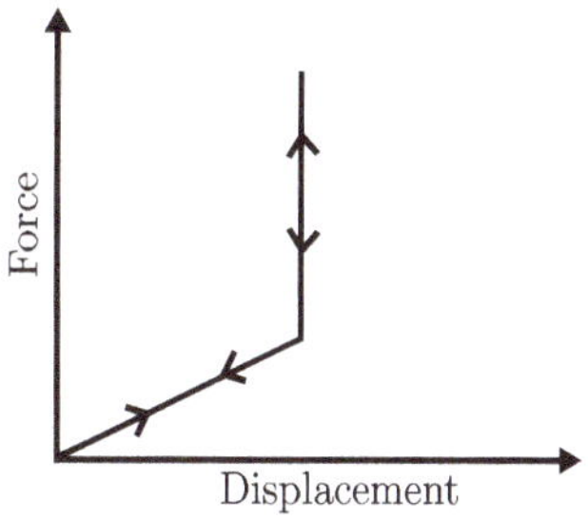

b: Force-displacement relation. Force is not a function of displacement.

c: Fully stretched string and stretched spring.

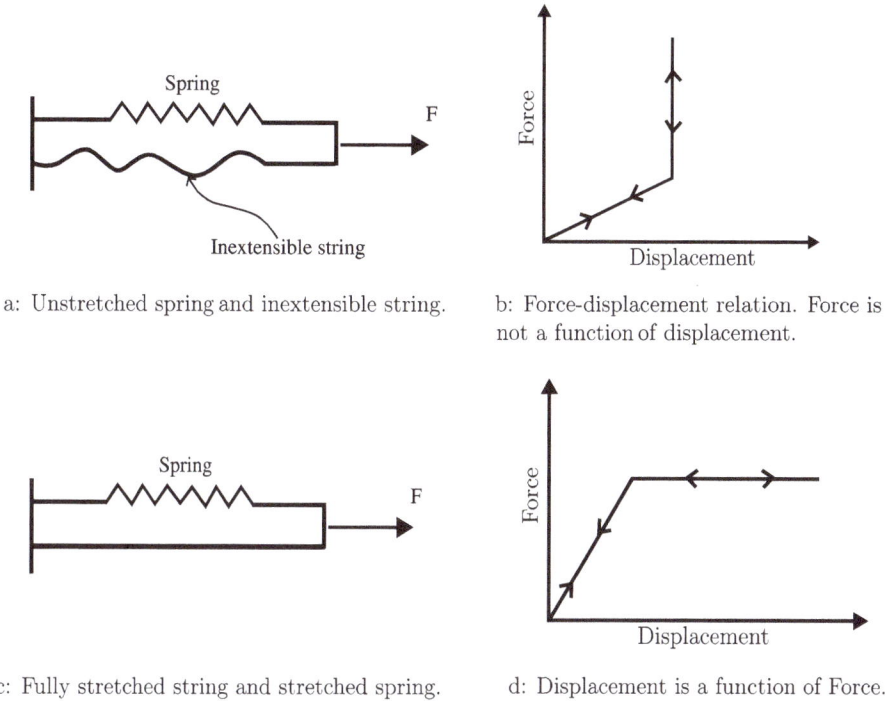

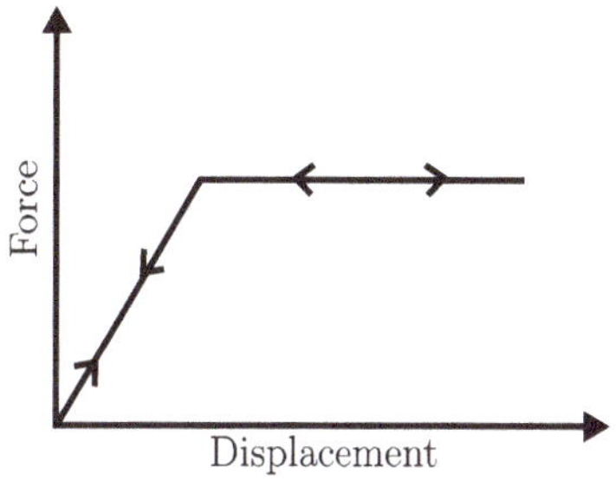

d: Displacement is a function of Force.

Figure 17.1: Stretching of string and spring and its force-displacement relationship.

and the stress is nonlinear. Within the construct of such an approximation, problems wherein the strains are singular within the context of classical, linearized elasticity as in the case of cracks, concentrated loads, doublets, and at corners would now have no such singularity as the strains are limited a'priori. Consider the implicit constitutive relation,

$$\mathbf{g}(\mathbf{X}, \rho, \mathbf{T}, \mathbf{F}) = \mathbf{0}. \tag{17.1.1}$$

We notice that the above implicit relation includes Cauchy elasticity as a special case. Next, let us consider the above restricted to isotropic implicit elastic bodies, in which case (17.1.1) would reduce to

$$\mathbf{h}(\mathbf{X}, \rho, \mathbf{T}, \mathbf{B}) = 0 \tag{17.1.2}$$

A special subclass of isotropic bodies is that wherein the left Cauchy-Green tensor is expressed explicitly in terms of the Cauchy stress **T**. Let us consider the constitutive relation[1]

$$\mathbf{B} = \mathbf{f}(\mathbf{X}, \rho, \mathbf{T}). \tag{17.1.3}$$

1 This constitutive relation is not an implicit relation, it is an explicit expression for **B** in terms of **T**. However, the constitutive relation is a sub-class of the general implicit relation in (17.1.2).

Suppose that the relation cannot be inverted to provide an expression for the stress as a function of the density and the Cauchy Green tensor.

Let us further suppose the body is homogeneous. Then, if **f** is an isotropic function, it follows that (see [5]):

$$\mathbf{B} = \alpha_0 \mathbf{1} + \alpha_1 \mathbf{T} + \alpha_2 \mathbf{T}^2, \tag{17.1.4}$$

where $\mathbf{B} = \mathbf{F}\mathbf{F}^T$ is the left Cauchy-Green tensor, and the α_i^s, $i = 0, 1, 2$ depend on the principal invariants of **T** and the density ρ. If instead of (17.1.3), we were to start with

$$\mathbf{C} = \mathbf{f}(\rho, \mathbf{T}) \tag{17.1.5}$$

we can describe anisotropic bodies. Frame indifference would require that

$$\mathbf{C} = \mathbf{f}(\rho, \mathbf{T}) = \mathbf{f}(\rho, \mathbf{Q}\mathbf{T}\mathbf{Q}^T) \quad \forall \mathbf{Q} \in \mathcal{O}. \tag{17.1.6}$$

Depending on the anisotropic body that we wish to study we need to use the appropriate representation for such bodies. We shall not get into a discussion of these issues here. We shall restrict our discussion to the response of isotropic bodies. Since the linearization of the constitutive relations (17.1.3) allows us to develop approximations wherein the linearized strain is a nonlinear function of the stress, we are in a position to describe the response of modern metallic alloys such as Gum metal, as well as the behavior of many geological materials, cement concrete, rocks, bones, etc., whose material moduli depend on density, even when the deformation is small. On requiring that the displacement gradient be small in the sense of maximum of the Frobenius norm of the displacement gradient is small for all points belonging to the body and for all times, we find that the response relation (17.1.4) reduces to

$$\varepsilon = \beta_0 \mathbf{1} + \beta_1 \mathbf{T} + \beta_2 \mathbf{T}^2, \tag{17.1.7}$$

and β_i, $i = 0, 1, 2$ are functions of the principal invariants of **T** and the density ρ. The approximation (17.1.7) would allow the constitutive relations proposed by Bernoulli, Ricatti, Gerstner, Wertheim and others to describe the elastic response of bodies undergoing small deformations, wherein the linearized strain is a nonlinear function of the stress, within its ambit. Such constitutive relations cannot be derived within the scope of Cauchy elasticity. It is also now possible in virtue of (17.1.2) to have a constitutive relation of the form

$$\varepsilon = E_1(1 + \lambda_1 \operatorname{tr} \varepsilon)\mathbf{T} + E_2(1 + \lambda_2 \operatorname{tr} \varepsilon)(\operatorname{tr} \mathbf{T})\mathbf{1}. \tag{17.1.8}$$

Notice that in the above constitutive relation ε and **T** appear linearly, but the relation is not bilinear. One cannot obtain such a constitutive relation within the context of linearization of Cauchy elasticity. When λ_1 and λ_2 are zero in (17.1.7), the constitutive relation reduces to that for classical linearized elastic constitutive approximation. Since in virtue of the balance of mass

$$\rho(\det \mathbf{F}) = \rho_R, \tag{17.1.9}$$

and in virtue of linearization

$$\det \mathbf{F} \approx 1 + \operatorname{tr} \varepsilon \tag{17.1.10}$$

we can express $\operatorname{tr} \varepsilon$ in terms of the current and reference densities. Thus, we can have a linearized theory in which we can define a generalized Young's modulus and generalized Poisson's ratio that depend on the density. Such approximations are not possible within the context of Cauchy elasticity. However, this has not deterred their usage without a proper basis. It is only within the context of the implicit constitutive theory can such approximations be justified. When we restrict the constitutive relation (17.1.7) to one-dimension, we obtain

$$\varepsilon = f(\sigma) \tag{17.1.11}$$

where ε, σ are the linearized strain and stress in one-dimension. Thus, if one has an experimental result in one-dimension as shown in Figure 17.2 which is the response of a Titanium alloy, it would be perfectly reasonable to express it as elastic response expressed by (17.1.11), as long as the displacement gradient and hence the strain are sufficiently small. However, if we are to use classical linearized elasticity, we would have to restrict ourselves to the domain wherein the stress is related to the strain in a linear fashion. Figure 17.2 depicts the response of Gum metal and the results obtained by using the one dimensional constitutive relation (see [1])

$$\varepsilon = 1.57\text{x}10^{-5}\sigma \exp\left(3.22\text{x}10^{-4}\sigma\right) \tag{17.1.12}$$

Let us consider the class of isotropic bodies defined through

$$\mathbf{f}(\rho, \mathbf{T}, \mathbf{B}) = \mathbf{0}. \tag{17.1.13}$$

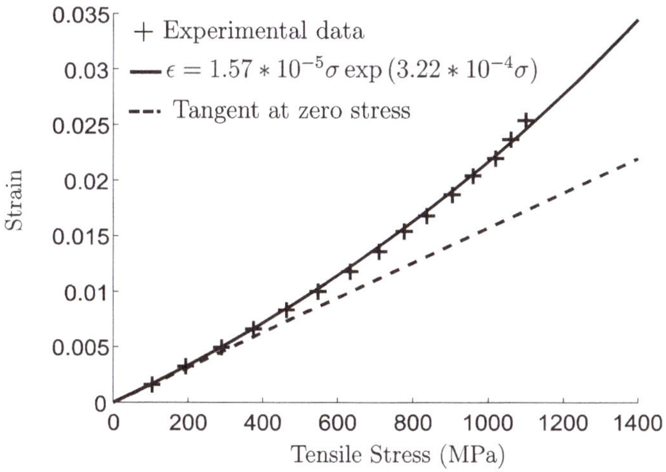

Figure 17.2: Strain-stress relation for a Gum metal alloy.

We have suppressed the dependence on $\mathbf{X}$ for the sake of simplicity. Let us suppose $\mathbf{f}$ is an isotropic function of the variables. We have not yet defined what we mean by a material that is defined by an implicit constitutive relation being isotropic. We shall say that a body described by an implicit constitutive relation is isotropic if the implicit function is an isotropic function of the variables (see [7] for a discussion of anisotropy of bodies described by implicit constitutive relations). In this case we find that $\mathbf{f}$ needs to obey

$$\mathbf{f}(\mathbf{Q}\mathbf{T}\mathbf{Q}^T, \mathbf{Q}\mathbf{B}\mathbf{Q}^T) = \mathbf{Q}\mathbf{f}(\mathbf{T}, \mathbf{B})\mathbf{Q}^T \quad \forall\, \mathbf{Q} \in \mathcal{O}, \tag{17.1.14}$$

where $\mathcal{O}$ denotes the orthogonal group. It then follows that (see Spencer (1975))

$$\begin{aligned} &\alpha_0\mathbf{1} + \alpha_1\mathbf{T} + \alpha_2\mathbf{B} + \alpha_3\mathbf{T}^2 + \alpha_4\mathbf{B}^2 + \alpha_5(\mathbf{BT} + \mathbf{TB}) + \alpha_6(\mathbf{BT}^2 + \mathbf{T}^2\mathbf{B}) \\ &\quad + \alpha_7(\mathbf{B}^2\mathbf{T} + \mathbf{TB}^2) + \alpha_8(\mathbf{B}^2\mathbf{T}^2 + \mathbf{T}^2\mathbf{B}^2) = 0, \end{aligned} \tag{17.1.15}$$

where the material moduli $\alpha_i, i = 0, \dots 8$ depend upon ρ, $\operatorname{tr}\mathbf{T}$, $\operatorname{tr}\mathbf{B}$, $\operatorname{tr}\mathbf{T}^2$, $\operatorname{tr}\mathbf{B}^2$, $\operatorname{tr}\mathbf{T}^3$, $\operatorname{tr}\mathbf{B}^3$, $\operatorname{tr}\mathbf{TB}$, $\operatorname{tr}\mathbf{T}^2\mathbf{B}$, $\operatorname{tr}\mathbf{TB}^2$, $\operatorname{tr}\mathbf{T}^2\mathbf{B}^2$. Once again if were to require that the body can only undergo small displacement gradients, we find that (17.1.15) reduces to

$$\varepsilon + \hat{\alpha}_0\mathbf{1} + \hat{\alpha}_1\mathbf{T} + \hat{\alpha}_2\mathbf{T}^2 + \hat{\alpha}_3\mathbf{T}^2 + \hat{\alpha}_4[\mathbf{T}\varepsilon + \varepsilon\mathbf{T}] + \hat{\alpha}_5[\mathbf{T}^2\varepsilon + \varepsilon\mathbf{T}^2] = \mathbf{0} \tag{17.1.16}$$

where the function $\hat{\alpha}_i$, $i = 0, 1, 2, 3$ are scalar valued functions that can at most depend linearly on ε, but arbitrarily on the invariants of $\mathbf{T}$ while $\hat{\alpha}_4$ and $\hat{\alpha}_5$ are functions of $\operatorname{tr}\mathbf{T}$, $\operatorname{tr}\mathbf{T}^2$ and $\operatorname{tr}\mathbf{T}^3$. The expression (17.1.16) would only be valid if the coefficients $\hat{\alpha}_i$, $i = 0, \dots 5$ and ε and $\mathbf{T}$ are such that each of the terms that appear in (17.1.16) are of $0(\delta)$, $\delta \ll 1$.

The constitutive relation (17.1.16) wherein the material moduli are functions of the density and the various invariants described above is just untenable to use as even the determination of one function is the equivalent of infinity of constants. Thus, in order to put such constitutive relations to use we have to simplify them drastically. Thus, we shall now consider some simple but useful constitutive relations which have been used to describe the response of intermetallic alloys, cement, concrete, rocks, etc. As mentioned earlier, an interesting feature that models of the form (17.1.16) is that they admit strain limiting constitutive relations, the strain being always small and below a certain value. Numerous ad hoc models have been introduced in the literature that exhibit such a feature, but none of them are within a fully three dimensional approach and more importantly none of them stem from a proper thermodynamic basis. We will consider few simple models that belong to the class defined by (17.1.16). First, let us consider the constitutive relation

$$\varepsilon = \alpha\left\{\left[1 - \exp\left(\frac{-\beta(\operatorname{tr}\mathbf{T})}{1 + \delta(\operatorname{tr}\mathbf{T}^2)^{1/2}}\right)\right]\mathbf{1} + \frac{\gamma}{[1 + (\operatorname{tr}\mathbf{T}^2)^{1/2}]}\mathbf{T}\right\} \tag{17.1.17}$$

which is not an implicit model but provides an explicit relationship for the linearized strain in terms of the stress. However, such a constitutive relation can only be justified by starting with a constitutive relation of the form (17.1.17) and linearizing. In the above equation, $\alpha, \beta, \delta, \gamma$ and n are constants. We note that when $\mathbf{T} = \mathbf{0}$, $\varepsilon = \mathbf{0}$ that is the stress-free state is also the unstrained state. Also, if we require linearity in $\mathbf{T}$, the above model reduces to

$$\varepsilon = \alpha\beta(\operatorname{tr}\mathbf{T})\mathbf{1} + \alpha\gamma\mathbf{T}, \tag{17.1.18}$$

and thus if

$$\alpha\beta = \frac{-(\nu)}{E}, \quad \alpha\gamma = \frac{(1+\nu)}{E} \tag{17.1.19}$$

the constitutive relation reduces to the classical linearized elastic constitutive relation. We shall assume that α and γ are negative, and β and δ are positive. It follows from (17.1.17) that the model exhibits limited strain as the stress increases indefinitely provided the material constants are chosen appropriately. The fact that the strains remain bounded makes them possible candidates for the study of cracks in brittle elastic bodies undergoing small strains. Yet another constitutive relation that exhibits strain limiting response is

$$\varepsilon = \alpha\left[1 - \frac{1}{1 + \operatorname{tr}\mathbf{T}}\right]\mathbf{1} + \beta\left[\frac{1}{(1 + \gamma(\operatorname{tr}\mathbf{T}^2))}\right]^n \mathbf{T}, \tag{17.1.20}$$

where α, β, γ and n are constants when the material constants are appropriately chosen. We shall not discuss the numerous special models that arise within the class of models (17.1.16) that predict small strains in the presence of large stresses.[2]

Let us consider some very simple homogeneous deformations of bodies whose constitutive relation is given by (17.1.7). By a state of pure shear stress we mean a stress of the form

$$\mathbf{T} = T_{xy}(\mathbf{e}_x \otimes \mathbf{e}_y + \mathbf{e}_y \otimes \mathbf{e}_x), \tag{17.1.21}$$

where T_{xy} is the shear stress and $\mathbf{e}_x$ and $\mathbf{e}_y$ denote the unit vectors along the x and y directions. It follows from (17.1.7) and (17.1.21) that the linearized strain is given by

$$\varepsilon = \beta_0\mathbf{1} + \beta_1 T_{xy}(\mathbf{e}_x \otimes \mathbf{e}_y + \mathbf{e}_y \otimes \mathbf{e}_x) + (\beta_2 T_{xy}^2)(\mathbf{e}_x \otimes \mathbf{e}_x + \mathbf{e}_y \otimes \mathbf{e}_y), \tag{17.1.22}$$

and thus it immediately follows that

$$\operatorname{tr}\varepsilon = 3\beta_0 + 2\beta_2 T_{xy}^2 \neq 0. \tag{17.1.23}$$

2 Since stresses are dimensional quantities, it does not make sense to talk about large stresses unless we are discussing stresses that have been rendered non-dimensional.

In the case of the classical linearized elastic solid a state of pure shear stress implies a state of pure shear strain, i. e., the tr $\varepsilon = 0$. It follows from (17.1.23) that is not the case unless the material moduli β_0 and β_2 are identically zero. The constitutive relation (17.1.7) also exhibits response characteristics that is typical of nonlinear elastic solids. It follows for the state of simple shear stress, (17.1.7) implies that

$$\varepsilon_{xx} - \varepsilon_{zz} = \beta_2 T_{xy}^2, \tag{17.1.24}$$

$$\varepsilon_{yy} - \varepsilon_{xx} = 0, \tag{17.1.25}$$

$$\varepsilon_{yy} - \varepsilon_{zz} = \beta_2 T_{xy}^2. \tag{17.1.26}$$

Normal strain differences develop and two of the normal strain differences are the same.

Thus far, we were discussing implicit constitutive theories and their special subclasses when the stress is the Cauchy stress (stress in the current deformed configuration). One could also define implicit constitutive relations for elastic solids in terms of

$$\mathbf{f}(\rho, \mathbf{S}, \mathbf{E}) = \mathbf{0}, \tag{17.1.27}$$

where **S** is the Piola-Kirchoff stress and **E** the Green-St. Venant strain given by

$$\mathbf{E} = \frac{1}{2}\left\{\left(\frac{\partial \mathbf{u}}{\partial \mathbf{X}}\right) + \left(\frac{\partial \mathbf{u}}{\partial \mathbf{X}}\right)^T + \left(\frac{\partial \mathbf{u}}{\partial \mathbf{X}}\right)^T\left(\frac{\partial \mathbf{u}}{\partial \mathbf{X}}\right)\right\} \tag{17.1.28}$$

Yet another class of implicit constitutive relations to describe the response of elastic solids takes the form

$$\mathbf{A}(\mathbf{S}, \mathbf{E})\dot{\mathbf{S}} + \mathbf{B}(\mathbf{S}, \mathbf{E})\dot{\mathbf{E}} = \mathbf{0} \tag{17.1.29}$$

The two definitions (17.1.27) and (17.1.29) are not equivalent. It is possible that (17.1.27) is not differentiable and so cannot be cast into the form (17.1.29), while (17.1.29) might not be integrable to the form (17.1.27).

We shall not discuss implicit theories for the response of elastic bodies in any further detail. Our aim was to merely show that even within the context of small strains, there is a lot that can be said about response of elastic bodies that is not governed by the classical linearized elastic response. These new constitutive theories add a whole new dimension to the description of the elastic response of bodies.

References

[1] K. R. Rajagopal. On the nonlinear elastic response of bodies in the small strain range. Acta Mechanica, 225(6):1545–1553, 2014.

[2] K. R. Rajagopal and A. R. Srinivasa. On the response of non-dissipative solids. Proceedings of the Royal Society A: Mathematical, Physical and Engineering Sciences, 463(2078):357–367, 2007.

[3] K. R. Rajagopal and A. R. Srinivasa. On the thermodynamics of fluids defined by implicit constitutive relations. Zeitschrift für angewandte Mathematik und Physik, 59(4):715–729, 2007.
[4] K. R. Rajagopal and A. R. Srinivasa. A Gibbs-potential-based formulation for obtaining the response functions for a class of viscoelastic materials. Proceedings of the Royal Society A: Mathematical, Physical and Engineering Sciences, 467(2125):39–58, 2010.
[5] K. R. Rajagopal. The elasticity of elasticity. Zeitschrift für angewandte Mathematik und Physik, 58:309–317, 2007.
[6] K. R. Rajagopal. On implicit constitutive theories. Applications of Mathematics, 48(4):279–319, 2003.
[7] K. R. Rajagopal. A note on the classification of the anisotropy of bodies defined by implicit constitutive equations. Mechanics Research Communications, 64:38–41, 2015.

A Finite dimensional vector spaces

We shall provide a very brief introduction to finite dimensional vector spaces. A detailed discussion of the material in this appendix can be found in Halmos [1], MacLane and Birkhoff [2], and Nickerson, Spencer and Steenrod [3]. The material in this appendix is based on material taken from the above sources. A vector space $\mathcal{V}$ defined over the field[1] $\mathbb{R}$ of real numbers is a set endowed with two operations, namely multiplication $*$ of the members of the set by real numbers and an addition $+$ of the members such that for all $\mathbf{u}, \mathbf{v}, \mathbf{w} \in \mathcal{V}$ and for all $\alpha, \beta \in \mathbb{R}$

1. $\mathbf{u} + (\mathbf{v} + \mathbf{w}) = (\mathbf{u} + \mathbf{v}) + \mathbf{w}$
2. $\mathbf{u} + \mathbf{v} = \mathbf{v} + \mathbf{u}$
3. $\exists$ an element of $\mathcal{V}$ called zero, $\mathbf{0}$, such that $\mathbf{u} + \mathbf{0} = \mathbf{u}$.
4. $\exists$ an element $-\mathbf{u}$ in $\mathcal{V} \ni \mathbf{u} + (-\mathbf{u}) = \mathbf{0}$.
5. $\alpha * (\beta * \mathbf{u}) = (\alpha\beta) * \mathbf{u}$
6. $1 * \mathbf{u} = \mathbf{u}$
7. $(\alpha + \beta) * \mathbf{u} = \alpha * \mathbf{u} + \beta * \mathbf{u}$
8. $\alpha * (\mathbf{u} + \mathbf{v}) = \alpha * \mathbf{u} + \alpha * \mathbf{v}$.

Henceforth, we shall not use $*$ and merely express $\alpha * \mathbf{u}$ as $\alpha\mathbf{u}$, etc. Also, we shall confine ourselves to finite dimensional vector spaces. We have not defined what we mean by "finite dimension" as yet.

A finite set of vectors $\{\mathbf{u}_i\}, i = 1, \ldots, N$ is linearly dependent if there exists a set of scalars $\{\alpha_i\}, i = 1, \ldots, N$, not all zero, such that

$$\sum_{i=1}^{N} \alpha_i \mathbf{u}_i = \mathbf{0}. \tag{A.0.1}$$

If $\sum_{i=1}^{N} \alpha_i \mathbf{u}_i = \mathbf{0}$ implies that all $\alpha_i = 0$, then the set $\{\mathbf{u}_i\}$ is linearly independent. If a vector

$$\mathbf{u} = \sum_{i=1}^{N} \alpha_i \mathbf{u}_i, \tag{A.0.2}$$

then $\mathbf{u}$ is said to be a linear combination of $\{\mathbf{u}_i\}$. A set of linearly independent vectors $\{\mathbf{b}_i\}, i = 1, \ldots, N$ is a basis if every vector in the vector space $\mathcal{V}$ is a linear combination of the elements of the basis. A vector space $\mathcal{V}$ is finite dimensional if it has a finite basis. Every linearly independent set in $\mathcal{V}$ can be extended to be a basis for $\mathcal{V}$. An important

1 We shall not define the notion of a field. For our purpose, knowledge of what is meant by real numbers suffices. A rigorous development of vector spaces would start with the definition of a Magma, followed by Semi-Group, Monoid, Group, Ring, Field, Left and Right Module, and then a Vector Space. However such a development adds little to our understanding of mechanics.

https://doi.org/10.1515/9783110789515-018

result is that the number of elements in any basis of a finite dimensional vector space is the same as in any other basis.

A.1 Subspaces and affine spaces

A non-empty subset $\mathcal{V}'$ of a vector space $\mathcal{V}$ is called a subspace if along with every pair $\mathbf{x}$ and $\mathbf{y}$ of vectors contained in $\mathcal{V}'$, every linear combination $\alpha\mathbf{x} + \beta\mathbf{y}$, $\alpha, \beta \in \mathbb{R}$, is also contained in $\mathcal{V}'$.

Suppose $\mathcal{U}$ is a subspace (linear subspace) of $\mathcal{V}$, and suppose $\mathbf{v} \in \mathcal{V}$. Let us define a set $\mathbf{v} + \mathcal{U}$ through

$$\mathbf{v} + \mathcal{U} = \{\mathbf{v} + \mathbf{u} \mid \mathbf{u} \in \mathcal{U}\}. \tag{A.1.1}$$

The set $\mathbf{v} + \mathcal{U}$ is a translation of $\mathcal{U}$ by the vector $\mathbf{v}$.

An inner product space $\mathcal{H}$ is a vector space with an inner product, denoted by $(\cdot)$, such that for all $\mathbf{u}, \mathbf{v}, \mathbf{w} \in \mathcal{H}$

1. $\mathbf{u} \cdot \mathbf{v} = \mathbf{v} \cdot \mathbf{u}$
2. $\mathbf{u} \cdot (\mathbf{v} + \mathbf{w}) = \mathbf{u} \cdot \mathbf{v} + \mathbf{u} \cdot \mathbf{w}$
3. $\alpha(\mathbf{u} \cdot \mathbf{v}) = \alpha\mathbf{u} \cdot \mathbf{v}$
4. $\mathbf{u} \cdot \mathbf{u} \geq 0$, equality holding if and only if $\mathbf{u} \equiv \mathbf{0}$.

We use the notation

$$\|\mathbf{u}\| = (\mathbf{u} \cdot \mathbf{u})^{1/2}. \tag{A.1.2}$$

The operator $\|\cdot\|$ is referred to as the norm induced by the scalar product.

A set of vectors $\mathcal{W}$ is orthonormal if whenever $\mathbf{u}, \mathbf{v} \in \mathcal{W}$, it follows that $(\mathbf{u} \cdot \mathbf{v}) = 0$ whenever $\mathbf{u} \neq \mathbf{v}$, and $(\mathbf{u} \cdot \mathbf{u}) = 1$. If the field over which the vector space is defined is the set of complex numbers, then

$$(\mathbf{u} \cdot \mathbf{v}) = \overline{(\mathbf{u} \cdot \mathbf{v})}, \tag{A.1.3}$$

where the overbar denotes conjugation. As we shall only consider the field of real numbers, we shall not discuss this issue further. Not all norms are induced by an inner product.

One can define the notion of an angle between two vectors $\mathbf{u}$ and $\mathbf{v}$ through

$$\cos\theta = \frac{(\mathbf{u} \cdot \mathbf{v})}{\|\mathbf{u}\|\|\mathbf{v}\|}. \tag{A.1.4}$$

We call an orthonormal set of vectors complete if it is not contained in a larger orthonormal set. If $\mathcal{V}$ is an n-dimensional inner product space, then there exists complete orthonormal sets in $\mathcal{V}$ and every complete orthonormal set in $\mathcal{V}$ contains exactly n elements.

A.2 Gram-Schmidt procedure

Let $\{\mathbf{b}_i\}, i = 1, \ldots, n$ be any basis in $\mathbf{V}$. We shall construct a complete orthonormal set. First, $\mathbf{b}_1 \neq 0$. Since $\{\mathbf{b}_1, \ldots, \mathbf{b}_n\}$ is a basis, they are linearly independent. Let

$$\mathbf{e}_1 = \frac{\mathbf{b}_1}{||\mathbf{b}_1||}, \tag{A.2.1}$$

where $||\mathbf{b}_1||^2 = (\mathbf{b}_1 \cdot \mathbf{b}_1)$. Next, let

$$\mathbf{e}_2 = \frac{\mathbf{b}_2 - \alpha_1 \mathbf{e}_1}{||\mathbf{b}_2 - \alpha_1 \mathbf{e}_1||}, \quad \alpha_1 = (\mathbf{b}_2 \cdot \mathbf{e}_1). \tag{A.2.2}$$

Thus, $||\mathbf{e}_1|| = 1$ and $\mathbf{e}_1 \cdot \mathbf{e}_2 = 0$. Suppose $\mathbf{e}_1, \ldots \mathbf{e}_r$ have been found. Then, we define

$$\mathbf{e}_{r+1} = \frac{\mathbf{b}_{r+1} - (\alpha_1 \mathbf{b}_1 + \cdots + \alpha_r \mathbf{b}_r)}{||\mathbf{b}_{r+1} - (\alpha_1 \mathbf{b}_1 + \cdots + \alpha_r \mathbf{b}_r)||}. \tag{A.2.3}$$

Again, $||\mathbf{e}_{r+1}|| = 1$ and $\mathbf{e}_{r+1} \cdot \mathbf{e}_i = 0 \; \forall \; i = 1, 2, \ldots, r$, if the α_j values are so chosen that

$$\alpha_r = (\mathbf{b}_{r+1} \cdot \mathbf{e}_r). \tag{A.2.4}$$

Thus, $\mathbf{e}_{r+1}$ are different from zero, and $\mathbf{e}_{r+1} \cdot \mathbf{e}_j = 0, j = 1, \ldots, r$. We continue this process until we get a complete orthonormal set.

A.3 Euclidean space, distance and inequalities

If $\mathbf{u}, \boldsymbol{\upsilon} \in \mathcal{V}$ is an inner product space, then

$$|(\mathbf{u} \cdot \boldsymbol{\upsilon})| \leq \big| ||\mathbf{u}|| \, ||\boldsymbol{\upsilon}|| \big|. \tag{A.3.1}$$

This is known as the Cauchy-Schwarz inequality. If $\mathbf{u}, \boldsymbol{\upsilon} \in \mathcal{V}$, where $\mathcal{V}$ is an inner product space, then

$$||\mathbf{u} + \boldsymbol{\upsilon}|| \leq \big| ||\mathbf{u}|| \, ||\boldsymbol{\upsilon}|| \big|. \tag{A.3.2}$$

The distance between two points $\hat{\mathbf{u}}, \hat{\boldsymbol{\upsilon}} \in \mathcal{E}$, where $\mathcal{E}$ is an Euclidean space and $\mathcal{V}$ is the translation space of $\mathcal{E}$ on which an inner product is defined, expressed as $d(\hat{\mathbf{u}}, \hat{\boldsymbol{\upsilon}})$, is the norm of the vector $\boldsymbol{\upsilon} - \mathbf{u}$, where $\mathbf{u}$ and $\boldsymbol{\upsilon}$ are the vectors $\mathbf{u} = \hat{\mathbf{u}} - \hat{\mathbf{0}}, \boldsymbol{\upsilon} = \hat{\boldsymbol{\upsilon}} - \hat{\mathbf{0}}$, where $\hat{\mathbf{0}}, \hat{\mathbf{u}}, \hat{\mathbf{u}} \in \mathcal{E}$. The distance so defined satisfies

1. For each pair $\mathbf{u}, \boldsymbol{\upsilon} \in \mathcal{V}$, $||\mathbf{u} - \boldsymbol{\upsilon}|| = ||\boldsymbol{\upsilon} - \mathbf{u}||$
2. For each pair $\mathbf{u}, \boldsymbol{\upsilon} \in \mathcal{V}$, $||\mathbf{u} - \boldsymbol{\upsilon}|| \geq 0$, equality holding if and only if $\mathbf{u} = \boldsymbol{\upsilon}$.
3. For all $\mathbf{u}, \boldsymbol{\upsilon}, \mathbf{w} \in \mathcal{V}$, $||\mathbf{u} - \mathbf{w}|| \leq ||\mathbf{u} - \boldsymbol{\upsilon}|| + ||\boldsymbol{\upsilon} - \mathbf{w}||$.

All metrics are not necessarily derived from an inner product.

A.4 Vector product

A vector product defined on $\mathcal{V}$ is a function which assigns to each ordered pair of vectors $\mathbf{u}, \mathbf{v} \in \mathcal{V}$, a vector $\mathbf{u} \times \mathbf{v}$, which has the following properties:
1. $\forall\ \mathbf{u}, \mathbf{v} \in \mathcal{V}$, $\mathbf{u} \times \mathbf{v} = -\mathbf{v} \times \mathbf{u}$,
2. $\forall\ \mathbf{u}, \mathbf{v}, \mathbf{w} \in \mathcal{V}$, $\mathbf{u} \times (\mathbf{v} + \mathbf{w}) = \mathbf{u} \times \mathbf{v} + \mathbf{u} \times \mathbf{w}$,
3. $\forall\ \mathbf{u}, \mathbf{v} \in \mathcal{V}$ and $\alpha \in \mathbb{R}$, $\alpha(\mathbf{u} \times \mathbf{v}) = (\alpha\mathbf{u}) \times \mathbf{v}$,
4. $\forall\ \mathbf{u}, \mathbf{v} \in \mathcal{V}$, $\mathbf{u} \cdot (\mathbf{u} \times \mathbf{v}) = 0$,
5. $\forall$ non-zero $\mathbf{u}, \mathbf{v} \in \mathcal{V}$, $||\mathbf{u} \times \mathbf{v}|| = ||\mathbf{u}||\,||\mathbf{v}|| \sin\theta$, where $\cos\theta = \frac{(\mathbf{u} \cdot \mathbf{v})}{||\mathbf{u}||\,||\mathbf{v}||}$.

There are exactly two vector products that can be defined. One of these is called the right-handed product and the other the left-hand product. A vector product is a bilinear function. The magnitude of the vector product between two vectors $\mathbf{u}$ and $\mathbf{v}$ can be interpreted as the area of the parallelogram whose sides are $\mathbf{u}$ and $\mathbf{v}$.

A vector product can only be defined in three dimensional vector spaces. However, a generalized notion of a vector product, referred to as exterior product can be defined in vector spaces whose dimension is not three (see Cartan [4], Edelen [5]).

A.5 The scalar triple product

Let us suppose that we have a scalar product and a vector product defined on $\mathcal{V}$. If $\mathbf{u}, \mathbf{v}$, and $\mathbf{w} \in \mathcal{V}$, their triple product expressed as $[\mathbf{u}, \mathbf{v}, \mathbf{w}]$ is defined through

$$[\mathbf{u}, \mathbf{v}, \mathbf{w}] = \mathbf{u} \cdot (\mathbf{v} \times \mathbf{w}). \tag{A.5.1}$$

The scalar triple product can be interpreted as the volume of a parallelepiped whose edges are $\mathbf{u}$, $\mathbf{v}$, and $\mathbf{w}$. The definition of the scalar triple product, Eq. (A.5.1), implies that it is skew-symmetric. That is, interchanging any two vectors leads to a reversal of the sign of the expression. It is also trilinear and is zero if and only if the three vectors are linearly dependent.

A.6 Linear transformations

Let $\mathcal{U}$ and $\mathcal{V}$ be two vector spaces and let $\mathbf{T}$ be a function $\mathbf{T} : \mathcal{U} \longrightarrow \mathcal{V}$ such that
1. $\mathbf{T}(\mathbf{u} + \mathbf{v}) = \mathbf{T}\mathbf{u} + \mathbf{T}\mathbf{v}\ \ \forall\ \ \mathbf{u}, \mathbf{v} \in \mathcal{U}$,
2. $\mathbf{T}(\alpha\mathbf{u}) = \alpha\mathbf{T}(\mathbf{u})\ \ \forall\ \ \mathbf{u} \in \mathcal{V}, \alpha \in \mathbb{R}$.

Then, $\mathbf{T}$ is said to be a linear transformation.

A.7 Existence of linear transformations

Let $\mathcal{V}$ be a vector space of finite dimension n and let $\mathbf{b}_1, \mathbf{b}_2, \ldots, \mathbf{b}_n$ be a basis for $\mathcal{V}$. Let $\mathcal{W}$ be a vector space and let $\mathbf{a}_1, \mathbf{a}_2, \ldots, \mathbf{a}_n$ be a set of n vectors in $\mathcal{W}$. Then there exist one and only one linear transformation $\mathbf{T} : \mathcal{V} \longrightarrow \mathcal{W}$ such that $\mathbf{T}(\mathbf{b}_i) = \mathbf{a}_i$, $i = 1, 2, \ldots, n$. $\mathbf{T}$ is completely determined by $\mathbf{T}(\mathbf{b}_i)$, $i = 1, 2, \ldots, n$. Thus, the coefficients $\mathbf{T}_{ji}$, $i = 1, 2, \ldots, n$; $j = 1, 2, \ldots, k$ characterize $\mathbf{T}$. The array of k rows and n columns is referred to as the matrix associated with the linear transformation $\mathbf{T}$:

$$(\mathbf{T}) = \begin{pmatrix} T_{11} & T_{12} & \cdots & T_{1n} \\ T_{21} & T_{22} & \cdots & T_{2n} \\ \vdots & \vdots & \ddots & \vdots \\ T_{k1} & T_{k2} & \cdots & T_{kn} \end{pmatrix}. \tag{A.7.1}$$

The ith column are the components of $\mathbf{T}(\mathbf{b}_i)$ with respect to the basis $\mathbf{a}_k$.

The zero linear transformation, represented as $\mathbf{0}$, is the linear transformation that assigns the zero vector to all vectors $\mathbf{u}$ belonging to $\mathcal{U}$, i. e.,

$$\mathbf{0u} = \mathbf{o} \quad \forall \, \mathbf{u} \in \mathcal{U}. \tag{A.7.2}$$

Let us define the addition of two linear transformations $\mathbf{S}$ and $\mathbf{T}$ through

$$(\mathbf{S} + \mathbf{T})(\mathbf{u}) = \mathbf{Su} + \mathbf{Tu} \tag{A.7.3}$$

and

$$(a\mathbf{T})(\mathbf{u}) = a[\mathbf{T}(\mathbf{u})]. \tag{A.7.4}$$

It then follows that the set of all linear transformations, $\mathbf{T}; \mathcal{V} \longrightarrow \mathcal{W}$ where $\mathcal{V}$ and $\mathcal{W}$ are vector spaces, forms a vector space. That is

$$\mathrm{Lin}(\mathcal{V}, \mathcal{W}) := \{\mathbf{T} : \mathcal{V} \longrightarrow \mathcal{W} \mid \mathbf{T} \text{ being a linear transformation}\}, \tag{A.7.5}$$

is a vector space.

References

[1] P. R. Halmos. Finite Dimensional Vector Spaces. Undergraduate Texts in Mathematics. Springer-Verlag-Heidelberg , 1987.

[2] S. MacLane and G. Birkhoff. Algebra. The MacMillan Company, Collier-MacMillan, London, 1967.

[3] H. K. Nickerson, D. C. Spencer, and N. E. Steenrod. Van Nostrand Company, Inc., Princeton, 1959.

[4] H. Cartan. Differential Forms. Dover Publications Inc., Mineola, 2006.

[5] D. G. B. Edelen. Applied Exterior Calculus. Dover Publications Inc., Mineola, 2011.

B Tensor analysis

We shall not define the tensor product of two vector spaces $\mathcal{V}$ and $\mathcal{W}$ and the elements that belong to the tensor product of the two vector spaces $\mathcal{V}$ and $\mathcal{W}$. Instead, we define the tensor product of two vectors $\mathbf{u}$ and $\mathbf{v}$ belonging to $\mathcal{V}$ as that linear transformation $(\mathbf{u} \otimes \mathbf{v})$ belonging to $\mathrm{Lin}(\mathcal{V}, \mathcal{V})$, such that

$$(\mathbf{u} \otimes \mathbf{v})\mathbf{w} = \mathbf{u}(\mathbf{v} \cdot \mathbf{w}) \quad \forall\, \mathbf{w} \in \mathcal{W}. \tag{B.0.1}$$

We can define the tensor product of the two spaces $\mathcal{V} \otimes \mathcal{W}$ as well as $(\mathbf{v} \otimes \mathbf{w}) \in \mathcal{V} \otimes \mathcal{W}$, $\mathbf{v} \in \mathcal{V}$ and $\mathbf{w} \in \mathcal{W}$, in a more abstract manner (for a discussion of tensor product of vector spaces, see Halmos [1] and Nickerson, Spencer and Steenrod [2]). Note that

$$(\mathbf{a} \otimes \mathbf{b}) \neq (\mathbf{b} \otimes \mathbf{a}), \quad \mathbf{a}, \mathbf{b} \in \mathcal{V}. \tag{B.0.2}$$

B.1 Representation theorem

Any linear transformation $\mathbf{T} : \mathcal{V} \longrightarrow \mathcal{V}$ where the dimension of $\mathcal{V}$ is three has the representation

$$\mathbf{T}\mathbf{u} = (\mathbf{e}_1 \otimes \mathbf{t}_1)\mathbf{u} + (\mathbf{e}_2 \otimes \mathbf{t}_2)\mathbf{u} + (\mathbf{e}_3 \otimes \mathbf{t}_3)\mathbf{u}, \tag{B.1.1}$$

where $\mathbf{t}_i \in \mathcal{V}$, $i = 1, 2, 3$ are unique and independent of $\mathbf{u}$ and where $\mathbf{e}_i$, $i = 1, 2, 3$ is the orthonormal basis. The above representation, Eq. (B.1.1), can be easily generalized to a higher finite dimensional vector space. It also follows from Eq. (B.1.1) that any $\mathbf{T}$ can be expressed as

$$\mathbf{T} = T_{ij}\mathbf{e}_i \otimes \mathbf{e}_j, \; i = 1, 2, 3. \tag{B.1.2}$$

Moreover,

$$T_{ij} = \mathbf{e}_i \cdot \mathbf{T}\mathbf{e}_j. \tag{B.1.3}$$

B.2 Matrix algebra

Matrix algebra follows the following set of rules, namely,

1. $A_{ij} + B_{ij} = (\mathbf{A} + \mathbf{B})_{ij}$,
2. $(\alpha\mathbf{A})_{ij} = \alpha A_{ij}$,
3. $(\mathbf{AB})_{ij} = A_{ik}B_{kj}$,
4. $\mathbf{A} + \mathbf{B} = \mathbf{B} + \mathbf{A}$,
5. $\mathbf{A}(\mathbf{B} + \mathbf{C}) = \mathbf{AB} + \mathbf{AC}$.

https://doi.org/10.1515/9783110789515-019

We use the usual summation convention when an index is repeated. Note that $\mathbf{AB} \neq \mathbf{BA}$, i. e., matrix multiplication does not commute.

Exercise. Show that $(\mathbf{AB})_{ij} = A_{ik}B_{kj}$.

B.3 Determinant

Let $\mathcal{V}$ be a vector space of dimension three. Let $\mathbf{T}$ be a linear transformation such that $\mathbf{T} : \mathcal{V} \longrightarrow \mathcal{V}$. The ratio

$$\frac{(\mathbf{Tu}) \cdot (\mathbf{T}\mathbf{\upsilon}) \times (\mathbf{Tw})}{\mathbf{u} \cdot \mathbf{\upsilon} \times \mathbf{w}}, \tag{B.3.1}$$

where $\mathbf{u}, \mathbf{\upsilon}, \mathbf{w} \in \mathcal{V}$ are linearly independent, denotes the determinant of $\mathbf{T}$. Since we are using the vector product to introduce the notion of the determinant, this definition obviously holds only in three dimensions. Later, we provide a more general definition of a determinant.

Suppose $\mathbf{e}_1$, $\mathbf{e}_2$, and $\mathbf{e}_3$ are an orthonormal basis for $\mathcal{V}$. Then, it follows that

$$\det \mathbf{T} = (\mathbf{Te}_1) \cdot (\mathbf{Te}_2) \times (\mathbf{Te}_3). \tag{B.3.2}$$

$\mathbf{T}$ transforms the edges $\mathbf{e}_1$, $\mathbf{e}_2$, $\mathbf{e}_3$ of a parallelepiped to the edges $\mathbf{Te}_1$, $\mathbf{Te}_2$, and $\mathbf{Te}_3$ of another parallelepiped. The determinant of $\mathbf{T}$ denotes the ratio of the volume of these two parallelepipeds. The $\det \mathbf{T} = 0$ if and only if $\mathbf{T}$ is singular. That is, $\ker \mathbf{T} \neq 0$ (the Kernel of a linear transformation, also referred to as the null space of the linear transformation, expressed as Ker, is the set of all vectors that get mapped to the zero vector. The image of the linear transformation $\mathbf{T} : \mathcal{V} \to \mathcal{V}$ is the subspace defined through $\{\mathbf{T}(\mathbf{\upsilon}) : \mathbf{\upsilon} \in \mathcal{V}\}$). It follows from Eq. (B.3.1) that

$$\det(\mathbf{TS}) = (\det \mathbf{T})(\det \mathbf{S}). \tag{B.3.3}$$

Also,

$$\det \mathbf{I} = 1. \tag{B.3.4}$$

Let us consider $n \times n$ matrices (square). Associated with any square matrix of order n we define the determinant, $\det \mathbf{A}$, as a number obtained as the sum of all possible products in each of which there appears one and only one element from each row and each column, each such product being prefixed by a plus or minus sign according to the following rule. Let the elements involved in a given product be joined in pairs by line segments. If the total number of such segments moving upward to the right is even, prefix a plus sign to the product, and a negative sign if it is otherwise. This rule for the calculation for the determinant can be found in Hildebrand [3]. Thus, suppose we have a 3×3 matrix. Then according to the above rule, we have

$$\det \mathbf{A} = a_{11}a_{22}a_{33} - a_{11}a_{32}a_{23} - a_{12}a_{21}a_{33} + a_{12}a_{31}a_{23} - a_{13}a_{22}a_{31} + a_{13}a_{21}a_{32}, \quad \text{(B.3.5)}$$

where a_{ij}, $i, j = 1, 2, 3$ are the components of $\mathbf{A}$.

Properties of the determinant

1. In any row or column of a nxn matrix, if all the terms are zeros, then the determinant of the matrix is zero.
2. When the rows and columns of a nxn matrix are interchanged, the value of the determinant does not change.
3. Interchanging two rows (or two columns) of a nxn matrix only changes its sign.
4. If all the elements of one row (or one column) of a nxn matrix are multiplied by $\lambda \in \mathbb{R}$, the determinant is multiplied by a scalar λ.
5. The determinant of a nxn matrix is zero if the corresponding elements of two rows (or two columns) are equal or in a constant ratio.
6. If each element in one row (or one column) is expressed as the sum of two terms, then the determinant is equal to the sum of two determinants, in each of which one of the two terms is deleted in each element of that row (or column).
7. If to the elements of any row (or column) are added λ times the corresponding elements of any other row (or column), the determinant is unchanged.
8. The determinant of the product of two nxn matrices is the product of the determinants of the individual matrices.

We have defined a matrix in the context of a linear transformation and a basis. The square array of numbers can be thought of in terms of as rows and columns only if we assign a strict order to the basis elements, and it is important to realize that a change of the ordering of the basis elements implies a change in the square array.

The calculation of the determinant can be determined in terms of permutations of the elements of the matrix. The interested reader can find a detailed discussion the same and other related concepts concerning matrices in Smirnov [4]

Let

$$\mathbf{A} : \mathcal{V} \longrightarrow \mathcal{V}. \quad \text{(B.3.6)}$$

It may happen that

(i) If $\mathbf{x}_1 \neq \mathbf{x}_2$, then $\mathbf{A}\mathbf{x}_1 \neq \mathbf{A}\mathbf{x}_2$. This implies that for each vector in $\mathcal{V}$, $\mathbf{A}$ assigns a distinct vector in $\mathcal{V}$. The transformation is said to be one to one.

(ii) To every $\mathbf{y} \in \mathcal{V}$, there exists at least one vector $\mathbf{x} \in \mathcal{V}$ such that $\mathbf{A}\mathbf{x} = \mathbf{y}$.

If the linear transformation is such that both (i) and (ii) hold, then $\mathbf{A}$ is called invertible. It has an inverse. When (ii) holds, the mapping is *onto*. By (ii) there is at least one $\mathbf{x}$ which gets mapped onto $\mathbf{y}$. By (i), there cannot be two of them which can be mapped onto $\mathbf{y}$. Thus, there exists a unique $\mathbf{x}$ which gets mapped to $\mathbf{y}$. The transformation

$$\mathbf{B} : \mathcal{V} \longrightarrow \mathcal{V} \quad \text{(B.3.7)}$$

such that $\mathbf{By} = \mathbf{x}$ is clearly well defined. $\mathbf{B}$ is called the inverse of $\mathbf{A}$ and is denoted by $\mathbf{A}^{-1}$. Of course, we have to check and see that $\mathbf{B}$ is linear.

Now, to show

$$\mathbf{B}(\alpha_1\mathbf{y}_1 + \alpha_2\mathbf{y}_2) = \alpha_1\mathbf{By}_1 + \alpha_2\mathbf{By}_2, \tag{B.3.8}$$

let $\mathbf{Ax}_1 = \mathbf{y}_1$ and $\mathbf{Ax}_2 = \mathbf{y}_2$. Then,

$$\mathbf{B}(\alpha_1\mathbf{y}_1 + \alpha_2\mathbf{y}_2) = \mathbf{B}(\alpha_1\mathbf{Ax}_1 + \alpha_2\mathbf{Ax}_2) = \mathbf{BA}(\alpha_1\mathbf{x}_1 + \alpha_2\mathbf{x}_2) = \alpha_1\mathbf{x}_1 + \alpha_2\mathbf{x}_2, \tag{B.3.9}$$

and

$$\alpha_1\mathbf{By}_1 = \alpha_1\mathbf{BAx}_1 = \alpha_1\mathbf{x}_1, \quad \alpha_2\mathbf{By}_2 = \alpha_2\mathbf{BAx}_2 = \alpha_2\mathbf{x}_2. \tag{B.3.10}$$

Thus,

$$\mathbf{B}(\alpha_1\mathbf{y}_1 + \alpha_2\mathbf{y}_2) = \alpha_1\mathbf{By}_1 + \alpha_2\mathbf{By}_2. \tag{B.3.11}$$

B.4 Indicial notation and summation convention

Let $\mathcal{V}$ be a finite dimensional inner product space and let $\{\mathbf{e}_i\}$, $i = 1, 2, \ldots, n$, be an orthonormal basis for $\mathcal{V}$. Then, any vector $\mathbf{v} \in \mathcal{V}$ can be represented as

$$\mathbf{v} = v_i\mathbf{e}_i. \tag{B.4.1}$$

Whenever any index is repeated twice it is tacitly assumed that we sum with respect to that index, thus

$$v_i\mathbf{e}_i = v_1\mathbf{e}_1 + v_2\mathbf{e}_2 + \cdots + v_n\mathbf{e}_n. \tag{B.4.2}$$

Sometimes we shall specifically use the summation symbol, Σ. However, when there is no cause for confusion, we shall use the above convention.

Let $\mathbf{u}$ be another vector which belongs to $\mathcal{V}$. Then

$$\mathbf{u} = u_k\mathbf{e}_k. \tag{B.4.3}$$

Then,

$$\mathbf{u} \cdot \mathbf{v} = \big((v_k\mathbf{e}_k) \cdot (v_i\mathbf{e}_i)\big) = u_k v_i(\mathbf{e}_k \cdot \mathbf{e}_i). \tag{B.4.4}$$

We define the Kronecker delta through

$$\delta_{ij} = \begin{cases} 0 & \text{if } i \neq j \\ 1 & \text{if } i = j \end{cases}. \tag{B.4.5}$$

It then follows that

$$\mathbf{u} \cdot \mathbf{v} = u_k v_i \delta_{ki} = u_k v_k = u_i v_i. \tag{B.4.6}$$

Thus,

$$\mathbf{u} \cdot \mathbf{v} = u_i v_i. \tag{B.4.7}$$

Next, consider the representation for the vector $\mathbf{v}$ which is given through

$$\mathbf{v} = \mathbf{T}\mathbf{u}. \tag{B.4.8}$$

Since $\mathbf{u} \in \mathcal{V}$,

$$\mathbf{u} = u_i \mathbf{e}_i. \tag{B.4.9}$$

Thus,

$$\mathbf{T}\mathbf{u} = \mathbf{T}(u_i \mathbf{e}_i) = u_i \mathbf{T}(\mathbf{e}_i). \tag{B.4.10}$$

Since $\mathbf{T}\mathbf{e}_i$ belongs to $\mathcal{V}$, we can express

$$\mathbf{T}\mathbf{u} = u_i T_{ji} \mathbf{e}_j = T_{ij} u_j \mathbf{e}_i. \tag{B.4.11}$$

But,

$$\mathbf{v} = v_i \mathbf{e}_i. \tag{B.4.12}$$

Thus,

$$v_i = T_{ij} u_j. \tag{B.4.13}$$

Note, we have used

$$\mathbf{T}(\mathbf{e}_i) = T_{ji} \mathbf{e}_j. \tag{B.4.14}$$

B.5 Cramer's rule

Consider the $n \times n$ square matrix corresponding to a linear transformation, $\mathbf{A}$. If the row and column containing an element A_{ij} are deleted, we have a $(n-1) \times (n-1)$ matrix. The determinant of this matrix is called a *minor* of A_{ij}, and let us denote it by M_{ij}. The quantity

$$(-1)^{i+j} M_{ij} \equiv C_{ij} \quad (i - \text{no sum}), \tag{B.5.1}$$

is called the co-factor of A_{ij}. It then follows from the definition of the determinant that

$$\det \mathbf{A} = \sum_{k=1}^{n} A_{ik} C_{ik} \quad (i - \text{no sum}). \tag{B.5.2}$$

If the element A_{ik} is replaced by A_{rk}, then $\sum_{k=1}^{n} A_{rk} C_{ik}$ is the determinant of the matrix where the ith row has been replaced by the rth row. Thus, in this matrix, unless $i = r$, there are two rows which have identical elements. Then by property (5) listed for determinants, it follows that

$$\sum_{k=1}^{n} A_{rk} C_{ik} = 0 \quad (i - \text{no sum}). \tag{B.5.3}$$

Consider the set of n-linear equations

$$\mathbf{Ax} = \mathbf{y}, \tag{B.5.4}$$

or, in other words,

$$\sum_{j=1}^{n} A_{ij} x_j = y_i. \tag{B.5.5}$$

It follows from Eq. (B.5.5) that

$$\sum_{i=1}^{n} \sum_{j=1}^{n} C_{ir} A_{ij} x_j = \sum_{i=1}^{n} C_{ir} y_i. \tag{B.5.6}$$

By virtue of Eq. (B.5.3) if $r \neq j$,

$$C_{ir} A_{ij} = 0. \tag{B.5.7}$$

If $r = j$, then

$$\sum_{i=1}^{n} A_{ij} C_{ij} = \det \mathbf{A}. \tag{B.5.8}$$

Thus,

$$(\det \mathbf{A}) x_j = \sum_{i=1}^{n} C_{ir} y_i \implies x_j = \frac{\sum_{i=1}^{n} C_{ir} y_i}{\det \mathbf{A}}. \tag{B.5.9}$$

Equation (B.5.9) leads to the following assertion known as Cramer's rule:

When the $\det \mathbf{A}$ of the a matrix of coefficients in a set of n linear algebraic equations in n unknowns $x_1, x_2, \dots, x_n$ is not zero, that set of equations has a unique solution. The value of x_j can be expressed as the ratio of two determinants, the denominator being the determinant of the matrix **A** and the numerator being the determinant of a matrix **A** where the jth column has been replaced by **y**.

B.6 Transpose

Let $\mathcal{V}$ be a finite dimensional inner product space. For each linear transformation $\mathbf{T} : \mathcal{V} \longrightarrow \mathcal{V}$, we define the transpose of the transformation denoted by $\mathbf{T}^{\mathrm{T}}$ through

$$(\mathbf{Tu}, \mathbf{\upsilon}) = \left(\mathbf{u}, \mathbf{T}^{\mathrm{T}}\mathbf{\upsilon}\right) \quad \forall\, \mathbf{u}, \mathbf{\upsilon} \in \mathcal{V} \tag{B.6.1}$$

$$(\mathbf{Tu}, \mathbf{\upsilon}) = \left(\mathbf{T}^{\mathrm{T}}\mathbf{\upsilon}, \mathbf{u}\right) \quad \forall\, \mathbf{u}, \mathbf{\upsilon} \in \mathcal{V} \tag{B.6.2}$$

Now,

$$(\mathbf{Tu}, \mathbf{\upsilon}) = T_{ij}u_j\upsilon_i = T_{ji}u_i\upsilon_j \tag{B.6.3}$$

$$\left(\mathbf{T}^{\mathrm{T}}\mathbf{\upsilon}, \mathbf{u}\right) = (\mathbf{T}^{\mathrm{T}})_{ij}\upsilon_j u_i = (\mathbf{T}^{\mathrm{T}})_{ji}u_j\upsilon_i. \tag{B.6.4}$$

Thus,

$$T_{ij} = (\mathbf{T}^{\mathrm{T}})_{ji} \quad \text{or} \quad (\mathbf{T}^{\mathrm{T}})_{ij} = T_{ji}. \tag{B.6.5}$$

Thus, the matrix associated with $\mathbf{T}^{\mathrm{T}}$ is obtained from the matrix **T** by interchanging the rows for the columns in **T**.

A linear transformation (matrix) is said to be symmetric or self-adjoint when

$$\mathbf{T} = \mathbf{T}^{\mathrm{T}}, \quad (T_{ij} = T_{ji}). \tag{B.6.6}$$

A linear transformation (matrix) is said to be skew-symmetrix if

$$\mathbf{T} = -\mathbf{T}^{\mathrm{T}}, \quad (T_{ij} = (-\mathbf{T}^{\mathrm{T}})_{ij} = -T_{ji}). \tag{B.6.7}$$

Exercise. Let $\mathbf{T} : \mathcal{V} \longrightarrow \mathcal{V}$, show that

$$(\mathbf{TS})^{\mathrm{T}} = \mathbf{S}^{\mathrm{T}}\mathbf{T}^{\mathrm{T}}. \tag{B.6.8}$$

B.7 Range

If a linear transformation $\mathbf{T} : \mathcal{V} \longrightarrow \mathcal{V}$, and $\mathcal{V}'$ is a subspace of $\mathcal{V}$ the image of $\mathcal{V}'$ under **T** denoted by $\mathbf{T}\mathcal{V}'$ is defined through

$$\mathbf{T}\mathcal{V}' = \{\mathbf{y} \in \mathcal{V} \mid \mathbf{y} = \mathbf{Tx} \quad \text{for } \mathbf{x} \in \mathcal{V}'\}. \tag{B.7.1}$$

The range of $\mathbf{T}$ denoted by $\mathcal{R}(\mathbf{T}) = \mathbf{T}\mathcal{V}$. The null space of $\mathbf{T}$, denoted by $\eta(\mathbf{T})$, is defined through

$$\eta(\mathbf{T}) = \{\mathbf{x} \in \mathcal{V} \mid \mathbf{T}\mathbf{x} = 0\} \tag{B.7.2}$$

Note, the transformation $\mathbf{T}$ is invertible if and only if $\mathcal{R}(\mathbf{T}) = \mathcal{V}$ and $\eta(\mathbf{T}) = 0$.

Proof. Recall the definition of invertible:
(i) If $\mathbf{x}_1 \neq \mathbf{x}_2, \mathbf{T}\mathbf{x}_1 \neq \mathbf{T}\mathbf{x}_2$.
(ii) If $\mathbf{y} \in \mathcal{V}, \exists \mathbf{x} \in \mathcal{V} \ni \mathbf{T}\mathbf{x} = \mathbf{y}$.

Now, suppose $\mathbf{T}$ is invertible. It follows from (ii) that $\mathcal{R}(\mathbf{T}) = \mathcal{V}$. Next, $\mathbf{T0} = \mathbf{0}$. Suppose $\mathbf{T}\mathbf{x} = \mathbf{0}$ for some $\mathbf{x}$. Then such a transformation contradicts (i).

Next suppose $\eta(\mathbf{T}) = \mathbf{0}$, i. e.,

$$\mathbf{T}\mathbf{x} = \mathbf{0} \implies \mathbf{x} = \mathbf{0}. \tag{B.7.3}$$

Now, if $\mathbf{T}\mathbf{x}_1 = \mathbf{T}\mathbf{x}_2$, then $\mathbf{T}(\mathbf{x}_1 - \mathbf{x}_2) = \mathbf{0}$. By Eq. (B.7.3), $\mathbf{x}_1 - \mathbf{x}_2 = \mathbf{0}$. Thus,

$$\mathbf{x}_1 \neq \mathbf{x}_2 \implies \mathbf{T}\mathbf{x}_1 \neq \mathbf{T}\mathbf{x}_2 \tag{B.7.4}$$

Let $\{\mathbf{x}_i\}$, $i = 1, 2, \ldots, n$ be a basis in $\mathcal{V}$. Claim that $\mathbf{T}\mathbf{x}_i$ is a basis. Suppose that $\{\mathbf{T}\mathbf{x}_i\}$ is not a basis, then for some i,

$$\mathbf{T}\mathbf{x}_i = \sum \alpha_j \mathbf{T}\mathbf{x}_j = \mathbf{T} \sum \alpha_j \mathbf{x}_j, \quad j = 1, 2, \ldots, n, \ j \neq i \tag{B.7.5}$$

Thus,

$$\mathbf{T}\Big(\mathbf{x}_i - \sum \alpha_j \mathbf{x}_j\Big) = \mathbf{0}, \quad j = 1, 2, \ldots, n, \ j \neq i \tag{B.7.6}$$

Thus, since we have supposed that $\eta(\mathbf{T}) = \mathbf{0}$,

$$\mathbf{x}_i - \sum \alpha_j \mathbf{x}_j = \mathbf{0}, \quad j = 1, 2, \ldots, n, \ j \neq i. \tag{B.7.7}$$

Since $\{\mathbf{x}_i\}$ is a basis, it follows that Eq. (B.7.7) is not possible. Thus, $\{\mathbf{T}\mathbf{x}_i\}$ is a basis.
It trivially follows that any $\mathbf{y} \in \mathcal{V}$ can be expressed as

$$\mathbf{y} = \mathbf{T}\mathbf{x} \quad \text{for some } \mathbf{x} \in \mathcal{V}. \tag{B.7.8}$$

To show that

$$\mathbf{x}_1 \neq \mathbf{x}_2 \implies \mathbf{T}\mathbf{x}_1 \neq \mathbf{T}\mathbf{x}_2, \tag{B.7.9}$$

let $\mathbf{y}_1, \mathbf{y}_2, \ldots, \mathbf{y}_n$ be a basis in $\mathcal{V}$. Then, $\mathbf{x}_1, \mathbf{x}_2, \ldots, \mathbf{x}_n$ such that

$$\mathbf{T}\mathbf{x}_i = \mathbf{y}_i \tag{B.7.10}$$

is a basis in $\mathcal{V}$.

If $\sum \alpha_i \mathbf{x}_i = \mathbf{0}$, then

$$\mathbf{T}\Big(\sum \alpha_i \mathbf{x}_i\Big) = \sum \alpha_i \mathbf{y}_i = \mathbf{0} \implies \alpha_1 = \alpha_2 = \cdots = \alpha_n = 0 \tag{B.7.11}$$

Now, any $\mathbf{x} = \alpha_i \mathbf{x}_i$. Thus,

$$\mathbf{T}\mathbf{x} = \mathbf{0} \implies \mathbf{x} = \mathbf{0}. \tag{B.7.12}$$

Exercise. If $\mathbf{x}$, $\mathbf{y}$, and $\mathbf{z}$ are linearly independent, does it imply that $\mathbf{x} + \mathbf{y}$, $\mathbf{y} + \mathbf{z}$, and $\mathbf{z} + \mathbf{x}$ are linearly independent?

Exercise. Under what conditions on the scalar ζ are the vectors $(\zeta, 1, 0)$, $(1, \zeta, 1)$, and $(0, 1, \zeta)$ in $\mathbb{R}^3$ linearly dependent?

B.8 Inverse

Consider

$$\det \mathbf{A} = \sum_{k=1}^{n} A_{ik} C_{ik}, \quad \det \mathbf{A} = \sum_{k=1}^{n} A_{kj} C_{kj}. \tag{B.8.1}$$

Then,

$$\sum_{k=1}^{n} A_{ik} C_{jk} = (\det \mathbf{A}) \delta_{ij}, \tag{B.8.2}$$

$$\sum_{k=1}^{n} A_{kj} C_{ki} = (\det \mathbf{A}) \delta_{ij} \tag{B.8.3}$$

Let us assume that $\det \mathbf{A} \neq 0$. Define

$$M_{ij} = \frac{C_{ij}}{\det \mathbf{A}}. \tag{B.8.4}$$

Then,

$$\sum_{k=1}^{n} A_{ik} M_{kj} = \delta_{ij}, \tag{B.8.5}$$

$$\sum_{k=1}^{n} A_{kj} M_{ik} = \delta_{ij}, \tag{B.8.6}$$

with the matrix with M_{ij} as its elements by means of the usual matrix multiplication leads to the identity matrix:

$$[A_{ik}][M_{kj}] = [\delta_{ij}]. \tag{B.8.7}$$

Let $\mathbf{M}$ denote the linear transformation whose matrix entries are M_{ij}. Then it follows that

$$\mathbf{AM} = \mathbf{MA} = \mathbf{I}. \tag{B.8.8}$$

$\mathbf{M}$ is denoted by $\mathbf{A}^{-1}$. It follows from our analysis that

$$\mathbf{A}^{-1} = \frac{\mathbf{C}^{\mathrm{T}}}{\det \mathbf{A}}, \tag{B.8.9}$$

where $\mathbf{C}$ is the linear transformation corresponding to the adjoint matrix. Thus,

$$(\mathbf{A}^{-1})_{ij} = \frac{C_{ji}}{\det \mathbf{A}}. \tag{B.8.10}$$

B.9 Rank

The rank of a linear transformation $\mathbf{T} : \mathcal{V} \longrightarrow \mathcal{V}$ is the dimension of $\mathcal{R}(\mathbf{T})$. The nullity of $\mathbf{T}$ is the dimension of $\eta(\mathbf{T})$.

B.10 Eigen values

A scalar λ is an eigen value (or a proper value) and a non-zero vector $\mathbf{x}$ is an eigen vector (proper vector) of a linear transformation $\mathbf{T} : \mathcal{V} \longrightarrow \mathcal{V}$ if

$$\mathbf{Tx} = \lambda \mathbf{x}. \tag{B.10.1}$$

Suppose λ is a proper value of $\mathbf{T}$ and $\mathbf{x}$ is an eigen vector. It follows from Eq. (B.10.1) that $\alpha\mathbf{x}$ is an eigen vector for any scalar α. Also, if $\mathbf{x}$ and $\mathbf{y}$ are two distinct eigen vectors corresponding to λ, then by Eq. (B.10.1) $(\mathbf{x} + \mathbf{y})$ is also an eigen vector. Let us define the set $\mathcal{M}$ as the set of all eigen vectors corresponding to a particular eigen value λ and the zero vector. It then follows that $\mathcal{M}$ is a subspace of $\mathcal{V}$.

B.11 Geometric multiplicity of an eigen value

The geometric multiplicity of a eigen value is the dimension of $\mathcal{M}$. The set of proper values of $\mathbf{T}$ is called the spectrum of $\mathbf{T}$. Now,

$$\mathbf{Tx} - \lambda\mathbf{x} = \mathbf{0} \iff (\mathbf{T} - \lambda\mathbf{I})\mathbf{x} = \mathbf{0}. \tag{B.11.1}$$

The value λ as an eigen value of $\mathbf{T}$ is the same as the nullity of the transformation $(\mathbf{T}-\lambda\mathbf{I})$.

Next, since a necessary and sufficient condition that $(\mathbf{T} - \lambda\mathbf{I})$ have a non-trivial null space (i. e., λ have an eigen vector associated with it) is that $(\mathbf{T}-\lambda\mathbf{I})$ be singular, it follows that $\det(\mathbf{T} - \lambda\mathbf{I}) = 0$.

It follows from the definition of the determinant that $\det(\mathbf{T} - \lambda\mathbf{I}) = 0$ is a polynomial of degree n (where $n = \dim \mathcal{V}$) in λ. Thus, the eigen values are the roots of a polynomial equation of degree n. Since polynomial equations do not always have real roots, all linear transformations do not necessarily have real eigen values when dealing with a linear transformation that acts on a vector space defined over the field of real numbers. However, if the scalar field is the set of complex numbers, it then follows from the fundamental theorem of algebra that every linear transformation has at least one eigen value in the field under consideration.

B.12 Algebraic multiplicity of eigen values

Let $\mathbf{T} : \mathcal{V} \longrightarrow \mathcal{V}$. As before let λ be a proper value of the linear transformation $\mathbf{T}$. The multiplicity of λ as a root of the characteristic equation $\det(\mathbf{T} - \lambda\mathbf{I}) = 0$, is referred to as the algebraic mulltiplicity of $\mathbf{T}$.

We will note that

1. algebraic multiplicity is greater than or equal to geometric multiplicity.
2. it follows from the definition of the determinant that

$$\det \mathbf{T} = \prod_{j=1}^{n} (\lambda_j)^{m_j}, \tag{B.12.1}$$

where m_j is the algebraic multiplicity of λ_j.

A linear transformation is said to be real if $\mathbf{T} = \bar{\mathbf{T}}$ (conjugation). Let us next consider a real symmetric transformation. Let $\mathcal{V}$ be an inner product space and let $\mathbf{T} : \mathcal{V} \longrightarrow \mathcal{V}$ be such that $\mathbf{T} = \mathbf{T}^{\mathrm{T}}$. Let λ_1 and λ_2 be two distinct eigen values and let $\mathbf{x}_1$ and $\mathbf{x}_2$ be the eigen vectors corresponding to λ_1 and λ_2, respectively (the same eigen vector cannot correspond to two distinct eigen values). Thus,

$$\mathbf{Tx}_1 = \lambda_1\mathbf{x}_1, \quad \mathbf{Tx}_2 = \lambda_2\mathbf{x}_2. \tag{B.12.2}$$

Now

$$(\mathbf{Tx}_1 \cdot \mathbf{x}_2) = (\lambda_1 \mathbf{x}_1 \cdot \mathbf{x}_2) = \lambda_1 (\mathbf{x}_1 \cdot \mathbf{x}_2) \tag{B.12.3}$$

$$(\mathbf{x}_1 \cdot \mathbf{Tx}_2) = (\mathbf{x}_1 \cdot \lambda_2 \mathbf{x}_2) = \lambda_2 (\mathbf{x}_1 \cdot \mathbf{x}_2). \tag{B.12.4}$$

By definition,

$$(\mathbf{Tx}_1 \cdot \mathbf{x}_2) = (\mathbf{x}_1 \cdot \mathbf{T}^{\mathrm{T}} \mathbf{x}_2) = (\mathbf{x}_1 \cdot \mathbf{Tx}_2). \tag{B.12.5}$$

Thus, two eigen vectors of a real symmetric transformation corresponding to distinct eigen values are orthogonal.

We can also show that the eigen values of a self-adjoint transformation are real. The proof follows trivially from

$$(\mathbf{Tx}, \mathbf{Tx}) = (\mathbf{Tx}, \mathbf{T}^{\mathrm{T}} \mathbf{x}) = (\mathbf{T}^2 \mathbf{x}, \mathbf{x}) = (\lambda)^2 (\mathbf{x}, \mathbf{x}). \tag{B.12.6}$$

Next, consider the equation

$$\mathbf{Tx} - \lambda \mathbf{x} = \mathbf{c}, \tag{B.12.7}$$

where $\mathbf{T}$ is a real self-adjoint transformation. Let $\mathbf{e}_i$, $i = 1, 2, \ldots, n$ be the n eigen vectors corresponding to

$$\mathbf{Te}_i - \lambda_i \mathbf{e}_i = \mathbf{0}. \tag{B.12.8}$$

It follows that $\{\hat{\mathbf{e}}_i\}$, $i = 1, 2, \ldots, n$ forms an orthonormal basis. Thus, any $\mathbf{x} \in \mathcal{V}$ can be expressed as

$$\mathbf{x} = \alpha_i \hat{\mathbf{e}}_i. \tag{B.12.9}$$

It then follows from Eq. (B.12.7) that

$$\alpha_i \mathbf{T} \hat{\mathbf{e}}_i - \alpha_i \lambda \hat{\mathbf{e}}_i = \mathbf{c}, \tag{B.12.10}$$

i. e.,

$$\sum_{i=1}^{n} \alpha_i \lambda_i \hat{\mathbf{e}}_i - \alpha_i \lambda \hat{\mathbf{e}}_i = \mathbf{c} \tag{B.12.11}$$

Thus,

$$\sum_{i=1}^{n} \alpha_i (\lambda_i - \lambda) \hat{\mathbf{e}}_i = \mathbf{c}, \tag{B.12.12}$$

and

$$\left(\sum_{i=1}^{n} a_i(\lambda_i - \lambda)\hat{\mathbf{e}}_i \cdot \hat{\mathbf{e}}_k\right) = (\mathbf{c} \cdot \hat{\mathbf{e}}_k). \tag{B.12.13}$$

Hence,

$$a_k(\lambda_k - \lambda) = (\mathbf{c} \cdot \hat{\mathbf{e}}_k), \quad \text{no sum on } k. \tag{B.12.14}$$

The above analysis leads to the following assertion: the equation

$$\mathbf{Tx} - \lambda\mathbf{x} = \mathbf{c} \tag{B.12.15}$$

has no solution when λ is an eigen value, unless $\mathbf{c}$ is orthogonal to the corresponding eigen vectors.

We conclude this section by making the following assertions (without proof):

1. If $\mathbf{T}$ is a self-adjoint linear transformation on a finite dimensional inner product space, then the algebraic multiplicity of each eigen value is equal to its geometric multiplicity.
2. If an eigen value of a symmetric matrix is of multiplicity p, then the $\mathbf{x}$ correspond to p linearly independent eigen vectors.

References

[1] P. R. Halmos. Finite Dimensional Vector Spaces. Undergraduate Texts in Mathematics. Springer-Verlag-Heidelberg, 1987.
[2] H. K. Nickerson, D. C. Spencer, and N. E. Steenrod. Van Nostrand Company, Inc., Princeton, 1959.
[3] F. B. Hildebrandt, Methods of Applied Mathematics. Dover Publications, New York. 1992.
[4] V. I. Smirnov, Linear Algebra and Group theory. Dover Publications, New York. 1970.

Index

https://doi.org/10.1515/9783110789515-020

www.ingramcontent.com/pod-product-compliance
Lightning Source LLC
Chambersburg PA
CBHW061403210326
41598CB00035B/6082
* 9 7 8 3 1 1 0 7 8 9 4 2 3 *